Modern Microscopies

Techniques and Applications

Edited by

P. J. Duke

SERC Daresbury Laboratory
Warrington, United Kingdom

and

A. G. Michette

King's College
University of London
London, United Kingdom

Plenum Press • New York and London

Library of Congress Cataloging-in-Publication Data

Modern microscopies : techniques and applications / edited by P.J. Duke and A.G. Michette.
p. cm.
Includes bibliographical references (p.
ISBN 0-306-43288-9
1. Microscope and microscopy--Technique. 2. X-ray microscope--Technique. I. Duke, P. J. II. Michette, Alan G.
QH207.M63 1990
578'.4--dc20 90-6741
CIP

A Division of Plenum Publishing Corporation
233 Spring Street, New York, N.Y. 10013

Printed in the United States of America

Contributors

A. Boyde • Department of Anatomy and Developmental Biology, University College London, London WC1E 6BT, United Kingdom

C. J. Buckley • NSLS, Brookhaven National Laboratory, Upton, New York 11794. *Present address:* Physics Department, King's College, University of London, London WCR 2LS, United Kingdom

P. C. Cheng • Advanced Microscopy Laboratory, Department of Anatomical Sciences, School of Medicine and Biomedical Sciences/School of Engineering and Applied Sciences, State University of New York at Buffalo, Buffalo, New York 14214

P. J. Duke • SERC Daresbury Laboratory, Warrington WA4 4AD, United Kingdom

M. R. Howells • Center for X-Ray Optics, Lawrence Berkeley Laboratory, Berkeley, California 94720

I. S. Hwang • Department of Electrical and Computer Engineering, State University of New York at Buffalo, Buffalo, New York 14260

C. Jacobsen • Center for X-Ray Optics, Lawrence Berkeley Laboratory, Berkeley, California 94720

A. Jasinski • Institute of Nuclear Physics, 31-342 Krakow, Poland

H. G. Kim • Laboratory for Laser Energetics, University of Rochester, Rochester, New York 12306

J. Kirz • Department of Physics, State University of New York at Stony Brook, Stony Brook, New York 11794

J. R. Mallard • Department of Bio-Medical Physics and Bio-Engineering, University of Aberdeen, and Grampian Health Board, Aberdeen AB9 2ZD, United Kingdom

D. J. O. McIntyre • Department of Biochemistry, University of Cambridge, Cambridge CB2 1QW, United Kingdom

K. McQuaid • Schools of Medicine and Dentistry, University of California–San Francisco, San Francisco, California 94143

A. G. Michette • Physics Department, King's College, University of London, London WC2R 2LS, United Kingdom

P. G. Morris • Department of Biochemistry, University of Cambridge, Cambridge CB2 1QW, United Kingdom

S. P. Newberry • CBI Labs, Schenectady, New York 12306

B. Niemann • University of Göttingen, Forschungseinrichtung Röntgenphysik, D-3400 Göttingen, Federal Republic of Germany

H. Rarback • NSLS, Brookhaven National Laboratory, Upton, New York 11794

S. S. Rothman • Schools of Medicine and Dentistry, University of California–San Francisco, San Francisco, California 94143

D. Rudolph • University of Göttingen, Forschungseinrichtung Röntgenphysik, D-3400 Göttingen, Federal Republic of Germany

G. Schmahl • University of Göttingen, Forschungseinrichtung Röntgenphysik, D-3400 Göttingen, Federal Republic of Germany

M. G. Somekh • Department of Electrical and Electronic Engineering, University of Nottingham, Nottingham NG7 2RD, United Kingdom

Murray Stewart • MRC Laboratory of Molecular Biology, Cambridge CB2 2QH, United Kingdom

M. E. Taylor • Interdisciplinary Research Centre in Superconductivity, University of Cambridge, Cambridge CB3 0HE, United Kingdom

M. E. Welland • Department of Engineering, University of Cambridge, Cambridge CB2 1PZ, United Kingdom

Preface

For several decades the electron microscope has been the instrument of choice for the examination of biological structures at high resolution. Biologists have become familiar with the techniques and pitfalls of sample preparation and with the interpretation of the images obtained. The purpose of this book is to introduce the biologist to a number of new imaging techniques that are now becoming available to supplement and even extend the information that can be obtained from the now-traditional electron microscope. Some of these techniques are still at the experimental stage, while others, such as cryoelectron microscopy and confocal optical microscopy, are at advanced stages of development and are already available commercially.

This book represents a first attempt to quantify the progress made by bringing together, in one volume, an account of the technical bases and the future potentials of the various techniques. Although the content is primarily aimed at biologists, microscopists in other fields should also find the information of interest and use. All the chapters are written by leading experts who are at the forefront of these exciting developments. About half the book is concerned with x-ray microscopy; the editors make no apology for this since they are both intimately involved with developments associated with this field and therefore view it, perhaps with bias, as being of the utmost importance.

The application of the physical properties of electrons to the elucidation of biological structure has been one of the success stories of modern biophysics. It is our hope that this volume will help to focus attention on other properties of the inanimate world that can be used as probes of the material of life.

Much of the material prepared for this book was first presented at a conference organized by the editors at King's College, London, on behalf of the British Biophysical Society (BBS). However, the content does not in any sense represent a proceedings of that conference and should not be taken as reflecting the views of the BBS. Any errors and misleading statements are solely the responsibility of the editors.

The editors are very grateful to all the contributors for submitting their manuscripts in forms that did not require much editing, and (in most cases) not too long after the originally agreed deadlines. We would also like to thank our colleagues, in particular at King's College and Daresbury but also at other laboratories too numerous to mention, for many illuminating discussions and for their forbearance during the production of the final manuscripts.

P. J. Duke
A. G. Michette

Daresbury and London

Contents

1

Modern Microscopy

P. J. Duke and A. G. Michette

1.1. Introduction

The intention of this book is to introduce to both experienced and budding microscopists some of the advances in techniques that have taken place over the past few years and, indeed, are still underway. It is not meant to be a complete survey of the huge field of microscopy, and many topics will not be addressed. In particular, established microscopical techniques will not be discussed; a comprehensive account of these has been given by Rochow and Rochow,[1] who concentrate on optical and electron microscopies but also give brief introductions to x-ray and acoustic microscopies. Such is the rate of advance, however, that their discussions of modern methods, particularly in x-ray microscopy, are very much out of date even though their book was published only a decade ago.

Advances in technique invariably lead to increased ranges of application; this is true of course in fields other than microscopy. The rationale behind pursuing technical advances in microscopy, and indeed for developing new methods of microscopy, is in improving the quality of images (in terms of resolution and contrast), in extending the range of specimen types amenable to imaging, and, perhaps most importantly, in reducing the possibility of drawing false inferences from information contained in the images. The interpretation of microscope images is clearly of paramount importance if conclusions are to be drawn about the nature of the object (the specimen), and in this respect continuing

P. J. Duke • SERC Daresbury Laboratory, Warrington WA4 4AD, United Kingdom. **A. G. Michette** • Physics Department, King's College, University of London, London WC2R 2LS, United Kingdom.

improvements in all aspects of image processing[2–5] should be closely followed by all microscopists.

The quality of the images, and of the information they contain, are major factors in assessing the quality of the technique. Thus, as well as describing the modern methods of microscopy, examples of their applications will be considered. The range of applications covered is limited only by what has been tried so far, and should not be taken as an exhaustive survey of the possibilities.

The advances to be covered are nearly all related to the need to circumvent limitations of established techniques. For example, although living biological material can be studied in conventional optical microscopy the amount of detail that can be observed (i.e., the resolution) is limited by the wavelength of light, and also by aberrations introduced into off-axis image points. Some gain in resolution (particularly in depth) can be obtained by using confocal scanning systems[6,7] which also allow the reduction of aberrations because all image points are obtained on axis. To reach very high resolutions, however, it is necessary to use the electron microscope (or, recently, devices such as the scanning tunnelling microscope), which has been developed to a very high degree over the past 40 years or so, and which has provided most of the present knowledge in, for example, cell biology. In conventional transmission electron microscopy, however, specimens have to be dried (to prevent excessive electron scattering by water), thinned or sectioned (since electron penetration depths are small), and fixed and stained with heavy metals (since natural contrast in biological material is low). This means that quantitative information about mass distribution in the specimen, for example, is difficult, if not impossible, to obtain. Alternatively the surfaces of specimens can be imaged in scanning electron microscopy; this normally involves coating the surface with a heavy metal. All this preparation means that the specimen will not be in a state corresponding to its natural one, and the resulting images will therefore be subject to artifacts leading to possibly false interpretation.

Attempts to solve the problems of conventional electron microscopy by using high-voltage microscopes[8] and environmental specimen containment cells have largely been unsuccessful. More recent techniques include cryoelectron microscopy,[9] in which the water is left in specimens that are very rapidly frozen (the natural contrast between the material of interest and the surrounding vitreous ice then being large), and x-ray microscopy.[10–12] Imaging in x-ray microscopy normally relies on x-ray absorption by the specimen, and, because this can vary considerably between materials, natural contrast can be obtained from wet specimens, particularly in the so-called water-window x-ray wavelength range between the K_α absorption edges of oxygen (at 2.2 nm) and carbon (at 4.4 nm). X-ray absorption cross sections are also much less than electron scattering cross sections so that thicker specimens can be imaged. However, obtainable resolutions (probably 10–20 nm for most types of x-ray imaging[13]) are much poorer than in electron microscopy, although they are considerably better than in optical microscopy. For imaging surfaces at atomic resolutions, techniques such as scanning tunnelling microscopy can be used.

Both electron and x-ray microscopy are invasive techniques; the specimen is damaged by the radiation as it is being imaged. Although calculations indicate that, for comparable resolution in images of similar specimens, X rays cause less damage than electrons,[13] it is radiation damage that ultimately limits the resolution in most forms of x-ray microscopy. This is not the case in electron microscopy since the specimen, albeit modified, is protected by the preparation it has undergone. Thus, it is important also to consider the development of noninvasive forms of imaging such as nuclear magnetic resonance microscopy and acoustic microscopy, although these offer lower resolutions than most other imaging techniques.

1.2. Cryoelectron Microscopy

The techniques of cryoelectron microscopy are described in detail in Chapter 2, where its applications, in particular to the imaging of biological macromolecules, are also discussed. This type of imaging circumvents the need for specimen dehydration and staining, and therefore a specimen can be observed in a more natural state than in conventional electron microscopy. The primary requirement[9] is for extremely rapid freezing of the specimen (at a rate considerably greater than 10^4 K s^{-1}) to prevent the formation of ice crystals which would mask the sought-for detail. Instead, vitreous (i.e., glasslike) ice is formed which has no observable structure. Also in this chapter the techniques of image processing are introduced, particularly with reference to the electron microscopy of biological macromolecules, although the techniques are applicable to other types of imaging as well.

1.3. X-Ray Microscopy

Although the concept of x-ray microscopy is not new, original work on the subject being published by Goby early in the twentieth century,[14] until recently little progress was made in most forms of it because of the lack of sufficiently intense sources and of suitable optical systems. The advent of microfocus x-ray sources,[15] plasma sources,[16–22] and (especially) synchrotron sources,[23] however, has largely overcome the source problem, while advances in the manufacture of x-ray optical components such as grazing-incidence and multilayer reflectors and (especially) zone plates have largely overcome the optics problem.[24] Over the last few years these advances have allowed an approach to be made toward the main goal of x-ray microscopy, namely, high-resolution imaging of material in its natural state (up to several micrometers thick, wet, and, at least at the start of the imaging process, potentially living). The advances in, and the current state of, the various forms of x-ray microscopy (full-field imaging, scanning, contact, and holographic) are described in Chapters 3–7, and some future prospects (such as phase-contrast imaging and the formation of images via photoelectrons emitted when the x-ray photons are absorbed) are also discussed.

1.4. Imaging by Magnetic Resonance Techniques

Nuclear magnetic resonance (NMR) techniques are being used increasingly in the study of large (i.e., body-sized) samples; the history of and recent advances in this form of NMR imaging are described in Chapter 8. More recently, the possibility of obtaining much higher resolutions (on the order of micrometers) has been investigated, and NMR microscopy is now becoming a reality, as described in the latter part of Chapter 8 and in Chapter 9. The principal advantages of NMR microscopy are that no specimen preparation is required, it is noninvasive (i.e., it does not damage the specimen), and it can image material that is opaque to light. In addition, imaging of variously oriented slices and three-dimensional imaging are possible, and contrast and resolution are easily controllable. The resolution is limited, however, by available magnetic field gradients, and it is unlikely that resolutions significantly less than 1 μm will be achieved.

1.5. Confocal Optical Microscopy

In conventional optical microscopy confusion can arise in images due to the light interacting with the specimen over a large vertical range. This is because specimens are either translucent or, if images are formed by reflection from the surface, not flat. The basic idea behind the confocal scanning optical microscope is to eliminate this confusion by illuminating and imaging only one small spot at a time in the focal plane. This requires a high-power monochromatic light source, and thus the development of this form of microscopy has arisen from the use of lasers. The development and applications of confocal optical microscopes are described in Chapter 10.

1.6. Acoustic Microscopy

Another noninvasive technique is acoustic microscopy. Because the speed of sound is, depending on the material, about 10^5 times lower than the speed of light, short wavelengths (~1 μm), that is, potentially high resolutions in microscopy, can be obtained at ultrasonic frequencies (~1 GHz). In addition, ultrasound can penetrate optically opaque material, and its reflection is determined by the mechanical properties of the specimen—thus allowing natural contrast and perhaps obviating the need for staining. Developments in acoustic microscopy have taken place over a number of years,[25–27] and recent advances in techniques and applications are reviewed in Chapter 11.

1.7. Scanning Tunnelling Microscopy

Scanning tunnelling microscopy was first envisaged in the 1970s,[28] but for several years its potential was not realizable because of mechanical and electrical difficulties, and because the first attempts at making a usable instrument resulted in rather poor lateral resolutions. These problems were overcome in the early 1980s by making the probe smaller,[29] leading to horizontal resolutions of $\approx$0.2 nm and vertical (i.e., in the direction of the tunnelling current flow) resolutions of $\approx$0.01 nm. Since then the atomic resolution capabilities of scanning tunnelling microscopes have found a wide range of applications. A description of the development of the techniques and examples of its application are given in Chapter 12.

1.8. Summary

From the very brief introductions to the various forms of microscopy that have been given above, it is clear that each of them, to be described in detail in the subsequent chapters, can give clear advantages or disadvantages depending on the type of specimen and what detail it is that requires study. Before anyone embarks on the use of these microscopes, it is very important that they understand what the limitations of the techniques are, because these are what will ultimately define the information content of the images. Potential users are therefore urged to be critical about the material presented herein, particularly since in many cases rapid developments are still taking place.

References

1. T. G. Rochow and E. G. Rochow, *An Introduction to Microscopy by Means of Light, Electrons, X Rays, or Ultrasound,* Plenum, New York (1978).
2. R. C. Gonzalez and P. Wintz, *Digital Image Processing,* 2nd ed., Addison-Wesley, Reading, Mass. (1986).
3. A. Rosenfeld and A. C. Kak, *Digital Picture Processing,* 2nd. ed., Academic, Orlando, Fl. (1982).
4. U. Aebi, W. E. Fowler, E. L. Buhle, and P. R. Smith, "Electron microscopy and image

processing applied to the study of protein structure and protein–protein interactions," *J. Ultrastruct. Res.* **88,** 143–176 (1984).
5. M. Stewart, in: *Ultrastructure Techniques for Microorganisms* (H. C. Aldrich and W. J. Todd, eds.), pp. 333–364, Plenum, New York (1986).
6. C. J. R. Wilson and T. Wilson, "Fourier imaging of phase information in scanning and conventional optical microscopes," *Phil. Trans.* **A295,** 513–536 (1980).
7. M. Petráň, M. Hadravský, and A. Boyde, "The tandem scanning reflected light microscope," *Scanning* **7,** 97–108 (1985).
8. J. N. Turner, C. W. See, A. J. Ratkowski, B. B. Chang, and D. F. Parsons, "Design and operation of a differentially pumped environmental chamber for the HVEM," *Ultramicroscopy* **6,** 267–280 (1981).
9. W. Chiu, "Electron microscopy of frozen, hydrated biological specimens," *Ann. Rev. Biophys. Biophys. Chem.* **15,** 237–257 (1986).
10. G. Schmahl and D. Rudolph (eds.), *X-Ray Microscopy,* Springer Series in Optical Sciences, Vol. 43, Springer, Berlin (1984).
11. P. C. Cheng and G. J. Jan (eds.), *X-Ray Microscopy: Instrumentation and Biological Applications,* Springer, Berlin (1986).
12. D. Sayre, M. Howells, J. Kirz, and H. Rarback (eds.), *X-Ray Microscopy II,* Springer Series in Optical Sciences, Vol. 56, Springer, Berlin (1988).
13. D. Sayre, J. Kirz, R. Feder, D. M. Kim and E. Spiller, "Transmission microscopy of unmodified biological materials: Comparative radiation dosages with electrons and ultrasoft x-ray photons," *Ultramicroscopy* **2,** 337–341 (1977).
14. P. Goby, "A new application of Roentgen rays: microradiography," *J. Roy. Mic. Soc.,* August, 373–375 (1913).
15. D. J. Pugh and P. D. West, in: *Developments in Electron Microscopy and Analysis* (D. L. Misell, ed.), I.O.P. Conference Series, No. 36, pp. 29–32, The Institute of Physics, London (1977).
16. A. G. Michette, R. E. Burge, A. M. Rogoyski, F. O'Neill, and I. C. E. Turcu, "The potential of Laser Plasma Sources in Scanning X-ray Microscopy", in: *X-Ray Microscopy II* (D. Sayre, M. Howells, J. Kirz and H. Rarback, eds.), Springer Series in Optical Sciences, Vol. 56, pp. 59–62, Springer, Berlin (1988).
17. R. A. McKorkle, in: *Ultrasoft X-Ray Microscopy: Its Application to Biological and Physical Sciences* (D. F. Parsons, ed.), Annals of the New York Academy of Science, vol 342, pp. 53–64 (1980).
18. J. Bailey, Y. Ettinger, A. Fisher, and R. Feder, "Evaluation of the gas puff z-pinch as an x-ray lithography and microscopy source," *Appl. Phys. Lett.* **40,** 33–35 (1982).
19. P. Choi, A. E. Dangor, and C. Deeney, "Small gas puff z-pinch x-ray source," *Soft X-Ray Optics and Technology, Proc. SPIE* **733,** 52–57 (1986).
20. G. Herziger, in: *X-Ray Microscopy,* Springer Series in Optical Sciences (G. Schmahl and D. Rudolph, eds.), Vol 43, pp. 19–24, Springer, Berlin (1984).
21. J. L. Bourgade, C. Cavailler, J. de Mascureau, and J. J. Miquel, "Pulsed soft x-ray source for laser–plasma diagnostic calibrations," *Rev. Sci. Instrum.* **57,** 2165–2167 (1986).
22. Y. Kato and S. H. Be, "Generation of soft X rays using a rare gas–hydrogen plasma focus and its application to x-ray lithography," *Appl. Phys. Lett.* **48,** 686–688 (1986).
23. E.-E. Koch, D. E. Eastman, and Y. Farge, in: *Handbook on Synchrotron Radiation* (E.-E. Koch, ed.), Vol. 1, pp. 1–63, North-Holland, Amsterdam (1983).
24. A. G. Michette, *Optical Systems for Soft X Rays,* Plenum, New York (1986).
25. L. W. Kessler, A. Korpel, and P. R. Palermo, "Simultaneous acoustic and optical microscopy of biological specimens," *Nature* **239,** 111–112 (1972).
26. R. A. Lemons and C. F. Quate, "Acoustic microscopy: Biomedical applications," *Science* **188,** 905–911 (1975).

27. Special issue on acoustic microscopy: *IEEE Trans. Sonics and Ultrasonics* **Su-132**(2), 130–378 (1985).
28. T. Mulvey, "Tunnelling through scientific barriers?," *Phys. Bull.* **38**(1), 24–25 (1987).
29. G. Binnig, H. Rohrer, C. Gerber, and E. Weibel, "Surface studies by scanning tunnelling microscopy, *Phys. Rev. Lett.* **49**, 57–61 (1982).

2

Electron Microscopy of Biological Macromolecules

Frozen Hydrated Methods and Computer Image Processing

Murray Stewart

2.1. Introduction

Two recent advances have substantially improved the information that can be obtained from electron microscopy about the structure of biological macromolecules and the interactions between them. Computer image processing has enabled many of the intrinsic problems of electron microscopy of these materials to be circumvented. For example, many fine details are usually faint in electron micrographs of macromolecules, and they are further obscured by noise deriving from both the image formation mechanism of the microscope and also features of the methods used to contrast and support the materials. Filtering removes much of the background and so enables fine details to be seen more easily. In addition, structural information from different levels in the object is superimposed, and this can sometimes produce confusing moire patterns. These moire patterns can often be decomposed into their constituents by image processing. Low-dose conditions and electron diffraction usually need to be combined with image processing to obtain the highest resolution and, moreover, it is generally possible to combine different views of objects to generate three-dimensional structural models.

Murray Stewart • MRC Laboratory of Molecular Biology, Cambridge CB2 2QH, United Kingdom.

A second major shortcoming in conventional electron microscopy of biological material has been the changes in structure brought about by dehydration and staining. The development of methods for the examination of frozen hydrated biological specimens now offers the exciting possibility of investigating structure by transmission electron microscopy in material that closely approximates its native condition. By freezing specimens in thin films of vitreous ice, material can be preserved well and, moreover, can be examined unstained by using the intrinsic contrast difference between the water and biological material. Although there are sometimes some problems of interpretation arising from the phase-contrast nature of the images obtained, this method has led to a number of spectacular insights into biological structure, particularly with delicate specimens that are not well preserved by conventional methods and with crystals of membrane proteins.

2.2. Levels of Structure in Biological Material

Electron microscopes have the potential to resolve image details of the order of 0.1 nm, but with biological objects it is rare for this resolution to be approached. Problems associated with sample preparation, radiation damage, and the mechanism of contrast formation in the electron microscope usually impose a resolution limit of about 2 to 5 nm on this material. Resolutions of this order only enable the grossest features of biological macromolecules to be made out. Methods have now been devised to circumvent or at least reduce the effects of many of these problems, and these techniques have opened the way for the investigation of biological structures at higher resolution. Two important advances have been the development of methods of computer-based image processing—for extracting the maximum useful information from images—and techniques for examining frozen hydrated biological material. The application of these techniques has been generally most powerful when applied to ordered assemblies, such as helices and crystals, where sometimes quite remarkable insights into molecular structure and the interactions between molecules have been obtained. Before reviewing these developments, however, it is useful to consider the different levels of structure present in biological material and the effect of resolution on the information that can be obtained about each.

There are a number of different levels of structure present within biological material. Cells themselves range from about 1 μm to many centimeters in size and can usually be easily seen with the naked eye or with a light microscope. The next level of structure in eukaryotic cells is that of organelles, which have dimensions similar to those of procaryotic cells, that is, about 1–10 μm. These features also can usually be seen with light microscopes and often in the living state. Virus particles and most macromolecular assemblies within cells generally have dimensions of the order of 100 nm, and, although these particles can frequently be detected by high-contrast light microscopy methods (such as fluorescence), details of their substructure can only be obtained in conjunction with electron microscopy and, possibly, x-ray diffraction. With electron microscopy methods, it is necessary to dehydrate and stain specimens, although recent advances have made it possible to examine a broad range of specimens preserved in amorphous (vitreous) ice, which appears to give a close approximation to their usual hydrated environment. To detect substructure in these assemblies requires resolution better than about 50 nm. For example, the repeat distance in collagen (probably the longest spacing commonly found) is 63 nm, whereas the spacings in muscle are 43 nm and 14 nm. Most globular proteins have dimensions of the order of 5–10 nm.

It is important to realize, therefore, that there are few biological structures that can be investigated in any detail until resolutions of about 50 nm have been reached, and that most objects require a resolution of at least 10 nm. Any details

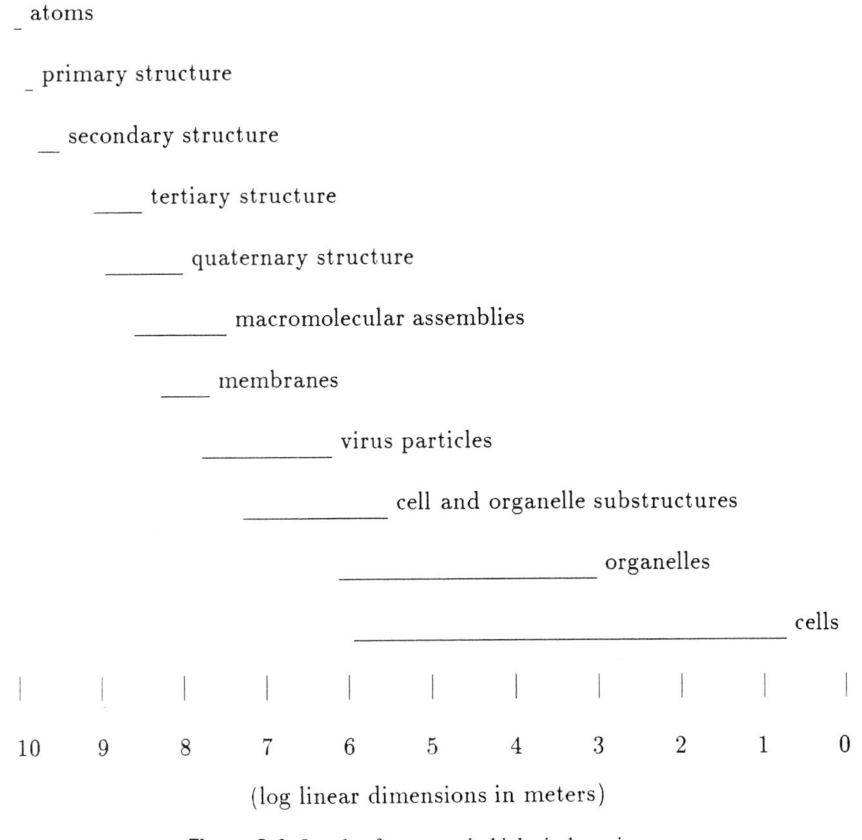

Figure 2.1. Levels of structure in biological specimens.

Table 2.1. Useful Test Objects

Resolution range	Objects
Better than 0.5 nm	Metal crystals such as evaporated gold Graphite Thallous chloride crystals
0.5–5 nm	Thin crystals of biological macromolecules such as catalase, purple membrane or long S-2 crystals
3–50 nm	Paracrystals of fibrous proteins such as collagen, paramyosin, and tropomyosin
10–2 μm	Striated muscle: spacing within filaments 14.3 nm and 38 nm; spacings between filaments of 40 nm, sarcomere spacing of 2.2 μm

about the fine structure of molecules generally require a resolution of better than 5 nm, and fine structure usually only becomes clear at about 2.5 nm. Resolutions of this order are adequate to locate subunits in most multiunit assemblies (such as virus particles, muscle filaments, and bacterial surface layers), although the orientation of subunits generally requires rather fine features to be discerned and may require resolutions of about 2.5 nm or better. Most biological macromolecules do not then yield a great deal more information until resolutions of about 0.7 nm are obtained, when elements of their secondary structure start to become visible. Atomic resolution, which is usually only obtainable by x-ray diffraction, generally means resolutions of about 0.3 nm or better. Figure 2.1 illustrates the structural levels in biological specimens, and Table 2.1 gives some test objects that can be used to assess resolution. Both image processing and frozen hydrated methods are very effective in extending resolution in biological material to the range 5–1 nm, and this can frequently yield significant new information about macromolecular assemblies and molecular fine structure.

2.3. Image Processing

Most biological material has little inherent contrast when examined by electron microscopy, and images are generally faint, even when stained with heavy metals. Contrast is also generated by the specimen support film and from the granular nature of most stains. This additional contrast tends to obscure the image of the biological specimen, and is compounded by the granularity introduced by the mechanism of image formation in the electron microscope. These effects all tend to be more severe at higher resolutions, and so fine structural features are often obscured in electron micrographs of biological macromolecules. Computer image processing can be used to remove many of these obscuring image features and so enhance the pattern due to the biological material. In its simplest form, usually referred to as "filtering," image processing improves the signal-to-noise ratio of the image by removing much of the obscuring background. However, these methods can be extended to average data from different areas, detect differences between similar objects, and to separate patterns from different structural levels that are superimposed in the micrograph. Such methods are particularly effective with crystalline objects or multisubunit arrays, such as helices, in which the subunit positions are related to one another in a specific manner. Frequently, higher-resolution information can be obtained by combining electron diffraction data with that obtained from low-dose images. Although electron microscope images are essentially two-dimensional (because they represent the density in the material projected in a direction parallel to the microscope's axis), three-dimensional models can be obtained by combining data from different views. For crystalline objects, three-dimensional models can be built up from a number of views taken from different angles, whereas with helical objects advantage can often be taken of their internal symmetry to generate a three-dimensional model from a single micrograph. There have been several comprehensive reviews of computer image processing of biological material[1–5] and so only an outline of the methods and the results that can be obtained will be given here.

2.3.1. Filtering

Filtering aims to enhance fine detail by improving the signal-to-noise ratio. Fourier transformation is generally used with regular objects, because this allows data to be manipulated conveniently and different components to be separated. Fourier methods depend on analyzing the density distribution of the object in terms of a sum of waves that can be combined to reconstruct the image. The basic principle of filtering can be understood by considering objects that are regular in one dimension; that is, regular objects in which the structure varies periodically in only one direction. The simplest regular mathematical function is a cosine wave, which has several properties that are useful when analyzing regular objects. The cosine wave has a wavelength or period (the distance between successive repeating features), an amplitude (its strength), and a phase term that species its position relative to an origin. A complicated, periodically varying pattern can be built up by adding waves together. Fourier transformation is a method of decomposing a function into its constituent waves or the inverse of this: synthesizing the function from its constituent waves. These waves are described in terms of their spatial frequency (the reciprocal of their wavelength). This analysis and synthesis is analogous to the action of a lens, where a diffraction pattern (corresponding to the object's Fourier transform) is produced in the back focal plane, and a final image is formed by recombining the waves from the diffraction pattern. This concept of Fourier transformation giving a diffraction pattern followed by Fourier synthesis to produce an image is central to filtering.

Fourier methods are useful in filtering because the transform of a regular object is concentrated in a small number of peaks. This is because only waves having the same period as the object, or fractions of this period, contribute to its density. Waves with other periods would make a different contribution in each repeat and so must be zero in an ideally regular object. Thus, it is possible to consider the density of a regular object as being analogous to a musical note that is built up from a fundamental frequency and its overtones. In contrast, the other contributions ("noise") are usually essentially random. A random function has a transform in which all frequencies are represented more or less equally, and so the transform of aperiodic noise will be distributed over the entire spectrum.

The difference between Fourier transforms of periodic and aperiodic structures can be used to separate one from the other. An image reconstructed from only the periodic signal peaks will have most of the noise removed, although periodic noise components will remain. Figure 2.2 illustrates this procedure. Filtering can be shown mathematically to correspond to superimposing all the repeating units (or "unit cells") in the image, and thus produces an average view.(6) Filtering is therefore most efficient when large numbers of repeats are used. These points are discussed in greater detail elsewhere.(3) An example of one-dimensional filtering is tropomyosin paracrystals,(7–10) as illustrated in Figure 2.4. Filtering is particularly useful in cases such as this where there are only small differences between similar objects (here the signal is produced by a mercurial label attached to the protein) which are difficult to evaluate in the original micrographs (Figure 2.3).

Images can be processed using either optical or digital methods. Although

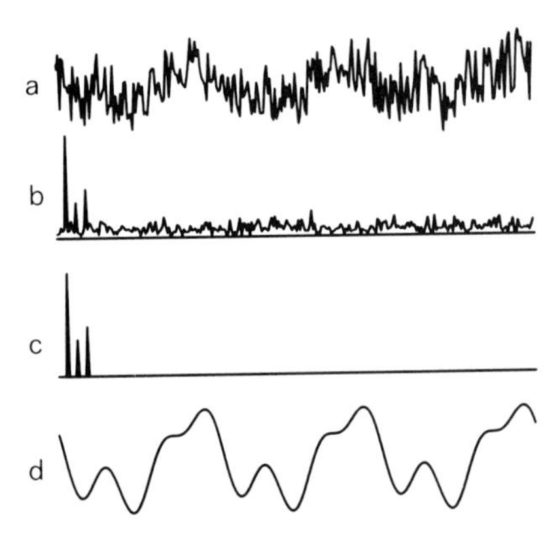

Figure 2.2. Filtering of a one-dimensional trace that contains both signal and noise. The initial trace (a) has its periodic pattern substantially masked by random noise. The Fourier transform (b) of this trace shows the peaks from the periodic pattern superimposed on the roughly constant background from the noise. To filter the trace, the transform is set to zero except at the peaks (c), and then an image is reconstructed by Fourier inversion (d) in which the periodic trace is much clearer.

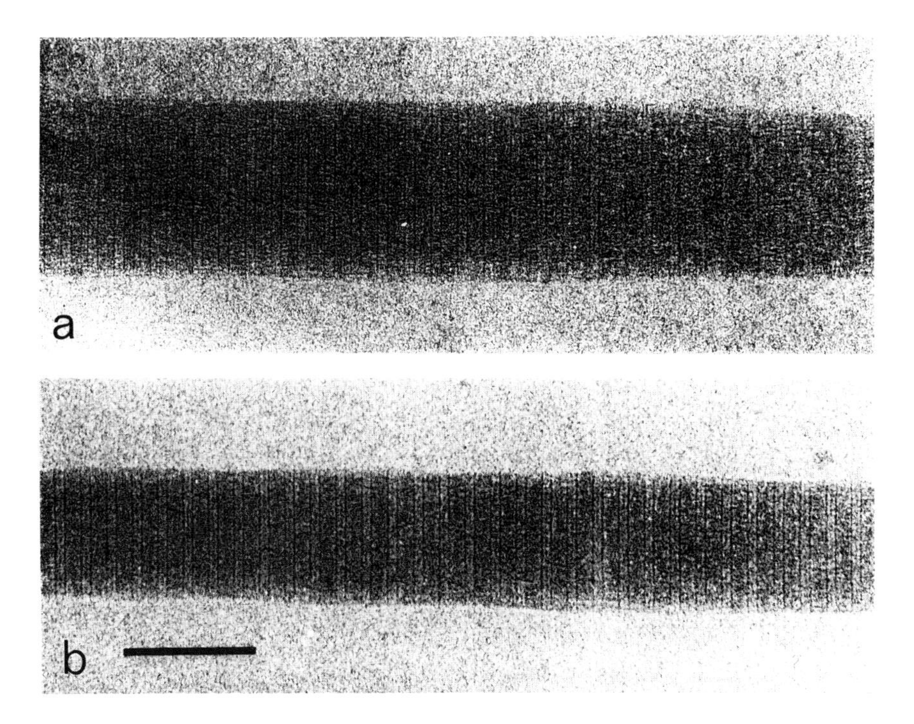

Figure 2.3. Electron micrographs of unstained, frozen hydrated tropomyosin magnesium paracrystals. These are an example of a one-dimensional regular object. The paracrystals in (b) have a mercurial label attached, but it is difficult to see any signal from this by eye and the pattern closely resembles that from unlabelled tropomyosin (a). Reproduced, with permission, from Ref. 10.

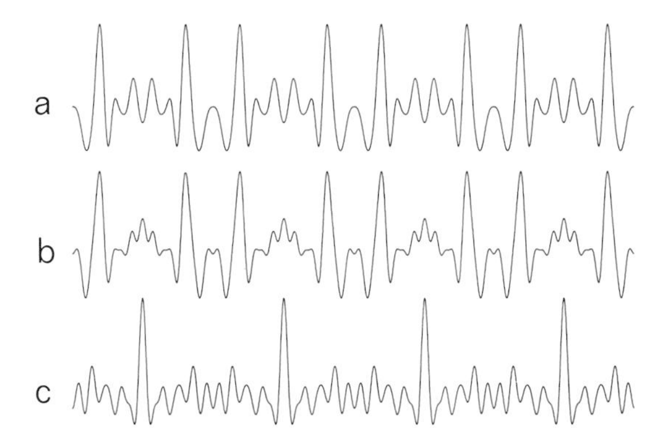

Figure 2.4. Filtered, axially projected density traces from Figure 2.3. The patterns from labelled (b) and unlabelled (a) tropomyosin are now clear and this allows the difference due to the mercury to be located. Both traces are similar, except for a single region. Subtracting (a) from (b) gives a difference trace (c) that corresponds to the signal from the mercury label. Reproduced from Ref. 3.

technically demanding, the former[11] require little expensive equipment. Computer methods, however, have many advantages, and so have now generally superseded optical methods of analysis. To process images by computer, they are digitized on a raster pattern. Processing is usually carried out along the lines suggested by DeRosier and Moore.[12] Fourier transforms are calculated[13] and displayed using either line printers or raster graphics devices. The required parts of the transform are then selected, averaged (internally according to the symmetry of the object, or with data from other micrographs) if required, and a filtered image is produced by Fourier synthesis. The filtered image is displayed either graphically or as a two-dimensional density array.

2.3.2. Regular Two-Dimensional Objects

Two-dimensionally ordered objects can be analyzed in an analogous manner. The object is conceived as a two-dimensional lattice on which a motif is reproduced at each lattice point. Mathematically, this represents convolution[14] of the motif with the lattice. The Fourier transform of a two-dimensional crystal is confined to a small number of peaks, corresponding to the limited number of waves that can contribute to the lattice. The peaks lie on a lattice (the "reciprocal lattice") related to the object lattice. Points on the reciprocal lattice are defined in terms of Miller indices, which are the number of unit translations along each lattice direction that are required to reach the point. Although the reciprocal lattice is evident in the computer Fourier transform of an image, it can be helpful to first obtain it by optical diffraction.[15] Moreover, optical diffraction gives a rapid and reliable indication of the degree of defocus and astigmatism in the micrograph.[16,17]

Figure 2.5 illustrates the effect of noise on image details. As the signal-to-noise ratio decreases, fine features are gradually lost, and eventually only the coarsest features remain. Enhancing the signal-to-noise ratio therefore maximizes the chance of observing fine structural detail. Figure 2.6 shows transforms of a crystal and noise, and a mixture of the two corresponding to a typical electron micrograph. The transform of the crystal is concentrated at lattice points, whereas that of the noise is spread over the entire spectrum. The mixture of the two has lattice peaks superimposed on a relatively continuous background. Reconstructing the image using only the data at the lattice points will clearly eliminate most of the noise.

It is important to ensure that the reciprocal lattice used for processing passes through all peaks in the transform. A common error is to miss weak spots that lie midway between strong reciprocal lattice points and to thereby assign a real-space lattice with half the true spacing. The consequence of this error is to miss subtle differences between adjacent subunits. Figures 2.7 and 2.8 show an example (the sheath of *Methanospirillum hungatei*) in which the difference between

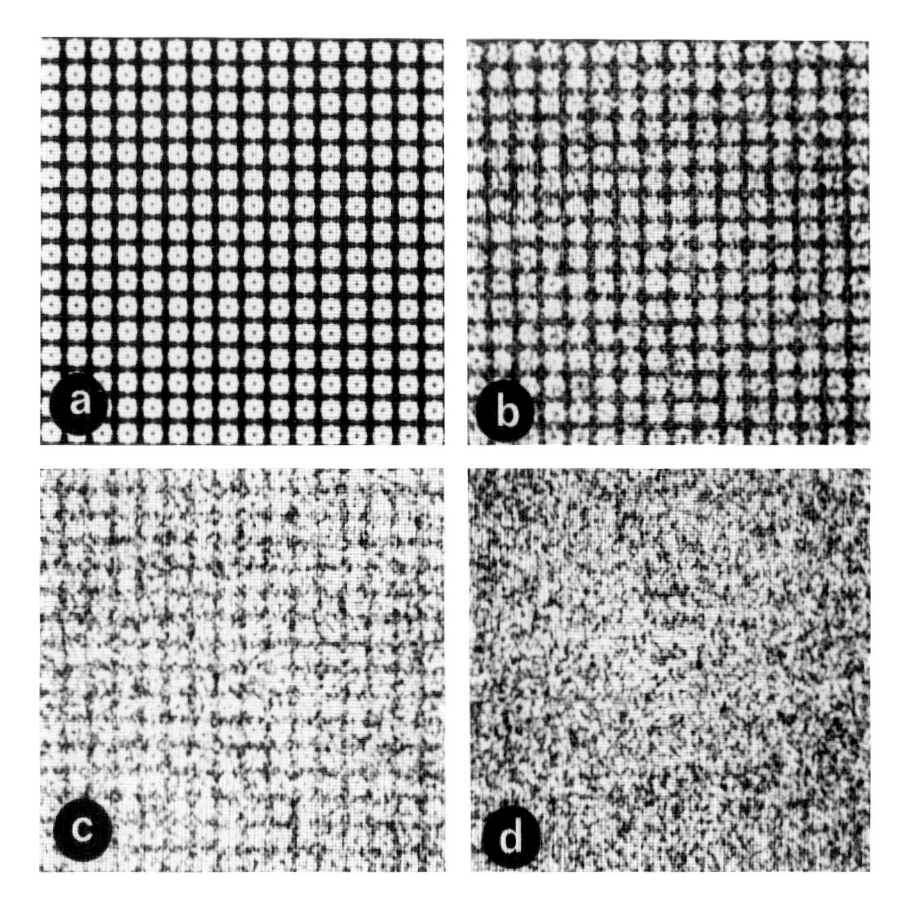

Figure 2.5. Effect of noise on fine structural detail: (a) a model two-dimensional crystal with an equal quantity of noise added (signal-to-noise ratio of 1). Fine details are easily made out. When the signal-to-noise ratio is decreased to 1/10 (b) most of the pattern is still visible, although fine details are harder to see. At a signal-to-noise ratio of 1/20 (c) only the position of the subunits is clear, and by a signal-to-noise ratio of 1/100 (d) even this feature cannot be made out. Electron micrographs usually resemble (b) and (c). Reproduced from Ref. 3.

alternate units is preserved only if the weak peaks that lie midway between the strong lattice peaks are included in the reconstructed image.[18]

2.3.3. Superposition Effects

The interference of patterns from different structural levels can sometimes make it difficult to interpret electron micrographs of regular biological arrays. Because the depth of focus in the electron microscope is large compared to

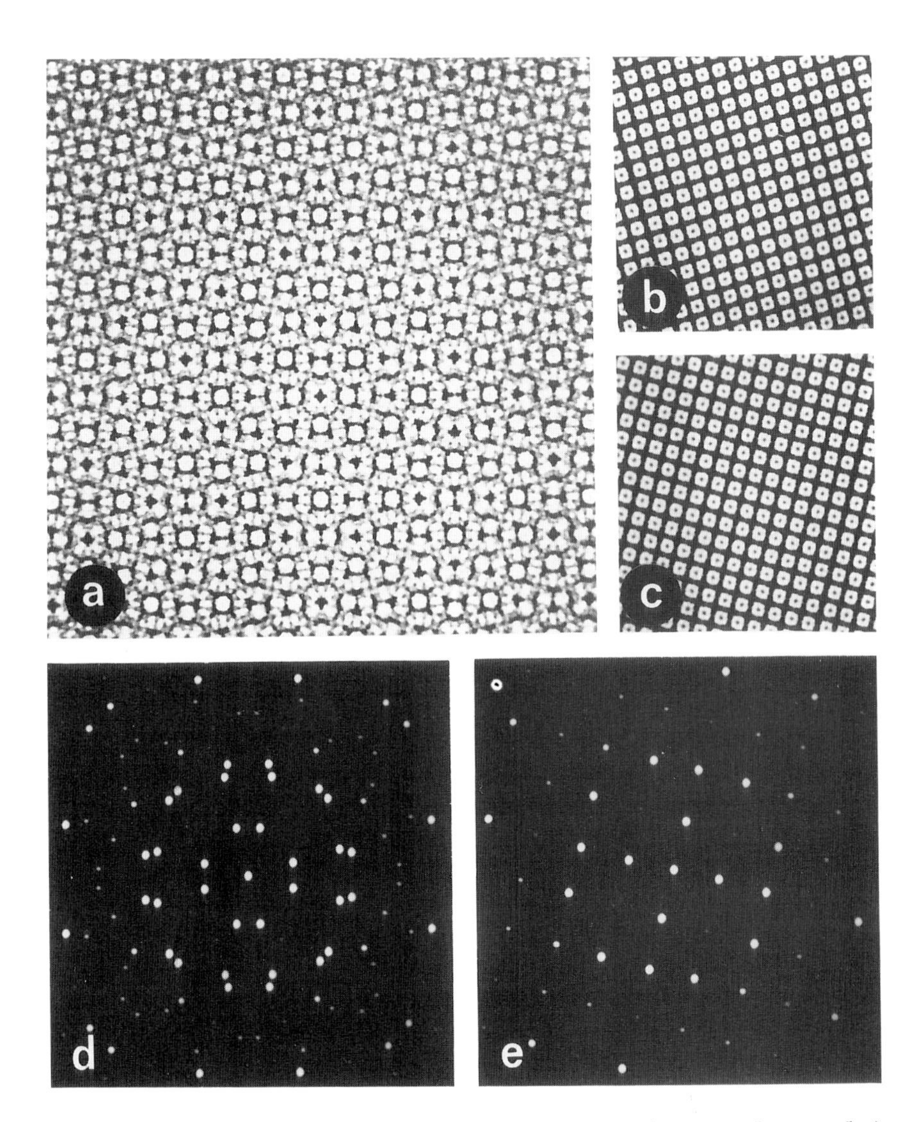

Figure 2.9. A rotational moire pattern (a) is produced by superimposing two regular arrays (b,c) that have been rotated relative to each other. The structure of the subunits can be easily made out in a single layer, but is confused in the moire pattern. The transform (d) of the superimposed layers contains spots from the transforms of each single layer, but these spots do not overlap. Thus, the transform of a single layer (e) can be identified and used to reconstruct a single layer [in this case, (c)]. Reproduced from Ref. 3.

noise been removed by filtering, but so has periodic information from another pattern. Examples of obtaining an image of a single sheet from the moire pattern formed from overlapping sheets included bacteriophage T4 polyheads,[19] the sheath of *Methanospirillum hungatei*,[18] tubular actin crystals,[20] acetylcholine receptors,[21] Ca-ATPase[22] and myosin subfragment-2.[23] It is sometimes possible to separate contributions from overlapping sheets that are related by translation rather than rotation. This requires some knowledge of the structure of the individual patterns, however, because the spots from each reciprocal lattice now overlap. Least squares analyses can be employed to show that the composite pattern is consistent with two overlapping layers.[24]

2.3.4. Contrast Transfer Functions

The mechanism of image formation in an electron microscope is not completely straightforward.[25,26] Overall, not all spatial frequencies in the object are represented with equal fidelity in the image. This is conveniently described by multiplying the Fourier transform of the object by a "transfer function." For most biological objects image contrast has two components, namely, phase contrast and amplitude contrast. Phase- and amplitude-contrast transfer functions describe how each alters the image. Broadly, amplitude contrast derives from electrons scattered outside the objective aperture and is more important at low spatial frequencies, whereas phase contrast derives from interference between scattered and unscattered electrons and is more important at higher spatial frequencies. Both phase- and amplitude-contrast transfer functions vary with microscope defocus, and this variation is responsible for the well-known changes seen on passing through focus.

The phase-contrast transfer function for a microscope can be calculated[25] and Figure 2.10 shows its appearance with different defocus values. The main effect of defocus on the contrast transfer function is to change the spatial frequency at which the first peak and subsequent oscillations occur. By changing the defocus, it is possible to emphasize different spatial frequencies. The oscillating nature of the contrast transfer function at high frequencies causes some terms to be present in the image with their phase reversed and, for this reason, naïve interpretation of high-resolution images is generally not possible. The effect of the phase-contrast transfer function can be seen in transforms of carbon foils (or indeed most biological objects) which show series of rings, known as Thon rings.[16] The position of these rings can be used to establish defocus and the degree to which they are deformed into ellipses defines the extent of axial astigmatism.[17]

For negatively stained biological material down to a resolution of about 2 nm, amplitude and phase contrast approximately complement each other for underfocus values of $\approx$300–500 nm.[25] For most unstained specimens, however, the contribution of amplitude contrast is much lower than the phase-con-

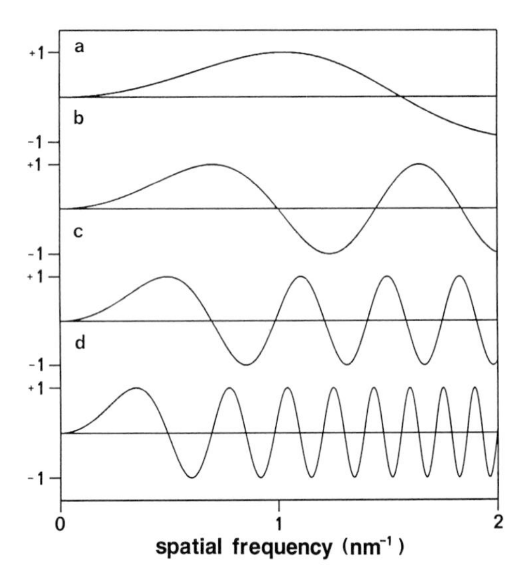

Figure 2.10. Typical variation of contrast transfer function with defocus in an electron microscope. (a)–(d) have defocus values of 125 nm, 250 nm, 500 nm, and 1000 nm respectively. The curves were calculated as described by Erickson and Klug[25] for an accelerating voltage of 100 kV and a microscope spherical aberration constant of 1.3 mm. These are typical values for modern microscopes. Closely related families of curves are produced with different values of the spherical aberration constant and accelerating voltage. Reproduced from Ref. 4.

trast contribution, and so some care is required in interpreting images.[27] Substantial correction is necessary for most higher-resolution work. Although the contrast transfer function alters considerably amplitudes in the Fourier transform, its effect on phase is quite different. The phase is either unaltered (when the transfer function is positive) or reversed (when it is negative). Thus, the phase information in the computed Fourier transform of the image will be easily corrected, provided the defocus is known accurately from the positions of the Thon rings. Rather than correct the computed amplitudes, it is usually simpler and more accurate to obtain amplitudes by electron diffraction, since these data are not modulated by the phase-contrast transfer function.[28]

2.3.5. Objects with Rotational Symmetry

Objects with rotational symmetry can be filtered in a manner analogous to that used for planar crystals. Signal-to-noise ratios can be enhanced by photographically superimposing a number of views of the repeating unit,[29] but determination of the rotational symmetry can be difficult. Moreover, it is not possible to verify objectively that the correct rotational symmetry has been assigned, and so simple rotational superposition is sometimes not very reliable. It

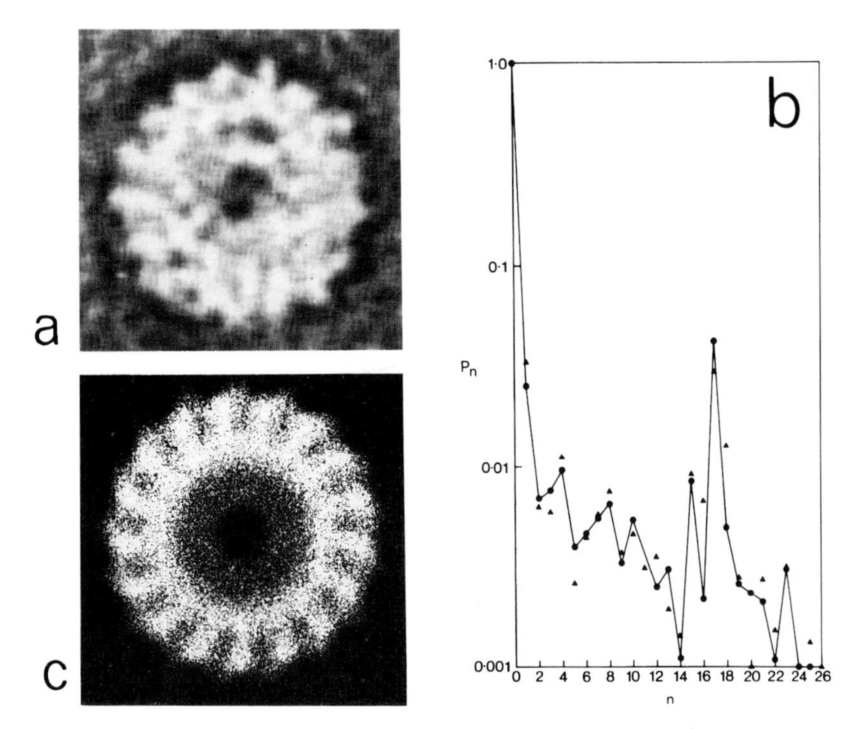

Figure 2.11. Analysis of the stacked-disk form of tobacco mosaic virus in terms of rotational harmonics: (a) electron micrograph of a negatively stained stacked disk, (b) its rotational power spectrum showing a strong 17-fold rotational harmonic, and (c) image reconstructed from the 17-fold rotational harmonics. Reproduced, with permission, from Ref. 30.

is more reliable to analyze the object in terms of a rotational power spectrum,[30] although in some instances a plot of the strength of different rotational frequency terms against radius can be helpful.[31] Figure 2.11 illustrates the application of rotational filtering to the analysis of stacked disks of tobacco mosaic virus.[30] The rotational symmetry of this particle had been controversial, but the rotational power spectrum (Figure 2.11b) shows a clear peak corresponding to 17-fold rotation. Figure 2.11c shows the reconstructed image in which the overall outline of the subunits is substantially enhanced.

2.3.6. Helical Objects

Helical objects have several unique properties that can be used to enhance the information present in their electron micrographs.[5,12] At the simplest level, an interference pattern is produced between the top and bottom of a helical particle, and so the structure and arrangement of subunits is usually not immedi-

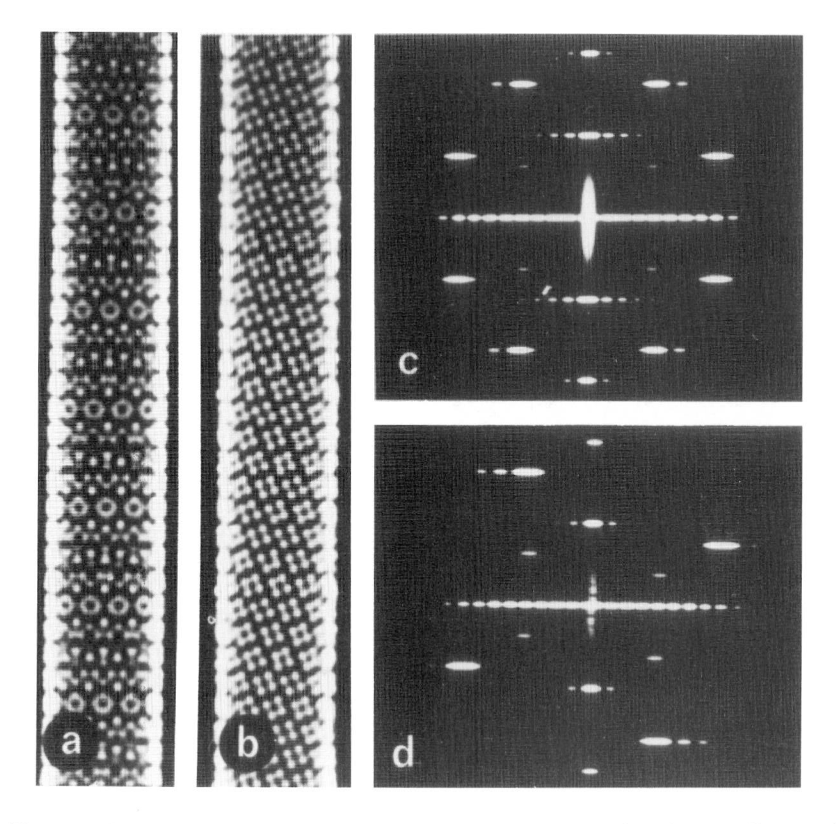

Figure 2.12. (a) Moire pattern produced by superposition of the patterns from the top and bottom of a helical structure. Views of one side (b) show the subunit structure clearly and can be produced by identifying components from one side (d) in the Fourier transform of the helix (c). Note that the helix transform has layer lines instead of the spots seen with crystals. Reproduced from Ref. 3.

ately clear from naïve inspection. Helical diffraction patterns (Figure 2.12) are composed of a number of layer lines,[32–34] and the contributions in Fourier transforms from each side can usually be identified unequivocally, although the analysis of helical diffraction patterns is not quite so straightforward as those from planar crystals. The distribution of intensity along layer lines can be described by sums of Bessel functions[5,32,34] that are related to helical waves in the object. Fraser and MacRae[34] give a good general introduction to the theory of diffraction by helical objects. Because some care is required when indexing helical diffraction patterns, it is wise to discuss these problems with someone who has some experience with this sort of object. Once the diffraction pattern has been successfully indexed, the layer line contributions corresponding to each side can be identified, and so images of a single side can then be reconstructed[11] as illustrated in Figure 2.12.

In three dimensions, the terms deriving from Bessel functions along layer lines in helical diffraction patterns have the very useful property that, at a given radius, their amplitudes do not change with azimuth, whereas their phases change in a predictable way.[32] In principle, therefore, if different Bessel-order terms do not overlap, it is possible to reduce the three-dimensional transform from the two-dimensional one, and thereby produce a three-dimensional model of the structure. Technically, this is most easily achieved using Fourier–Bessel transforms[5,12,32,34] that relate the Bessel terms in the diffraction pattern to helical waves in the object. Quite a range of helical objects has now been reconstructed in this way, and even when Bessel terms overlap, ways have been devised for separating them. A number of examples of this procedure have been recently reviewed.[5]

2.3.7. Three-Dimensional Reconstruction

It is often difficult to appreciate fully the structure of a biological macromolecule from the essentially two-dimensional views produced by electron microscopy. Much more information can be obtained by generating a three-dimensional model, which can be done with crystalline material (and some other ordered structures that have internal symmetry) by combining information from a number of tilted views.[2] In this way, three-dimensional models of structures such as the purple membrane of *Halobacterium halobium,*[35] acetylcholine receptors,[36] cytochrome *c* reductase,[37] actin,[20] gap junctions,[38] coated vesicles,[39,40] and bacterial surface layers, have been produced. The computational and methodological problems associated with this procedure are formidable. It is necessary to merge data from a large number of views and to compensate for focal changes across the micrograph.[2,41] There can also be problems in determining the precise tilt of views. The resolution is limited in a direction perpendicular to the plane of the crystal because data in an "excluded cone" cannot be accessed as the object cannot be tilted sufficiently to be viewed from the side. With particularly favorable specimens, however, such as the purple membrane, it is possible to obtain spectacular results that clearly show up protein secondary structure.[35] If the resolution is more limited, however, the result may not be so impressive, and if only data down to about 2–3 nm is available, the new information may sometimes not seem great compared with the effort required to obtain it.

2.3.8. Correction of Image Defects

Up to this point, image processing has been discussed in terms of essentially infinite and perfect crystals. This ideal object is seldom encountered in practice. Most biological crystals are rather limited in extent and generally are disordered in a number of ways. These effects degrade the Fourier transform, making it harder to extract high-resolution data.

Lattice disorder can be of two types.[34,42] Type I disorder is when each unit cell is randomly displaced from its ideal position, and so actual positions can be considered as distributed about the ideal ones in a Gaussian manner. Type I disorder progressively attenuates higher-frequency terms in the Fourier transform, but does not change the width of the peaks in the transform. Type II disorder involves a correlation between neighboring unit cells. The mean distance between unit cells is specified by a vector which has a randomly distributed error associated with it. Unit cell displacement errors are now cumulative to an extent and long-range order decreases. Type II disorder broadens peaks in the transform as well as attenuating higher spatial frequencies. This broadening can make it difficult to detect weak peaks and necessitates integration around peaks to obtain the true value of the transform. Some type II disorder is present in most electron micrographs of crystalline biological objects (see Figure 2.13).

Because both types of disorder decrease high-resolution information, it is vital to use only the most regular areas for processing. Often the most important determinant of the resolution obtained is the regularity of the objects processed. It is for this reason that so much effort should be put in to obtaining the best possible micrographs. However, it is sometimes impossible (or at least imprac-

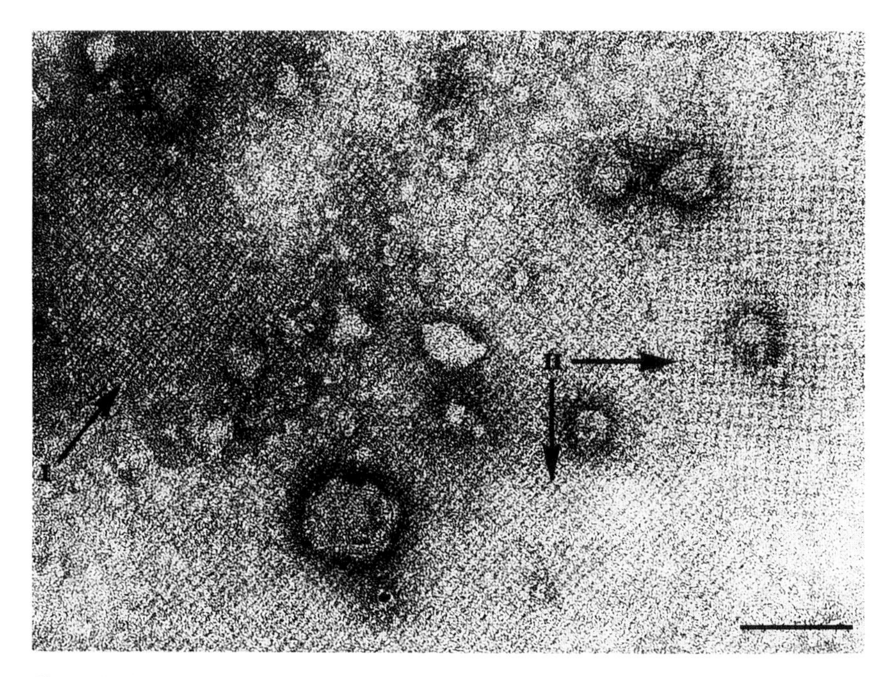

Figure 2.13. Electron micrograph of the A-layer of *Aeromonas salmonicida* negatively stained with phosphotungstate. Two types of pattern are present (I and II) but the lattice order, particularly of the type II patterns, is not very high, and this makes it difficult to obtain a filtered image using simple Fourier methods. Reproduced from Ref. 45.

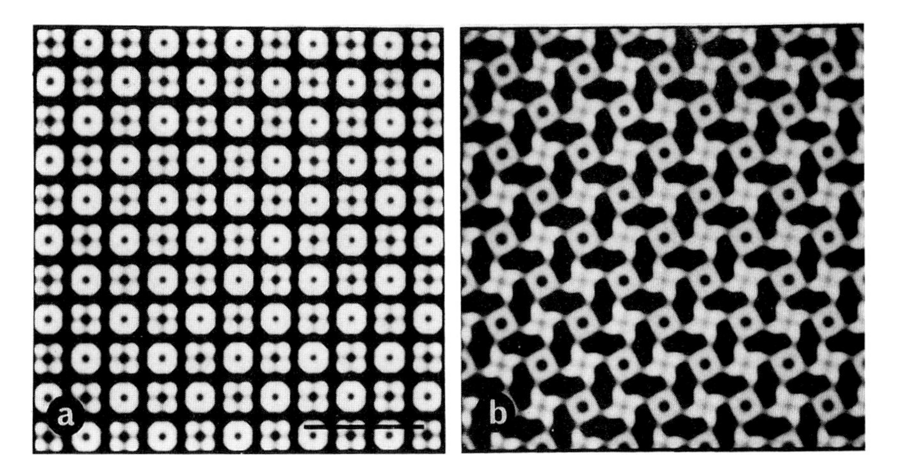

Figure 2.14. Comparison of the two patterns seen in the A-layer of *Aeromonas salmonicida* after correcting for lattice imperfections by locating the position of each unit cell and then forming an average by superposition. Each pattern is constructed from two types of morphological unit, which appear as an annulus and a square planar array of four dots, respectively. The change in the appearance of the pattern is caused by a rotation of the square planar unit. Reproduced from Ref. 45.

tical) to obtain perfectly crystalline arrays, and so computer methods have been devised to at least partially overcome disorder. These methods usually rely on identifying the position of each unit cell in the array and either superimposing them directly or interpolating a new scanning raster to "unbend" the lattice and so increase its perfection.[41,43,44] Figure 2.13 illustrates the application of these methods to disordered arrays of the A-layer of *Aeromonas salmonicida*.[45] This layer shows two types of pattern and filtered images, obtained by superimposing a large number of individual subunits, and show that these patterns are related (Figure 2.14) by rotation of one of the subunits. These cross-correlation methods can also be applied to noncrystalline specimens.[46,47]

2.4. Examination of Frozen Hydrated Material

Recently developed methods for examining frozen hydrated specimens have the potential to allow biological material to be examined by electron microscopy in conditions that closely resemble those in cells. This prospect has excited considerable interest as it offers the chance of eliminating many of the preparative artifacts associated with dehydration and heavy metal staining that have long bedevilled structural investigation of these rather delicate specimens. Moreover, because freezing takes place very rapidly, these methods also offer the chance to examine dynamic systems with high time resolution, and also to examine the effects of solvent composition on the structure and interactions of cellular components and macromolecules. Several recent reviews [27,48,49] have treated the development of methods for examining frozen hydrated specimens by transmission electron microscopy, and so only an outline of the methods will be given here, along with examples of the results that can be obtained with them and an indication of the problems that may be encountered.

2.4.1. Sample Preparation

The key to examining frozen hydrated biological material is to preserve the environment of the specimen by rapid freezing. If freezing is sufficiently rapid, the water in which the material is suspended is trapped in a vitreous, glasslike state.[50–53] Achieving this vitreous state prevents damage to the specimen by ice crystal formation (and associated changes in solvent composition due to eutectic formation) and is fundamental to the success of the method. The demonstration by Dubochet et al.[51] that thin films of vitreous ice could be produced by rapid freezing using simple, though elegant, apparatus transformed the field and opened the way for a whole range of specimens to be investigated. These workers produced thin (100–200 nm) films of vitreous ice containing a range of different biological specimens which, when examined by electron diffraction, gave no evidence for the presence of crystalline ice. Instead of the distinctive pattern of spots or sharp rings expected for crystalline ice, these films gave only broad diffuse rings (indicative of the average spacing between water molecules) characteristic of a vitreous state. Moreover, the films could be examined by electron microscopy and, provided their temperature was maintained sufficiently low, the water did not evaporate to a significant extent in the microscope's high-vacuum environment. As indicated by earlier studies in crystalline ice,[54] the intrinsic density difference between water and biological material provided sufficient contrast to record images without using heavy metal stains. Furthermore, the specimens preserved in theses films of vitreous ice retained a high degree of structural integrity.

The method generally used to produce specimens is based on the work of Dubochet et al.[51] although there have been a number of suggestions for minor improvements.[55–59] Thin aqueous films containing the material of interest are produced on electron microscope grids that have been coated with a fenestrated film. Generally, glow discharging of these grids in an atmosphere of aliphatic amine[51,60] seems to greatly assist the production of aqueous films that are sufficiently thin (100–200 nm). The grids containing films of the required thickness are then rapidly introduced into a suitable cryogen, such as supercooled liquid ethane or propane, which freezes the films sufficiently rapidly to prevent the formation of crystalline ice. The grids produced in this way are then stored under liquid nitrogen until required, when they are transferred to a cryoholder and introduced into the electron microscope at near liquid nitrogen temperature. It is important to maintain the sample temperature below about −140°C, because above this temperature the vitreous ice tends to crystallize.[51] Initially some problems were encountered with inadequate microscope cold holders and anti-contamination traps,[49] but simple modifications have usually enabled these difficulties to be circumvented. Some cryosections of biological specimens have also been examined,[55,61–64] but the freezing of bulk material and subsequent production of cryosections is generally difficult and requires considerable skill and experience. Consequently, most investigations have tended to concentrate on macromolecules and their assemblies frozen in thin films.

2.4.2. Virus Particles

A wide range of frozen hydrated biological specimens has now been examined, and it is illustrative to examine the results that have been obtained in order to evaluate the technique and also to identify those sorts of specimen for which the method seems most appropriate. The first results were obtained using fairly robust particles such as viruses. These had already been extensively characterized by electron microscopy and x-ray diffraction, and so served as useful test objects to demonstrate the potential of the method. Frozen suspensions of both icosahedral and helical viruses had dimensions close to those determined by x-ray studies, and the symmetry of their constituent subunits was well preserved.[65] A key advance was to record images at substantial defocus values (often several micrometers underfocus, or about an order of magnitude greater than is generally employed for biological material[25]), which greatly increased the rather low inherent contrast in these specimens. Under appropriate conditions, even fine structural features, such as adenovirus spikes, could be seen[65] and a three-dimensional reconstruction of the tail of bacteriophage T4 closely resembled that obtained by other methods.[66] Application of the method to more complex virus particles, such as the encapsulated influenza virus[67] and multi-layered structures such as Semliki forest virus[68] has given a number of important insights into their structure. Helical tobacco mosaic virus seems particularly well preserved in ice and it is possible to obtain data down to about 1 nm resolution.[69] Generally the structural information obtained on frozen hydrated virus particles has been at least as good as that obtained by other methods such as negative staining, and in some instances that from the frozen hydrated material has been superior. For example, frozen hydrated particles are generally not so distorted as by drying and this has enabled fine details to be more readily seen, particularly in more delicate examples such as influenza virus.[67] Moreover, some details of DNA packing can sometimes be seen.[48,65,70]

2.4.3. Crystals

Two-dimensional protein crystals, such as bacterial S-layers,[71–73] catalase,[54] and ribosomes[57] seem to be well preserved. Often, the preservation of

Figure 2.15. Frozen hydrated gap junctions, showing the hexagonal array of annular units. Micrograph obtained by E. Gogol and N. Unwin. Reproduced from Ref. 27.

closely packed crystals in vitreous ice seems only comparable to the degree of preservation obtained in crystalline ice, but crystal specimens with a more open structure are frequently superior when preserved in vitreous ice.[57] With most of these frozen hydrated specimens, however, the degree of preservation does not seem to be markedly superior to that obtained by negative staining. This may be because, in a crystal, adjacent molecules probably tend to support each other against the forces encountered in drying, and so preservation is superior to that seen for isolated particles. Much more impressive results have been obtained with frozen hydrated two-dimensional crystals of membrane proteins (Figure 2.15). These crystals appear to be not so well preserved by negative staining and, moreover, techniques using stains to produce contrast cannot give information about the part of the protein molecule that resides in the membrane. Thus, investigation of material such as gap junctions,[38] KATP synthase,[74] and acetylcholine receptors[21,75] in vitreous ice has given important new information about their structures. In the cases of gap junctions and acetylcholine receptors, it has been possible also to produce different physiological states in the material by manipulating the composition of the solution in which samples are frozen. Thus, with gap junctions, images have been obtained in the open and closed configurations that have led to a postulated mechanism relating structural changes to changes in porosity *in vivo*.[38] A similar study on acetylcholine receptors has indicated an analogous mechanism for controlling channel size and raises the possibility of a general mechanism for gating based on the tilting of protein subunits (P. N. T. Unwin, personal communication).

2.4.4. Cytoskeletal and Other Cellular Components

Cytoskeletal components have also been visualized by this method, and aspects of the structure of microtubules[76] and actin filaments[77] have been investigated. Figure 2.16 shows a micrograph of frozen hydrated actin magne-

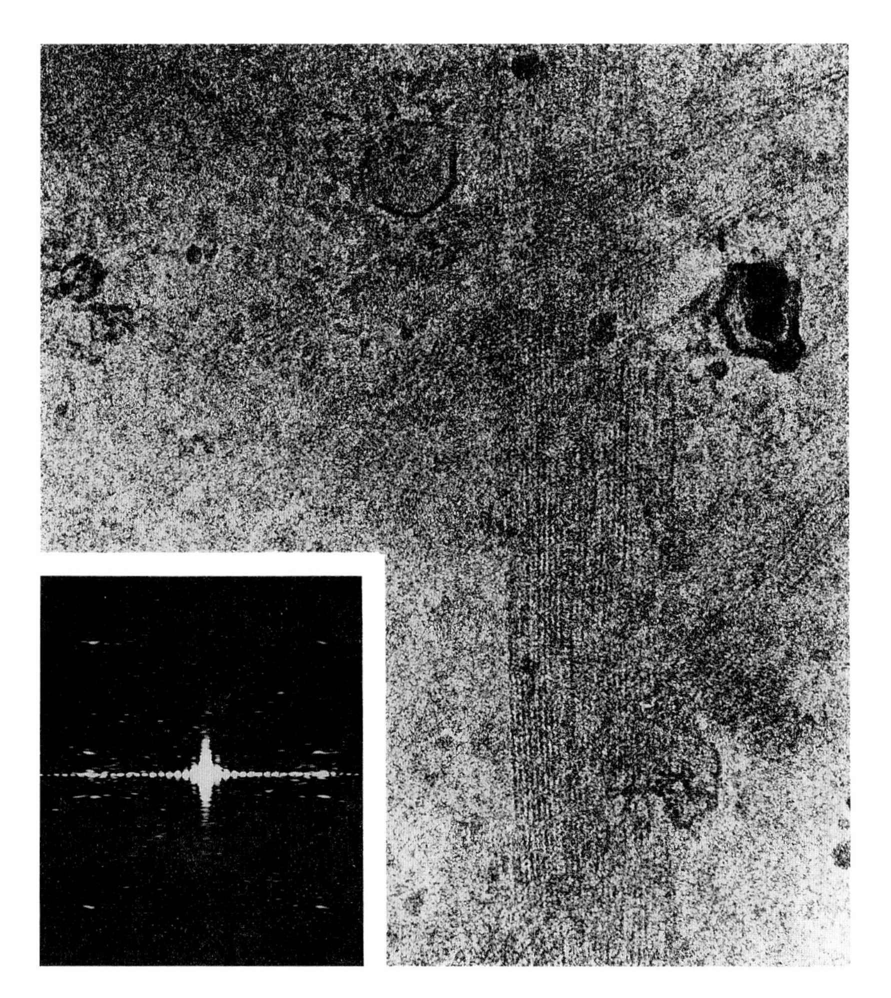

Figure 2.16. Frozen hydrated actin magnesium paracrystals. (Inset) Optical diffraction pattern indicating that the helical structure of the filaments has been well preserved. (Unpublished micrograph courtesy of J. Dubochet and M. Stewart.)

sium paracrystals in which optical diffraction (inset) indicated that the helical structure of the filaments had been well preserved and was comparable to that obtained by negative staining. Tropomyosin paracrystals (Figure 2.3) have also been investigated, where it was shown that it was possible to detect a single mercury atom bound to each molecular chain.[10] An impressive demonstration of the power of the frozen hydrated technique has recently been given by Milligan and Flicker[78] who were able to determine the arrangement of actin, tropomyosin, and myosin heads in decorated thin filaments of muscle. The structure of thick and thin filaments in cryosections of insect flight muscle also seems to be well preserved and offers an exciting possibility of correlating structural changes in this system with changes in x-ray diffraction patterns in contracting muscle.[63] Work is also progressing on the arrangement of microtubules and other components in eukaryotic flagella.[79]

Frozen hydrated specimens of delicate samples, such as chromosomes[64,80] have proved informative because this sort of material is usually poorly preserved by other methods. Similarly, cryosections of bacteria have indicated that the mesosome often observed in conventional plastic-embedded material is probably an artifact.[61] Some quite spectacular images of coated vesicles (Figure 2.17) have been obtained in vitreous ice,[39,40] and these have enabled three-dimen-

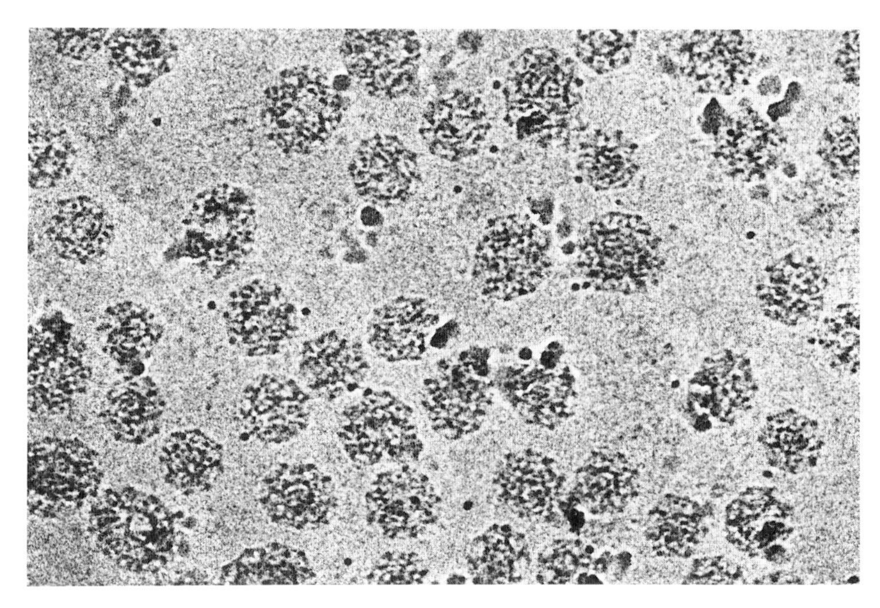

Figure 2.17. Frozen hydrated coated vesicles, in which the pattern of struts formed by clathrin arms and the membranous vesicle within them can be clearly seen. The pattern in individual particles varies according to their orientation. Reproduced from Ref. 27.

sional reconstructions to be produced and the location of different protein components to be established.

2.4.5. Problems of Interpretation

Most micrographs of frozen hydrated material are recorded at high defocus values. This is necessary to enhance the rather low intrinsic contrast between the biological material and surrounding vitreous ice. Such large defocus values can cause quite dramatic changes in image appearance, because of the way defocus alters contrast at different spatial frequencies in an electron microscope (see Section 2.3.4 and Refs. 25 and 26 for detailed discussions of the theory of image formation in electron micrographs of biological material and the effects of defocus on amplitude and phase contrast). Contrast in frozen hydrated objects results almost entirely from phase contrast, whereas most images from stained material contain substantial contributions from both phase and amplitude contrast. This can have important consequences for the interpretation of images. As discussed in Section 2.3.4, phase contrast contributes mainly to the higher spatial frequencies in the image, whereas amplitude contrast contributes mainly to the lower spatial frequencies. In a negatively stained object at optimal defocus, the low-frequency components due to amplitude contrast contributes mainly to the lower spatial frequencies. In a negatively stained object at optimal defocus, the low-frequency components due to amplitude contrast complement the higher-frequency components due to phase contrast. In frozen hydrated objects, however, the contribution due to amplitude contrast is markedly reduced,[81] and so all spatial frequencies are not represented in the final image with equal fidelity. Instead, a band of frequencies (depending on the defocus) is represented strongly, whereas frequencies above and below this band are represented weakly. Thus, those features of the object that lie in a fairly narrow spatial frequency band are represented well in the image, but those outside this band are not. Moreover, because the contrast transfer function of the electron microscope oscillates at high spatial frequencies,[25] some higher-frequency components of the image may be included with reversed contrast. The actual position of the band of maximum contrast transfer depends on the defocus and, for the values of several micrometers commonly employed for frozen hydrated specimens, usually corresponds to resolutions of about 3–6 nm (see Figure 2.10). Consequently, great care is sometimes needed in interpreting higher-resolution features of these objects.[27] Moreover, the lack of low spatial frequency information can cause difficulties in interpretation, particularly with respect to defining particle boundaries and the relative weights of different components of the image. These difficulties are not severe in most crystalline specimens, because there are usually not many spots at low spatial frequencies in the Fourier transforms of these objects. However, as discussed in detail elsewhere,[27] the appearances of bounded objects such as viruses, helices, and small cytoplasmic components, may be substantially altered by these phase contrast effects, and so considerable care is needed when making detailed interpretation of such data. Some attempts have been made to correct for these problems,[66,68,76] but the inherently small signals at low spatial frequencies can make accurate correction difficult.

2.4.6. Radiation Damage

Biological material in vitreous ice is rapidly degraded by electron irradiation (as is the vitreous ice), and so it is necessary to record images under low-dose conditions. Despite considerable work on the subject[82,83] the actual mechanisms giving rise to damage are not completely clear. Generally, severe radiation damage is seen after electron doses of 2000–5000 electrons per square nanometer (about an order of magnitude lower than the dose usually employed to record micrographs of biological material), which means that images recorded often have a rather poor signal-to-noise ratio because the exposure is only just above the fog level of the film. Thus, images of frozen hydrated specimens are usually not so well defined statistically (particularly at higher spatial frequencies) as material in negative stain. Consequently, for frozen hydrated work, it is often necessary to employ crystalline specimens so that Fourier-based image processing methods can be used to enhance the signal-to-noise ratio. This is particularly so when higher-resolution information (better than about 3 nm) is required. Moreover, the radiation sensitivity of frozen hydrated specimens can make it difficult to record tilt series that can be used to produce three-dimensional reconstructions. Consequently, frozen hydrated specimens can usually be used most effectively to obtain information about locations of subunits and changes in structure (either as a result of conformational change or labelling) rather than precise details of molecular outlines.

2.5. Conclusions

Both image processing and methods for examining frozen hydrated material have substantially enhanced the information that can be obtained by electron microscopy about the fine structure and interactions between biological macromolecules. Although considerable effort is often required to analyze regular biological objects by image processing methods, the results obtained can be spectacular. Sometimes insights into the structure are obtained that cannot be achieved by simple inspection. The best results are usually obtained by trying to make superb data even better, rather than to improve poor-quality images, and so these methods should be thought of as an adjunct to careful and precise microscopy rather than as a substitute for it.

Image processing is usually most powerful in detecting positions and orientations of subunits and is often less precise about particle outlines and relative densities. But, in spite of these difficulties, image analysis has proved to be a powerful tool in the analysis of many regular biological objects and has yielded a wealth of quantitative information about the structure, function, and assembly of many biological systems. Of course, precise image processing is of only limited value if the biological material has been altered by the methods employed to produce specimens for electron microscopy, and so it is of great importance that a wide range of biological specimens can be successfully examined in a frozen hydrated state. Because the environment of the material remains close to that of its native milieu, fine structural details are often preserved more faithfully in these specimens. This is particularly so for delicate material such as chromosomes or material that is easily distorted by the forces involved in the usual methods of drying specimens.

Although the contrast in most unstained frozen hydrated biological specimens is quite low, significant information can usually be extracted, particularly if Fourier-based image processing methods are employed. Generally, most impressive results are obtained when the specimens examined are poorly preserved by conventional methods or when the composition of the aqueous phase can be manipulated to introduce structural transformations. Frozen hydrated specimens are particularly useful for examining the structure of proteins in membranes because they enable the whole protein to be seen and not just that portion that projects outside the membrane. Another powerful use of these specimens is in detecting small changes in structure produced by labelling or adding or subtracting subunits. The phase nature of the contrast in electron micrographs of frozen hydrated material can produce problems of interpretation with bounded objects, and so crystalline objects may be an advantage. Certainly the examination of frozen hydrated biological material offers exciting prospects for determining the structure of macromolecules and the manner in which they interact in aqueous solution.

ACKNOWLEDGMENTS. I am most grateful to my colleagues in Cambridge for their many helpful comments, criticisms, and suggestions.

References

1. U. Aebi, W. E. Fowler, E. L. Buhle, and P. R. Smith, "Electron microscopy and image processing applied to the study of protein structure and protein–protein interactions," *J. Ultrastruct. Res.* **88,** 143–176 (1984).
2. L. A. Amos, R. Henderson, and P. N. T. Unwin, "Three-dimensional structure determination by electron microscopy of two-dimensional crystals," *Proc. Biophys. Mol. Biol.* **39,** 183–231 (1982).

3. M. Stewart, in: *Ultrastructure Techniques for Microorganisms* (H. C. Aldrich and W. J. Todd, eds), pp 333–364, Plenum, New York (1986).
4. M. Stewart, "An introduction to computer image processing of two-dimensionally ordered biological structures," *J. Electron Microsc. Tech.* **9,** 301–324 (1988).
5. M. Stewart, "Computer image processing of electron micrographs of biological structures that have helical symmetry," *J. Electron Microsc. Tech.* **9,** 325–358 (1988).
6. R. D. B. Fraser and G. R. Millward, "Image averaging by optical filtering," *J. Ultrastruct. Res.* **31,** 203–211 (1970).
7. M. Stewart, "Location of the troponin binding site on tropomyosin," *Proc. Roy. Soc. Lond. Ser. B.* **190,** 257–266 (1975).
8. M. Stewart and A. D. McLachlan, "Structure of magnesium paracrystals of α-tropomyosin, *J. Mol. Biol.* **103,** 251–269 (1976).
9. M. Stewart, "Structure of α-tropomyosin magnesium paracrystals," *J. Mol. Biol.* **148,** 411–425 (1981).
10. M. Stewart and J. Lepault, "Cryo-electron microscopy of tropomyosin magnesium paracrystals," *J. Microsc.* **138,** 53–60 (1985).
11. A. Klug and D. J. DeRosier, "Optical filtering of electron micrographs: reconstruction of one-sided images," *Nature* **212,** 29–32 (1966).
12. D. J. DeRosier and P. B. Moore, "Reconstruction of three-dimensional images from electron micrographs of structures with helical symmetry," *J. Mol. Biol.* **52,** 355–369 (1970).
13. J. W. Cooley and J. W. Tookey, "An algorithm for the machine calculation of the complex Fourier series," *Math. Comput.* **19,** 297–301 (1965).
14. S. G. Lipson and H. Lipson, *Optical Physics,* Cambridge University Press, Cambridge (1969).
15. A. Klug and J. E. Berger, "An optical method for the analysis of periodicites in electron micrographs, with some observations on the mechanism of negative staining," *J. Mol. Biol.* **10,** 565–569 (1964).
16. F. Thon, "Zur Defokussierungsabhängigkeit des Phasenkontrastes bei der elektronmikroskopischen Abbildung," *Z. Naturforschung* **21a,** 476–478 (1966).
17. F. Thon, in: *Electron Microscopy in Materials Science* (U. Valdre, ed.), pp. 571–625 Academic, London (1971).
18. M. Stewart, T. J. Beveridge, and D. G. Sprott, "Crystalline order to high resolution in the sheath of *Methanospirillum hungatei:* a cross-beta structure," *J. Mol. Biol.* **183,** 509–515 (1985).
19. A. C. Steven, E. Couture, U. Aebi, and M. K. Showe, "Structure of T4 polyheads. II. A pathway of polyhead transformations as a model for T4 capsid maturation," *J. Mol. Biol.* **106,** 187–221 (1976).
20. P. R. Smith, W. E. Fowler, T. D. Pollard, and U. Aebi, "Structure of the actin monomer determined from electron microscopy of crystalline sheets with a tentative alignment of the molecule in the actin filament," *J. Mol. Biol.* **167,** 641–660 (1983).
21. A. Brisson and P. N. T. Unwin, "Tubular crystals of acetylcholine receptors," *J. Cell Biol.* **99,** 1202–1211 (1984).
22. K. A. Taylor, L. Dux, and A. Martinosi, "Structure of vanadate-induced crystals of sarcoplasmic reticulum," *J. Mol. Biol.* **174,** 193–204 (1984).
23. R. A. Quinlan and M. Stewart, "Crystalline tubes of chicken myosin subfragment-2 showing the coiled-coil and molecular interaction geometry," *J. Cell Biol.* **105,** 403–415 (1987).
24. M. Stewart and R. G. E. Murray, "Structure of the regular surface layer of *Aquaspirillum serpens* MW5," *J. Bacteriol.* **150,** 348–357 (1982).
25. H. P. Erickson and A. Klug, "Measurement and compensation of defocusing and aberrating by Fourier processing of electron micrographs," *Phil. Trans. Roy. Soc. Lond. Ser. B* **261,** 105–118 (1971).
26. R. M. Glaeser, "Electron crystallography of biological macromolecules," *Ann. Rev. Phys. Chem.* **36,** 234–275 (1985).

27. M. Stewart and G. Vigers, "Electron microscopy of frozen hydrated biological material," *Nature* **319,** 631–636 (1986).
28. P. N. T. Unwin and R. Henderson, "Molecular structure determination of unstained crystalline specimens," *J. Mol. Biol.* **94,** 425–440 (1975).
29. R. Markham, S. Frey, and G. Hills, "Methods for the enhancement of detail and accentuation of structure in electron micrographs," *Virology* **20,** 88–102 (1963).
30. R. A. Crowther and L. A. Amos, "Harmonic analysis of electron microscope images with rotational symmetry," *J. Mol. Biol.* **60,** 123–130 (1971).
31. M. Stewart, F. T. Ashton, R. Lieberson, and F. A. Pepe, "The myosin filament. IX. Determination of subunit position by computer processing of electron micrographs, *J. Mol. Biol.* **153,** 381–392 (1981).
32. A. Klug, F. H. C. Crick, and W. Whykoff, "Diffraction by helical structures," *Acta Crystallogr.* **11,** 199–213 (1958).
33. D. J. DeRosier and A. Klug, "Reconstruction of three-dimensional structures from electron micrographs," *Nature* **217,** 130–134 (1968).
34. R. D. B. Fraser and T. P. MacRae, *Conformation in Fibrous Proteins,* Academic, New York (1973).
35. R. Henderson and P. N. T. Unwin, "Three-dimensional model of purple membrane obtained by electron microscopy," *Nature* **257,** 28–32 (1975).
36. A. Brisson and P. N. T. Unwin, "Quarternary structure of the acetylcholine receptor," *Nature* **315,** 474–477 (1985).
37. K. Leonard, P. Wingfield, T. Arad, and H. Weiss, "Three-dimensional reconstruction of ubiquinol: cytochrome *c* reductase from neurospora mitochondria determined by electron microscopy of membrane crystals," *J. Mol. Biol.* **149,** 259–274 (1981).
38. P. N. T. Unwin and P. D. Ennis, "Two configurations of a channel-forming membrane protein," *Nature* **307,** 609–613 (1984).
39. G. P. A. Vigers, R. A. Crowther, and B. F. M. Pearse, "Three-dimensional structure of clathrin cages in ice," *EMBO J.* **5,** 529–534 (1986).
40. G. P. A. Vigers, R. A. Crowther, and B. F. M. Pearse, "Location of the 100 kD accessory proteins in clathrin coats," *EMBO J.* **5,** 2079–2085 (1986).
41. R. Henderson, J. M. Baldwin, K. H. Downing, J. Lepault, and F. Zemlin, "Structure of purple membrane from *Halobacterium halobium:* recording, measurement, and evaluation of electron micrographs at 3.5 Å resolution," *Ultramicroscopy* **19,** 147–178 (1986).
42. B. K. Vainstein, *Diffraction of X Rays by Chain Molecules,* Elsevier, Amsterdam (1966).
43. R. H. Crepeau and E. K. Fram, "Reconstruction of imperfectly ordered, Zn-induced tubulin sheets using cross correlation and real space averaging," *Ultramicroscopy* **6,** 7–18 (1981).
44. W. Baumeister, M. Barth, R. Hegerl, R. Guckenberger, M. Hahn, and O. Saxton, "Three-dimensional structure of the regular surface layer (HPI layer) of *Deinococcus radiodurans,*" *J. Mol. Biol.* **187,** 241–253 (1986).
45. M. Stewart, T. J. Beveridge, and T. J. Truss, "Two patterns in the *Aeromonas salmonicida* A-layer that may reflect a structural transformation that alters permeability," *J. Bacteriol.* **166,** 120–127 (1986).
46. J-P. Bretaudiere and J. Frank, "Reconstruction of molecule images by correspondence analysis: a tool for structural interpretation," *J. Microscopy* **144,** 1–14 (1986).
47. J. Frank, "New methods for averaging nonperiodic objects and distorted crystals in biological electron microscopy," *Optik* **63,** 67–89 (1982).
48. J. Dubochet, M. Adrian, J. Lepault, and A. W. McDowall, "Cryo-electron microscopy of vitrified biological specimens," *Trends Biochem. Sci.* **10,** 143–146 (1985).
49. W. Chiu, "Electron microscopy of frozen hydrated biological specimens," *Ann. Rev. Biophys. Biophys. Chem.* **15,** 237–257 (1986).

50. J. Dubochet and A. W. McDowall, "Vitrification of pure water for electron microscopy," *J. Microscopy* **124,** RP3–4 (1981).
51. J. Dubochet, J. Lepault, R. Freeman, J. A. Berriman and J-C. Homo, "Electron microscopy of frozen water and aqueous solutions," *J. Microscopy* **128,** 219–237 (1982).
52. P. Bruggeller and E. Mayer, "Complete vitrification in pure liquid water and in dilute solutions," *Nature* **288,** 569–571 (1980).
53. C. A. Angell and Y. Choi, "Crystallization and vitrification in aqueous systems," *J. Microscopy* **141,** 251–261 (1986).
54. K. A. Taylor and R. M. Glaeser, "Electron microscopy of frozen hydrated biological specimens," *J. Ultrastruct. Res.* **55,** 448–456 (1976).
55. J-J. Chang, A. W. McDowall, J. Lepault, R. Freeman, C. A. Walter, and J. Dubochet, "Freezing, sectioning and observation artifacts of frozen hydrated sections for electron microscopy," *J. Microscopy* **132,** 109–123 (1983).
56. J. Jaffe and R. M. Glaeser, "Preparation of frozen hydrated specimens for high-resolution electron microscopy," *Ultramicroscopy* **13,** 373–378 (1984).
57. R. A. Milligan, A. Brisson, and P. N. T. Unwin, "Molecular structure determination of crystalline specimens in frozen aqueous solutions," *Ultramicroscopy* **13,** 1–10 (1984).
58. J. M. Murray and R. Ward, "Principles for the construction and operation of a device for rapidly freezing suspensions for cryo-electron microscopy," *J. Electron Micros. Tech.* **5,** 279–284 (1987).
59. J. M. Murray and R. Ward, "Preparation of holey carbon films suitable for cryo-electron microscopy," *J. Electron Micros. Tech.* **5,** 285–290 (1987).
60. U. Aebi and T. D. Pollard, "A glow discharge unit to render electron microscope grids and other surfaces hydrophobic," *J. Electron Micros. Tech.* **7,** 29–33 (1987).
61. J. Dubochet, A. W. McDowall, B. Menge, E. N. Schmid, and K. G. Lickfeld, "Electron microscopy of frozen hydrated bacteria," *J. Bacteriol.* **155,** 381–390 (1983).
62. A. W. McDowall, J-J. Chang, R. Freeman, J. Lepault, C. A. Walter, and J. Dubochet, "Electron microscopy of frozen hydrated sections of vitreous ice and vitrified biological samples," *J. Microscopy* **131,** 1–9 (1983).
63. A. W. McDowall, W. Hofmann, J. Lepault, M. Adrian, and J. Dubochet, "Cryo-electron microscopy of vitrified insect flight muscle," *J. Mol. Biol.* **178,** 105–111 (1984).
64. A. W. McDowall, J. M. Smith, and J. Dubochet, "Cryo-electron microscopy of vitrified chromosomes *in situ,*" *EMBO J.* **5,** 1395–1402 (1986).
65. M. Adrian, J. Dubochet, J. Lepault, and A. W. McDowall, "Cryo-electron microscopy of viruses," *Nature* **308,** 32–36 (1984).
66. J. Lepault and K. Leonard, "Three-dimensional structure of unstained, frozen hydrated extended tails of bacteriophage T4," *J. Mol. Biol.* **182,** 431–441 (1985).
67. F. P. Booy, R. W. H. Ruigrok, and E. F. J. van Bruggen, "Electron microscopy of influenza virus," *J. Mol. Biol.* **184,** 667–676 (1985).
68. R. H. Vogel, S. W. Provencher, C-H. von Bunsdorff, M. Adrian, and J. Dubochet, "Envelope structure of Semliki forest virus reconstructed from cryo-electron micrographs," *Nature* **320,** 533–535 (1986).
69. J. Lepault, "Cryo-electron microscopy of helical particles: TMV and T4 polyheads," *J. Microscopy* **140,** 73–80 (1985).
70. J. Lepault, F. P. Booy, and J. Dubochet, "Electron microscopy of frozen biological suspensions," *J. Microscopy* **129,** 89–102 (1983).
71. J. Lepault and T. Pitt, "Projected structure of unstained, frozen hydrated T-layer of *Bacillus brevis,*" *EMBO J.* **3,** 101–105 (1984).
72. R. M. Glaeser and K. A. Taylor, "Radiation damage related to transmission electron microscopy of biological specimens at low temperature: A review," *J. Microscopy* **112,** 127–138 (1978).
73. R. Rachel, U. Jakubowski, H. Teitz, R. Hegerl, and W. Baumeister, "Projected structure of the

surface protein of *Deinococcus radiodurans* determined to 8 Å resolution by cryomicroscopy," *Ultramicroscopy* **20,** 305–316 (1986).
74. E. Gogol, U. Lucken, and R. Capaldi, "The stalk connecting the F1 and F0 domains of ATP synthase visualized by electron microscopy of unstained specimens," *FEBS Lett.* **219,** 274–278 (1987).
75. E. Kubalek, S. Ralston, J. Lindstrom, and N. Unwin, "Location of subunits within the acetylcholine receptor by electron image analysis of tubular crystals from *Torpedo marmorata,*" *J. Cell Biol.* **105,** 9–18 (1987).
76. E-M. Mandelkow and E. Mandelkow, "Unstained microtubules studied by cryo-electron microscopy," *J. Mol. Biol.* **181,** 123–135 (1985).
77. J. Trinick, J. Cooper, J. Seymour, and E. H. Egelman, "Cryo-electron microscopy and three-dimensional reconstruction of actin filaments," *J. Microscopy* **141,** 349–360 (1985).
78. R. A. Milligan and P. Flicker, "Structural relationships of actin, myosin and tropomyosin revealed by cryo-electron microscopy," *J. Cell Biol.* **105,** 29–39 (1987).
79. J. M. Murray, "Electron microscopy of frozen hydrated eukaryotic flagella," *J. Ultrastruct. Res.* **95,** 196–209 (1986).
80. J. Dubochet, M. Adrian, P. Schultz, and P. Oudet, "Cryo-electron microscopy of vitrified SV40 minichromosomes: The liquid drop method," *EMBO J.* **5,** 519–528 (1986).
81. R. Eusemann, H. Rose, and J. Dubochet, "Electron scattering in ice and organic materials," *J. Microscopy* **128,** 239–249 (1982).
82. Y. Talmon, M. Adrian, and J. Dubochet, "Electron beam radiation damage to organic inclusions in vitreous, cubic, and hexagonal ice," *J. Microscopy* **141,** 375–384 (1986).
83. E. Zeitler, *Cryomicroscopy and Radiation Damage,* North Holland, Amsterdam (1982).
84. M. Stewart, "Transmission electron microscopy of frozen-hydrated biological material," *Electron Microscr. Revs* **2**, 117–121 (1989).

3

Radiation Sources for X-Ray Microscopy

A. G. Michette

3.1. Introduction

In this chapter we describe sources that have been used or are being developed for soft x-ray microscopy. The most important of these are synchrotron radiation, plasma sources, and microfocus sources. There are some further processes, such as transition radiation[1] and channelling radiation[2] which cause emission of soft X rays but, as yet, sources that use these have not been constructed. First, however, the difficulties in using conventional (electron-impact) sources are discussed.

3.2. Electron-Impact Sources

The conventional way of obtaining a high-intensity beam of X rays is to bombard a solid target with high-energy electrons, causing x-ray emission via deceleration of the incident electrons through Coulomb interactions with the electrons and nuclei of the material (Bremsstrahlung), and via removal of an inner level atomic electron followed by repopulation of the level. The former gives a continuous spectrum characterized by the incident electron energy, and the latter gives a discrete line spectrum characterized by the target material.

A. G. Michette • Physics Department, King's College, University of London, London WC2R 2LS, United Kingdom.

In Bremsstrahlung the acceleration a is, on average, against the direction of electron motion, and the angular distribution of the radiation emitted into a solid angle $d\Omega$ is, for the electron being brought completely to rest,[3]

$$\frac{dP}{d\Omega}(\theta) = \frac{e^2a^2}{64\pi^2\epsilon_0 c^3} \frac{\sin^2\theta}{\cos\theta} \left\{ \frac{1}{[1 - (v/c)\cos\theta]^4} - 1 \right\} \tag{3.1}$$

Thus, there is no radiation directly forwards ($\theta = 0$), but as the electron energy increases the radiation peaks toward the forward direction. Experimentally, the intensity does not drop to zero at $\theta = 0$ because individual electrons may experience decelerations with components transverse to the direction of motion. This means, in addition, that the polarization is only partial instead of being completely in the plane of the incident electrons. The shape of the continuous spectrum depends primarily on the electron kinetic energy T, the nature of the target having only a small effect. The short-wavelength cut-off

$$\lambda_{\min} = \frac{hc}{T} \tag{3.2}$$

corresponds to an electron losing all its energy in one collision. Most electrons undergo several collisions before coming to rest, leading to a distribution of x-ray energies,

$$I(\lambda) = \frac{AcZ}{\lambda^2} \left[\frac{c(\lambda - \lambda_{\min})}{\lambda\lambda_{\min}} + BZ \right] \tag{3.3}$$

where A and B are constants, and Z is the atomic number of the target material.[4] The wavelength λ_m of maximum intensity is thus

$$\lambda_m = \frac{(3/2)\lambda_{\min}}{1 + (BZ/c)\lambda_{\min}} \simeq \frac{3}{2}\lambda_{\min} \tag{3.4}$$

where the approximation holds for $BZ \ll c(\lambda - \lambda_{\min})/\lambda\lambda_{\min}$. This condition is not satisfied for small T and high Z, when equation (3.3) does not accurately represent the intensity distribution. However, equation (3.4) may still be used to give an estimate of the electron kinetic energies ($\approx$150–3000 eV) needed to give peak wavelengths in the soft x-ray region. The correspondingly low accelerating voltages ($\approx$150–3000 V) can cause operational problems in this type of source.

The conversion efficiency ϵ of incident electron kinetic energy into continuum x-ray energy is given by[5]

$$\epsilon = 1.1 \times 10^{-9}\, ZV \tag{3.5}$$

where V is the accelerating voltage. For high-Z targets (e.g., tungsten, $Z = 74$), operated in the voltage range 150–3000 V, $\epsilon \approx 0.001$–0.025%. The useful efficiency is even lower than this for imaging experiments because only a small angular range can be accepted and, in most cases, only a small wavelength range is required. Even if electron currents of a few amperes could be obtained, giving a target loading of a few kilowatts, only a few hundred usable photons per second would be emitted, resulting in impossibly long imaging times. Practically, target loadings of a few kilowatts can be obtained in rotating-anode sources, but these are for higher operating voltages (leading to peak emission at higher x-ray energies) and correspondingly lower electron currents. It is clear that the continuous x-ray emission from electron-impact sources is not suitable for x-ray microscopy.

A similar conclusion follows from consideration of the x-ray line spectra from such sources, although it has been possible, in some cases, to use line-emission sources for contact imaging (see Chapter 6). An x-ray line is produced when the electron impact causes a vacancy in an inner atomic level, which is subsequently filled by an electron from a higher level that loses energy by radiating a photon. A line cannot appear in the spectrum unless the incident electron kinetic energy T is greater than the ionization energy E_I, but apart from this the wavelengths of the spectral lines are independent of T and depend only on the nature of the target. The total number of X rays emitted in a particular line, however, increases with increasing kinetic energy. The number of K_α x-ray photons emitted per steradian per incident electron from a target of atomic number Z and atomic weight A is given, approximately, by[6,7]

$$N = \frac{Z^4}{Z^4 + 7.5 \times 10^5} \left[\frac{2.8 \times 10^{-2} T^{-0.37}}{A} + \Delta \right] \left(\frac{T}{E_I} - 1 \right)^{1.65} f(\chi) \qquad (3.6)$$

for T and E_I in kiloelectron volts. Δ, a term due to reabsorption of continuum X rays by the target, causing subsequent emission of K_α X rays, is negligible for low-Z targets. The factor $f(\chi) = f(\mu \operatorname{cosec} \theta)$ is to take account of reabsorption of the X rays in the target material and depends on the electron energy, the target material, the x-ray take-off angle θ, and the mass absorption coefficient μ. For a carbon target ($\lambda = 4.47$ nm) and $\theta = 20°$, $f(\chi) \approx 0.5$.[8]

The x-ray flux increases rapidly for T just greater than E_I, but the increase starts to fall off when $T \approx 3E_I$ due to increasing penetration of the electrons into the target which results in greater reabsorption of the X rays. For a carbon target, an accelerating voltage of 10 kV and a take-off angle of 20°, $N \approx 6 \times 10^{-4}$ x-ray photons per steradian per incident electron. Thus, the usable fluxes for x-ray microscopy are again low, due in part to the small apertures of x-ray optical components and in part to the inherently low x-ray production efficiency. A similar conclusion follows from considering the L emission lines of calcium, scandium, titanium, and vanadium (which fall in the water-window range).

3.3. Synchrotron Radiation

Although a full derivation of the properties of synchrotron radiation involves the use of relativistic quantum mechanics,[9] the principal features important for the present discussion may be obtained from semiclassical considerations. A magnetic field applied perpendicularly to the direction of motion of a particle of electric charge e causes it to move in a circular path, so that it is accelerated centripetally and emits radiation. The angular distribution of the radiation emitted into a solid angle $d\Omega$ about the direction $\mathbf{r}$ is given by

$$\frac{dP}{d\Omega}(\theta,\psi) = \frac{e^2a^2}{16\pi^2\epsilon_0c^3}\,\frac{1}{[1-(v/c)\cos\theta]^3}\left\{1-\frac{\sin^2\theta\cos^2\psi}{\gamma^2[1-(v/c)\cos\theta]^2}\right\} \tag{3.7}$$

where γ is the usual relativistic factor, θ is the angle between the particle velocity $\mathbf{v}$ and $\mathbf{r}$, and ψ is the angle between the acceleration $\mathbf{a}$ and the projection of $\mathbf{r}$ onto the plane containing $\mathbf{a}$. This equation shows some of the important properties of synchrotron radiation. The radiation is peaked in the direction of motion ($\theta = 0$), with

$$\frac{dP}{d\Omega}(0,\psi) = \frac{e^2a^2}{16\pi^2\epsilon_0c^3}\,\frac{1}{[1-(v/c)]^3} \tag{3.8}$$

and the radiation intensity falls to half the peak value for $\theta \simeq \gamma^{-1}$. Thus, the radiation is concentrated in a very narrow cone about the forward direction, and

$$\Delta\theta = \gamma^{-1} = \frac{m_0c^2}{E} \tag{3.9}$$

(where m_0 is the rest mass of the particle) is a measure of its angular divergence.

An expression for the total power P radiated in one revolution is obtained by integrating equation (3.7) over all angles ($0 \leq \psi \leq 2\pi$, $0 \leq \theta \leq \pi$),

$$P = \frac{e^2a^2}{6\pi\epsilon_0c^3}\gamma^4 = \frac{e^2a^2}{6\pi\epsilon_0c^3}\left(\frac{E}{m_0c^2}\right)^4 \tag{3.10}$$

i.e., the rate of energy loss is much higher for high energies and for light particles. Synchrotron radiation is usually only significant for electrons and positrons. Purpose-built synchrotron sources (storage rings) almost always use stored circulating beams of electrons, although in some cases positrons can be used[10] because the lifetime of the stored beam is increased—one of the limiting factors on electron-beam lifetime being interactions of the electrons with positive ions trapped by the beam.

The spectral distribution of the radiation may be obtained from the relativistic gyrofrequency ν_g of an orbiting particle in a magnetic field B perpendicular to the particle velocity

$$\nu_g = \frac{eB}{2\pi\gamma m_0} \tag{3.11}$$

Due to relativistic time dilation the particle will appear to return to the same point in its orbit at a frequency

$$\nu_B = \gamma^3\nu_g = \gamma^2 \frac{eB}{2\pi m_0} \tag{3.12}$$

This is the frequency at which a pulse of radiation, of angular width γ^{-1} rad, is observed. The Fourier components of this pulse give the frequency spectrum of synchrotron radiation; since $\gamma \gg 1$ the harmonics involved are very large multiples of ν_g, and the spectrum merges into a continuum. The energy radiated by an electron of energy E over the frequency range $d\nu$ about ν in one revolution is[9,11,12]

$$I(\nu,E)d\nu = \frac{3e^3 B}{15\pi^2\epsilon_0 m_0 c^2}\left[\frac{\nu}{\nu_c}\int_{\nu/\nu_c}^{\infty} K_{5/3}(\eta)d\eta\right] d\nu \tag{3.13}$$

where $K_{5/3}$ is a modified Bessel function.[13] The characteristic frequency ν_c is defined by

$$\nu_c = \frac{3}{2}\nu_B = \frac{3\gamma^2 eB}{4\pi m_0} \tag{3.14}$$

with a corresponding characteristic wavelength

$$\lambda_c = \frac{c}{\nu_c} \tag{3.15}$$

Half the total power is emitted at wavelengths longer, and half at wavelengths shorter, than λ_c. The radiated energy $I(\nu,E)$ is plotted as a universal function of ν/ν_c in Figure 3.1, showing the very large spectral range of synchrotron radiation. The peak emission is at $\nu \simeq 0.3\ \nu_c$, with asymptotic variation $\sim(\nu/\nu_c)^{1/3}$ for $\nu \simeq 0.01\ \nu_c$ and $\sim(\nu/\nu_c)^{1/2}\exp(-\nu/\nu_c)$ for $\nu \geq 10\nu_c$.

Synchrotron radiation is elliptically polarized,[14] except in the orbital plane where it is 100% linearly polarized, but this property has not yet been used in x-ray microscopy. Synchrotron radiation sources, which are now mainly purpose

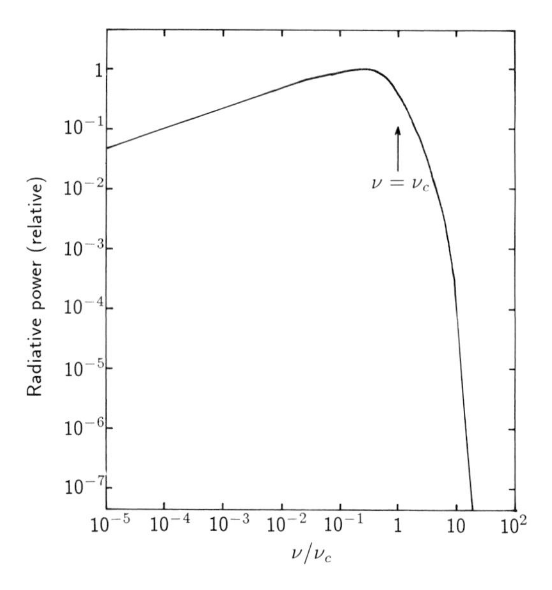

Figure 3.1. The synchrotron radiation spectrum.

built, are described by their characteristics frequencies given by, from equation (3.14),

$$\nu_c = 1.608 \times 10^{17} E^2 B \text{ Hz} \tag{3.16}$$

or their corresponding characteristic wavelengths

$$\lambda_c = \frac{1.864}{E^2 B} \text{ nm} = 0.559\, RE^{-3} \text{ nm} \tag{3.17}$$

where the electron energy E is in gigaelectron volts, the magnetic flux density B is in tesla, and the radius R of the electron orbit

$$R = \frac{(E^2 - m_0^2 c^4)^{1/2}}{eBc} \simeq \frac{E}{eBc} \tag{3.18}$$

is in meters.

The electrons are kept in their orbit by dipole (bending) magnets to give the required orbit radii and quadrupole (focusing) magnets (higher-order configurations can also be used) to prevent beam divergence. To lessen losses through interactions with residual gas molecules, the beam circulates in an ultrahigh vacuum ($< 10^{-9}$ T), and then $1/e$ lifetimes can be several hours (at the SRS Daresbury, $1/e$ lifetimes in excess of one day have been achieved). The energy losses through synchrotron radiation are replaced in one or more radiofrequency (rf) cavities and, because these can only accelerate the electrons when the field is in the correct direction, the electron beam has to circulate in a series of bunches, giving the radiation a pulsed structure with pulse lengths of $\approx$50 ps–1 ns and spacings $\approx$2–20 ns. Because each bunch radiates independently, synchrotron radiation is temporally incoherent for wavelengths smaller than the bunch length, but there is some degree of angular coherence.[15] The rf cavities are also used to accelerate the electrons to their full energy after injection, which takes place at intervals determined by the beam lifetime via a linear accelerator and, perhaps, a smaller (booster) synchrotron.

Higher fluxes of X rays can be obtained from synchrotrons by the use of insertion devices. Wigglers[12] are magnetic structures which, by use of local magnetic fields, cause the electron beam to move in a path with different radius of curvature to that in the dipole magnets. The wiggler field B_W is usually larger than the dipole-magnet field, giving a smaller radius of curvature and a shorter characteristic wavelength. The simplest type of wiggler uses a three-pole magnet and causes a single deviation of the electron beam. The intensity of radiation may be increased by using a multipole wiggler since an experiment can receive radiation from the same point on each of the wiggles. The spectral distribution is the same as that from a dipole magnet except that, at long wavelengths, interference effects become noticeable. Because λ_c is typically much shorter than for dipole magnets, these devices do not normally offer any advantages for soft x-ray microscopy.

Increasing the number of poles further causes interference effects to become dominant, and the device is then called an undulator, of which there are two types. The most common, planar undulators, cause the electrons to oscillate transversely in a plane (Figure 3.2), while helical undulators cause the electrons to describe a helix about the mean path. The wavelengths λ_n of the quasimonochromatic lines are given by

$$\lambda_n = \frac{\lambda_0}{2n\gamma^2}\left(1 + \frac{K^2}{2}\right)\left[1 + \frac{\gamma^2\theta^2}{1 + (K^2/2)}\right] \tag{3.19}$$

where λ_0 is the magnet period, $n = 1,3,5, \ldots$, is the order of interference, θ is the angle of observation with respect to the electron-beam axis, K is the deflection parameter,

$$K = \frac{\epsilon B_0 \lambda_0}{2\pi m_0 c} \tag{3.20}$$

such that $K\gamma^{-1}$ is the maximum angle through which the beam is deflected, and B_0 is the peak field. The spectral width of the peak emission on axis is

$$\frac{\Delta\lambda}{\lambda} \sim \frac{1}{nN} \tag{3.21}$$

where N is the number of undulator periods. By suitable choice of field and period, radiation may be obtained in a desired wavelength range so that undulators can be suitable sources of soft X rays, and photon fluxes several times higher than those from dipole magnets can be obtained, as shown in Figure 3.3.

A schematic diagram of a synchrotron radiation source is shown in Figure 3.4. Beam ports at the locations of dipole magnets or immediately downstream of insertion devices allow the radiation to be directed along beamlines toward monochromators (to select the require wavelength) and experimental stations. Although in principle beamlines and monochromators are separate systems, in practice they should be considered together because, particularly for microscopy where any losses of x-ray flux must be minimized, the beamline optics should be designed to match the acceptance of the monochromator to the source emittances σ_x and σ_y, respectively, in and perpendicular to the plane of the electron orbit.[16,17] The emittance is defined as the product of source size and angular divergence, and is determined by the design and operating conditions of the synchrotron; normally σ_x is limited by apertures in the storage ring or in the part of the beam line closest to the storage ring (the front end). The emittances of the photon beam must also be taken into consideration; these add in quadrature with the source emittances.

A comprehensive account of beamline design has been given by West and Padmore,[18] and it is only necessary here to briefly discuss some monochromators that have been used or are planned for use for x-ray microscopy. Most soft x-ray monochromators use a combination of a reflection grating (for wavelength resolution and perhaps focusing) and a grazing-incidence mirror (for focusing onto an exit slit), and most also have another grazing-incidence mirror prior to the grating to remove hard X rays from the beam.

A schematic design of a toroidal grating monochromator (TGM) is shown in Figure 3.5. In TGMs wavelength scanning is achieved simply by rotating the grating about its central axis, but they quickly go out of focus during scanning. They have poor spectral resolution, but this is not a disadvantage in x-ray microscopy. The scanning microscopy at the National Synchrotron Light Source (NSLS) has been operated on beamlines utilizing a TGM.[19] This monochromator was initially operated with the electron beam in the storage ring defining the source, that is, without an entrance slit. The holographic laminar grating had 600 lines mm^{-1} with grooves 12 nm deep over an area of 15×50

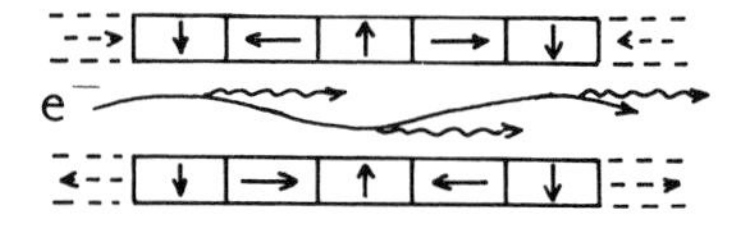

Figure 3.2. A schematic diagram of a planar undulator. Arrows indicate field directions.

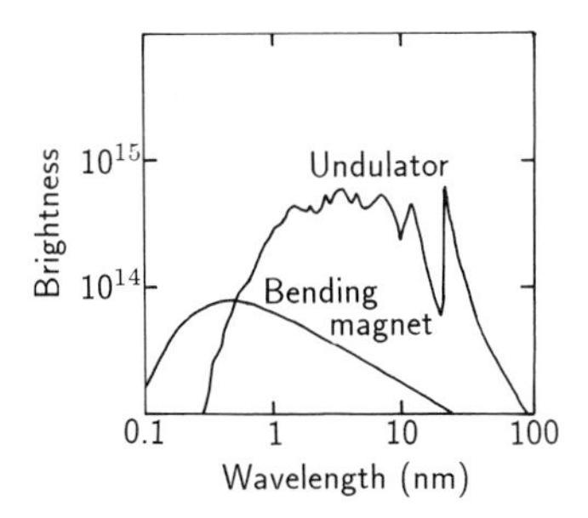

Figure 3.3. Spectral brightnesses of an SRS bending magnet and the undulator. Brightness is defined as the number of photons^{-1} Å^{-1} mrad^{-2} in 0.1% bandwidth.

mm^2; the two radii of curvature were 0.19 m and 88 m, respectively, and the total deviation was 5.72°. This gave a spectral resolution $\lambda/\Delta\lambda \geq 500$, which is sufficiently good for a zone plate microscope, but the efficiency, about 15% at 3.2 nm, reduced with time due to carbon contamination, even though the TGM was housed at a base pressure of about 10^{-10} T. The grating was large enough

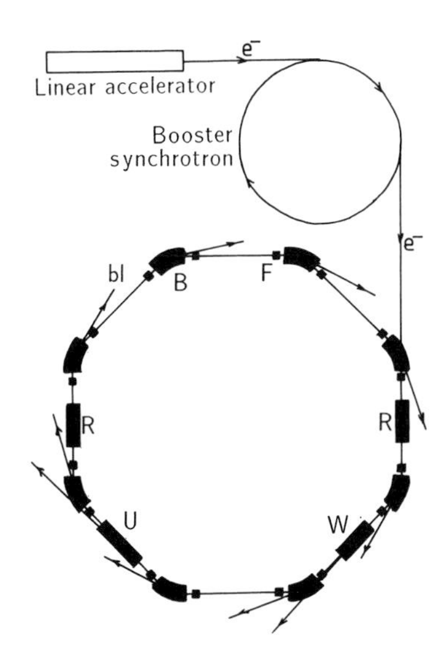

Figure 3.4. A schematic diagram of a synchrotron radiation source. KEY: B: bending magnet; F: focusing magnet; W: wiggler; U: undulator; R: radio frequency cavity; bl: beamline.

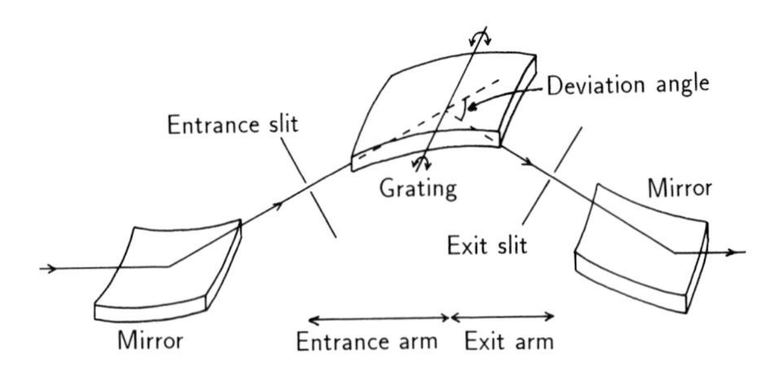

Figure 3.5. A schematic diagram of a toroidal grating monochromator.

however, that uncontaminated areas could be exposed to the beam when necessary, so that frequent refurbishment was not needed. A problem with the operation of this monochromator was the large high-order spectral contamination of about 15% at 3 nm in the output. A similar monochromator has been used with a later version of the same microscope installed on an undulator on the x-ray ring at the NSLS.[20] The principal difference was that a 300 μm pinhole was used as the entrance slit to ensure spatial coherence of the source.

The monochromator used to date for King's College/Daresbury scanning microscope[21] is similar to the SX-700 plane grating monochromator originally developed for spectroscopy at BESSY, Berlin[22] (Figure 3.6). In the Daresbury implementation a spherical mirror has been used in place of the ellipsoidal one for cost reasons. This is not a problem in the use of this monochromator for a zone plate microscope because the beam emittance can still be conserved. The movements necessary for wavelength scanning are rotation of the grating and rotation and translation of the plane mirror and, using a 1200 lines mm^{-1} grating, the wavelength range of the monochromator is 1.2–70 nm. The two major problems with the use of this monochromator for a microscope, both of which lead to significant flux losses, are the relatively high resolution (> 1000 over the water-window range), which is not matched to the bandpass acceptance of the zone plate, and the use of three optical components, giving (at best) an overall efficiency of 2%. Because of these problems another type of monochromator, which should have an efficiency of about 7%, is under construction for this microscope[23] (Figure 3.7). This will use a cylindrical mirror to provide horizontal focusing and spherical diffraction grating for vertical focusing and dispersion. In principle, a cylindrical grating should be used but, since the sagittal focal length of a spherical grating is very much greater than the meridian focal length,[24] this is not critical. The grating can be prepared with areas of different lines density so that by moving it through the x-ray beam optimum

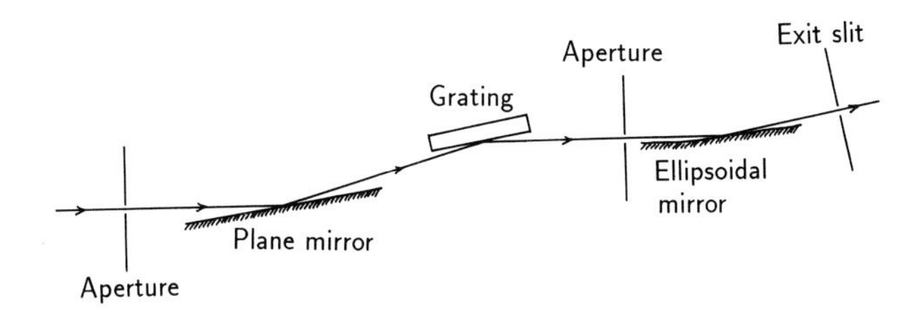

Figure 3.6. A schematic diagram of a SX-700 type plane grating monochromator.

performance, in terms of matching the monochromator and microscope optics, can be obtained for a range of wavelengths.

None of the monochromators discussed above are suitable for full-field imaging microscopy, because this needs a large aperture beamline to allow the simultaneous imaging of the whole field. Large apertures can be provided by zone plate monochromators as shown in Figure 3.8.[25] These rely on the wavelength dependence of the zone plate focal length ($f \propto \lambda^{-1}$), and different wavelengths are selected by moving the aperture along the axis. The central obstruction of the zone plate removes contamination by the zeroeth order, while the aperture removes that from higher orders. The main contamination is then due to higher

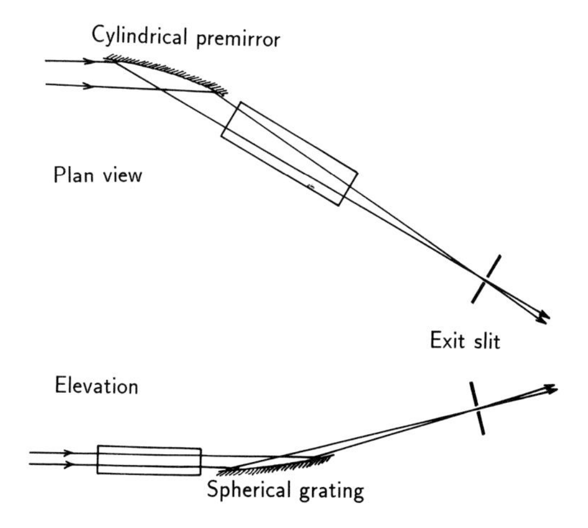

Figure 3.7. A schematic of a spherical grating monochromator.

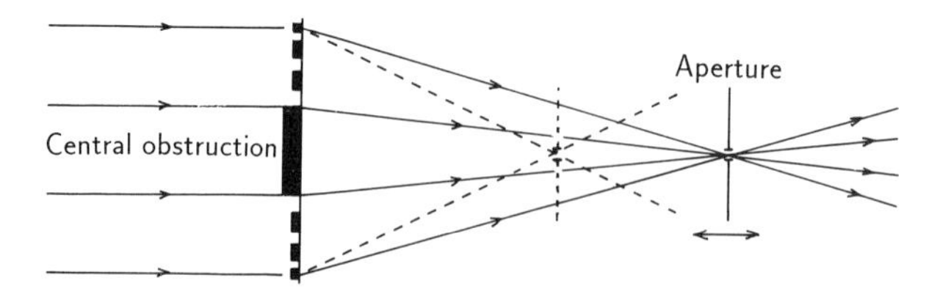

Figure 3.8. A zone plate monochromator.

orders of fractions of the selected wavelength, the third order of the wavelength $\lambda/3$ being focused at the same point at the first order of λ. The spectral resolution is given by[26]

$$\frac{\lambda}{\Delta\lambda} = \frac{r_N}{2\sqrt{2}r_a} \tag{3.22}$$

where r_N is the radius of the zone plate and r_a that of the aperture. For high resolution the aperture must be small. This type of monochromator is used for the microscopes at BESSY.[27]

An alternative type of high-aperture monochromator, based on a multilayer grating, has been proposed.[23] The reflectivities of mirrors and gratings are enhanced at nongrazing angles by the use of multilayer coatings, and the possibility of using reflective optics at larger glancing angles clearly increases the aperture without needing to increase the mirror or grating size. However, monochromators based on these principles have not yet been implemented.

The use of transmission gratings has also been suggested.[28,29] These have not been used in a practical system, and it is not clear whether they will have any significant advantages over reflection gratings. They may, however, be easier to use for microscopes based on laser plasma rather than synchrotron sources (see Section 3.4).

The next generation of synchrotron sources, such as the ESRF in Grenoble, the ALS in Berkeley, and Bessy II in Berlin, will have much higher brilliancies than present sources, and so will allow much shorter imaging times for x-ray microscopy. Despite these high brilliancies, however, it can be argued that x-ray microscopy will never become routine like optical and electron microscopies if synchrotrons remain as the only viable type of source. This is because these sources are normally large, expensive, multiuser facilities remote from an individual user's base. Even the compact synchrotrons now under development[30] are prohibitively expensive compared to sources for other types of microscopy, and thus research into smaller and cheaper sources forms an important part of the current development of x-ray microscopy.

3.4. Plasma Sources

Dense, high-temperature plasmas emit X rays in continua, due to Bremsstrahlung or to an ion capturing a free electron, and in lines, due to transitions between the energy levels of an ion. For sources suitable for microscopy, much of the development has been concerned with using pulsed lasers to produce the plasma.(31–34) Sources suitable for some types of imaging can also be produced by, for example, electron beam–plasma interactions,(35) gas-puff *z*-pinches,(36,37) and dense plasma focus devices.(38–40)

Laser-produced plasmas have been used to produce some striking images by contact x-ray microscopy.(41,42) For this type of imaging all the x-ray photons needed to form the image should be delivered in one pulse (this allows the radiation damage limit to the resolution to be circumvented), and this implies using a large laser, which is likely to be a national facility and suffer from the same types of disadvantage as synchrotron sources do. At soft x-ray energies photoresists have sensitivities in the 10–100 mJ cm^{-2} range,(43) and typically the specimen and protective window will absorb about 75% of the radiation. To protect the specimen and window from target debris, recording has to be done at a distance of about 5 cm or greater so that the corresponding x-ray intensity must be in the range 1–10 J sr^{-1}. Conversion efficiencies(44) of laser intensity into water-window x-ray intensity can be ~10% sr^{-1} for irradiances of $\sim 10^{13}$ W cm^{-2} so that laser pulses with energies in the range 10–100 J are required. Lasers with ≈40 J in 1 ns or 50 ns pulses, focused to spot diameters of ≈300 μm or ≈80 μm, respectively, with corresponding irradiances of $\approx 5 \times 10^{13}$ W cm^{-2} and $\approx 1.5 \times 10^{13}$ W cm^{-2}, respectively, have been used to form images.(41) The requirements for a laser-generated plasma source for full-field imaging x-ray microscopy are similar.

Smaller lasers can, in principle, be used for scanning microscopy.(32–34) Now, the ideal requirements are sufficient photons in a given bandpass to allow imaging of one pixel per pulse, and a high repetition rate to give short imaging times. Measurements using currently available lasers show that ~1 J per pulse at a repetition rate of ~1 kHz should be ideal.(34,45) Such lasers are expected to become available within a few years. (Currently, repetition rates of a few hundred Hertz are available, and even these will be usable.)

A source based on the interaction of an intense electron beam with a dense plasma has been used for contact microscopy.(35,46) In this the plasma was formed in a vacuum of 10^{-6}–10^{-7} torr by surface erosion of polyethylene or Teflon, giving a carbon-rich plasma with strong emission in the water-window region. The plasma density was (2–5) $\times 10^{25}$ m^{-3} and its temperature was about 40 eV. The tightly self-focused 25–50 kV electron beam had a power density of 10^{17} W m^{-2}, giving an energy storage of 40–80 J and an x-ray pulse duration of 60–100 ns. The source size as viewed from the specimen was about 200 μm.

The gas-puff z-pinch has also been suggested as a potential source for contact microscopy.[36,37] Gas is injected via a fast puff valve through a nozzle between discharge electrodes, and switching spark gaps are fired before the gas spreads and before it reaches an insulating surface inside the vacuum chamber. The amount of gas between the electrodes is governed by the delay between opening of the puff valve and triggering the spark gaps, and at optimum the pinch occurs at the peak current, thereby maximizing the energy delivered to the gas and therefore the x-ray intensity from the resultant plasma. Although different gases and compounds can be used, this type of source is limited to materials that can be obtained in gaseous form.

For full-field imaging x-ray microscopy the best potential laboratory source is the dense plasma focus discharge.[38] This consists of a coaxial electrode structure powered by a capacitor bank and produces a high-temperature plasma by cylindrically symmetric pinch compression of a gas. The plasma is strongly dependent on the filling gas (typically, neon, argon, or krypton), and the soft x-ray emission can be enhanced, and the source made more reproducible, by the addition of about 40–50% of hydrogen.[40] Several percent of the energy stored in the capacitor bank can be converted into X rays giving possible pulse energies of several joules (over 4π sr) into a single x-ray line.

3.5. Microfocus Sources

The limitations of conventional electron-impact sources for x-ray microscopy can be overcome in part by the use of microfocus sources.[47] A source of this type been built and tested[48] for use with a Wolter scanning x-ray microscope under construction at the National Physical Laboratory (UK).[49] The source uses a tungsten filament electron gun delivering a current of about 1 μA, followed by stigmator coils, beam-locating deflector coils, aperture-searching scanning coils, and a water-cooled Mulvey lens[50] for fine focusing to give a spot diameter of about 1 μm on an aperture-mounted thin-foil target. A beam of X rays is emitted from the other side of the target. As for rotating-anode sources, the x-ray intensity in line emission is proportional to $(V_e - V_t)^n$, where V_e is the electron gun anode voltage, V_t is the threshold voltage for production of the x-ray line, and (empirically) $n \approx 1.8$. For a 2-μm thick, 300-nm diameter carbon target, with $V_t = 0.285$ kV for the K_α line at a wavelength of 4.47 nm, about 3.5×10^9 photons $\mu A^{-1}\ s^{-1}\ sr^{-1}$ are emitted for $V_e = 20$ kV. The main drawback of this type of source for biological x-ray microscopy is in obtaining a sufficiently intense line in the water-window region (i.e., at a wavelength *shorter* than 4.47 nm). This is not necessarily a disadvantage for the imaging of other types of material.

3.6. Choice of Source

It is clear that each of the types of source discussed here have relative advantages and disadvantages so far as x-ray microscopy is concerned. It is too early in the development of laboratory sources for x-ray imaging to be able to decide which, if any, will allow the techniques to become routine, but it appears likely that laser–plasma sources offer the best possibility of this for at least some forms of imaging. Until such sources have been fully characterized and have become widely available most experiments in x-ray microscopy will continue to take place at synchrotron radiation sources, and it may be that this will always be so for the imaging of certain types of specimen.

References

1. A. N. Chu, M. A. Piestrup, T. W. Barbee, R. H. Pantell, and F. R. Buskirk, "Observation of soft x-ray transition radiation from medium-energy electrons," *Rev. Sci. Instrum.* **51,** 597–601 (1980).
2. Y. A. Bazylev and N. K. Zhevago, "Intense electromagnetic radiation from relativistic electrons," *Sov. Phys. Usp.* **25,** 565–595 (1982).
3. A. H. Compton and S. K. Allison, *X Rays in Theory and Experiment,* 2nd ed., pp. 97–115, Van Nostrand, Princeton, N.J. (1935).
4. B. K. Agarwal, *X-Ray Spectroscopy,* Springer Series in Optical Sciences, Vol. 15, pp. 35–46, Springer, Berlin (1979).
5. A. H. Compton and S. K. Allison, *X Rays in Theory and Experiment,* 2nd ed., pp. 89–90, Van Nostrand, Princeton, N.J. (1935).
6. M. Green and V. E. Cosslett, "The efficiency of production of characteristic x-radiation in thick targets of a pure element," *Proc. Phys. Soc. London* **78,** 1206–1214 (1961).
7. M. Green, in: *X-Ray Optics and Microanalysis* (A. Engström, V. Cosslett, and H. Pattee, eds.), pp. 185–192, Academic, London (1963).
8. M. Green, in: *X-Ray Optics and Microanalysis* (A. Engström, V. Cosslett, and H. Pattee, eds.), pp. 361–377, Academic, London (1963).
9. A. A. Sokolov and J. M. Ternov, *Synchrotron Radiation,* Pergamon, Oxford (1968).
10. Y. Petroff, in: *X-Ray Microscopy,* Springer Series in Optical Sciences (G. Schmahl and D. Rudolph, eds.), Vol. 43, pp. 11–18, Springer, Berlin (1984).
11. J. D. Jackson, *Classical Electrodynamics,* 2nd ed., pp. 672–679, Wiley, New York (1975).
12. S. Krinsky, M. L. Perlman and R. E. Watson, in: *Handbook on Synchrotron Radiation* (E-E. Koch, ed.), Vol. 1, pp. 65–171, North-Holland, Amsterdam (1983).
13. G. N. Watson, *A Treatise on the Theory of Bessel Functions,* 2nd ed., University Press, Cambridge (1944).
14. V. L. Ginzburg and S. I. Syrovatskii, "Developments in the theory of synchrotron radiation," *Ann. Rev. Astron. Astrophys.* **7,** 375–420 (1969).
15. E-E. Koch, D. E. Eastman, and Y. Farge, in: *Handbook on Synchrotron Radiation* (E-E. Koch, ed.), Vol. 1, pp. 1–63, North-Holland, Amsterdam (1983).
16. R. L. Johnson, in: *Handbook on Synchrotron Radiation* (E-E. Koch, ed.), Vol. 1, pp. 173–260, North-Holland, Amsterdam (1983).

17. V. Saile and J. B. West, "VUV and soft x-ray monochromators for use with synchrotron radiation," *Nucl. Instrum. Methods* **208,** 199–213 (1983).
18. J. B. West and H. A. Padmore in: *Handbook on Synchrotron* (G. V. Marr, ed.), Vol. 2, pp. 21–120, Elsevier, Amsterdam (1988).
19. H. Rarback, J. M. Kenney, J. Kirz, M. R. Howells, P. Chang, P. J. Coane, R. Feder, P. J. Houzego, D. P. Kern, and D. Sayre D, in: *X-Ray Microscopy,* Springer Series in Optical Sciences (G. Schmahl and D. Rudolph, eds.), Vol. 43, pp. 203–216, Springer, Berlin (1984).
20. H. Rarback, D. Shu, S. C. Feng, H. Ade, J. Kirz, I. McNulty, D. P. Kern, T. H. P. Chang, Y. Vladimirsky, N. Iskander, D. Attwood, K. McQuaid, and S. Rothman, "Scanning x-ray microscope with 75-nm resolution," *Rev. Sci. Instrum.* **59,** 52–59 (1988).
21. G. R. Morrison, M. T. Browne, C. J. Buckley, R. E. Burge, R. C. Cave, P. Charalambous, P. J. Duke, A. R. Hare, C. P. B. Hills, J. M. Kenney, A. G. Michette, K. Ogawa, A. M. Rogoyski, and T. Taguchi, in: *X-Ray Microscopy II* (D. Sayre, M. Howells, J. Kirz, and H. Rarback, eds.), Springer Series in Optical Sciences, pp. 201–208, Springer, Berlin (1988) (in press).
22. H. Petersen and H. Baumgartel, "BESSY SX/700: A monochromator system covering the spectral range 3 eV $< h\nu <$ 700 eV," *Nucl. Instrum. Methods* **172,** 191–193 (1980).
23. H. A. Padmore, P. J. Duke, R. E. Burge, and A. G. Michette, in: *X-Ray Microscopy II* (D. Sayre, M. Howells, J. Kirz, and H. Rarback, eds.), Springer Series in Optical Sciences, pp. 63–67, Springer, Berlin (1988).
24. A. G. Michette, *Optical Systems for Soft X Rays,* p. 55, Plenum, New York (1986).
25. E. Spiller in: *Workshop on X-ray Instrumentation for Synchrotron Radiation Research* (H. Winick and G. Brown, eds.), SSRL Report No. 78/04, pp. V144–V149 (1978).
26. A. G. Michette, *Optical Systems for Soft X Rays,* p. 253, Plenum, New York (1986).
27. D. Rudolph, B. Niemann, G. Schmahl, and O. Christ, in: *X-Ray Microscopy,* Springer Series in Optical Sciences (G. Schmahl and D. Rudolph, eds.), Vol. 43, pp. 192–202, Springer, Berlin (1984).
28. R. Tatchyn and I. Lindau, "New monochromator designs for the soft x-ray range," *Nucl. Instrum. Methods Phys. Res.* **195,** 163–173 (1982).
29. R. Tatchyn, I. Lindau, and P. L. Csonka, "Optimization of rectangular transmission gratings: Applications to new monochromator design," *Nucl. Instrum. Methods Phys. Res.* **195,** 239–243 (1982).
30. G. P. Williams, "Commercial synchrotron storage rings," *Synchrotron Radiation News* **1**(2), 21–27 (1988).
31. M. L. Ginter, in: *X-Ray Microscopy,* Springer Series in Optical Sciences (G. Schmahl and D. Rudolph, eds.), Vol. 43, pp. 25–29, Springer, Berlin (1984).
32. J. A. Trail and R. L. Byer, "X-ray microscopy using multilayer optics with a laser-produced plasma source," *Applications of Thin-Film Multilayered Structures to Figured X-Ray Optics, Proc. SPIE* **563,** 90–97 (1985).
33. J. A. Trail and R. L. Byer, and J. B. Kortright, in: *X-Ray Microscopy II* (D. Sayre, M. Howells, J. Kirz, and H. Rarback, eds.), Springer Series in Optical Sciences, pp. 310–315, Springer, Berlin (1988).
34. A. G. Michette, R. E. Burge, A. M. Rogoyski, F. O'Neill, and I. C. E. Turcu, in: *X-Ray Microscopy II* (D. Sayre, M. Howells, J. Kirz, and H. Rarback, eds.), Springer Series in Optical Sciences, pp. 59–62, Springer, Berlin (1988).
35. R. A. McKorkle, in: *Ultrasoft X-Ray Microscopy: Its Application to Biological and Physical Sciences* (D. F. Parsons, ed.), Annals of the New York Academy of Science, Vol 342, pp. 53–64 (1980).
36. J. Bailey, Y. Ettinger, A. Fisher, and R. Feder, "Evaluation of the gas-puff z-pinch as an x-ray lithography and microscopy source," *Appl. Phys. Lett.* **40,** 33–35 (1982).
37. P. Choi, A. E. Dangor, and C. Deeney, "Small gas-puff z-pinch x-ray source," *Soft X-Ray Optics and Technology, Proc. SPIE* **733,** 52–57 (1986).

38. G. Herziger, in: *X-Ray Microscopy,* Springer Series in Optical Sciences (G. Schmahl and D. Rudolph, eds.), Vol. 43, pp. 19–24, Springer, Berlin (1984).
39. J. L. Bourgade, C. Cavailler, J. de Mascureau, and J. J. Miquel, "Pulsed soft x-ray source for laser–plasma diagnostic calibrations," *Rev. Sci. Instrum.* **57,** 2165–2167 (1986).
40. Y. Kato and S. H. Be, "Generation of soft X rays using a rare gas–hydrogen plasma focus and its application to x-ray lithography," *Appl. Phys. Lett.* **48,** 686–688 (1986).
41. A. G. Michette, P. C. Cheng, R. W. Eason, R. Feder, F. O'Neill, Y. Owadano, R. J. Rosser, P. T. Rumsby, and M. J. Shaw, "Soft x-ray contact microscopy using laser–plasma sources," *J. Phys. D: Appl. Phys.* **19,** 363–372 (1986).
42. P. C. Cheng, R. Feder, D. M. Shinozaki, K. H. Tan, R. W. Eason, A. Michette, and R. J. Rosser, "Soft x-ray contact microscopy," *Nucl. Instrum. Methods Phys. Res.* **A246,** 668–674 (1986).
43. C. G. Willson in: *Introduction to Microlithography, A.C.S. Symposium Series,* Vol. 219, pp. 87–159 (1983).
44. R. Kodama, K. Okada, N. Ikeda, M. Mineo, K. A. Tanaka, T. Mochizuki, and C. Yamanaka, "Soft x-ray emission from ω_0, $2\omega_0$, and $4\omega_0$ laser-produced plasmas," *J. Appl. Phys.* **59,** 3050–3052 (1986).
45. A. G. Michette, C. P. B. Hills, A. M. Rogoyski, and P. Charalambous, "Laser–plasma sources for scanning x-ray microscopy," *Soft X-Ray Optics and Technology, Proc. SPIE* **733,** 28–33 (1986).
46. B. J. Panessa, R. A. McKorkle, P. Hoffman, J. B. Warren, and G. Coleman, "Ultrastructure of hydrated proteoglycans using a pulsed plasma source," *Ultramicroscopy* **6,** 139–148 (1981).
47. D. J. Pugh and P. D. West, in: *Developments in Electron Microscopy and Analysis* (D. L. Misell, ed.), I.O.P. Conference Series No. 36, pp. 29–32, The Institute of Physics, London (1977).
48. J. Anderson, *X-Ray Microscopy Progress Report,* National Physical Laboratory (UK) Report (1988).
49. A. Franks and B. Gale, in: *X-Ray Microscopy,* Springer Series in Optical Sciences (G. Schmahl and D. Rudolph, eds.), Vol. 43, pp. 129–138, Springer, Berlin (1984).
50. T. Mulvey and M. J. Wallington, "The focal properties and aberrations of magnetic electron lenses," *J. Phys. E: J. Sci. Instrum.* **2,** 466–472 (1969).

4

Amplitude and Phase Contrast in X-Ray Microscopy

D. Rudolph, G. Schmahl, and B. Niemann

4.1. Introduction

When a material object is placed in the path of an electromagnetic wave, the latter undergoes a change in both its amplitude and its phase. In particular, calculations have been performed previously for two extremes[1] to obtain the contrast caused purely by the change in amplitude produced when an object is imaged by soft X rays (an amplitude object) or, alternatively, by the change in phase (a phase object) when imaged in a phase-contrast microscope with coherent illumination. These calculations show that phase contrast exceeded amplitude contrast. In fact, phase-contrast microradiography[2] and phase-contrast topography using x-ray interferometers[3] had already been suggested several years prior to the work reported here. To confirm these calculations, preliminary phase-contrast x-ray microscopy experiments were performed with the x-ray microscope at the BESSY electron storage ring.[4,5]

In this chapter the general case of an object that causes absorption as well as phase shift of the incident radiation in the wavelength range 0.62–5.74 nm is considered.

D. Rudolph, G. Schmahl, and B. Niemann • University of Göttingen, Forschungseinrichtung Röntgenphysik, D-3400 Göttingen, Federal Republic of Germany. This chapter is dedicated to Ulrich Bonse on the occasion of his 60th birthday.

4.2. Amplitude and Phase Contrast

The effect of a material object on the incident electromagnetic wave can be described using the complex refractive index $\tilde{n} = 1 - \delta - i\beta$. The real and imaginary parts of this expression can be related to the linear absorption coefficient μ and the phase shift per unit length η by

$$\mu = \frac{4\pi}{\lambda}\beta \tag{4.1}$$

and

$$\eta = \frac{2\pi}{\lambda}\delta \tag{4.2}$$

where λ is the wavelength of the radiation. The optical constants δ and β can be calculated from the atomic scattering factors $f = f_1 + if_2$, published by Henke et al.[6] using the relations

$$\delta = \frac{r_0\lambda^2}{2\pi} nf_1 \tag{4.3}$$

and

$$\beta = \frac{r_0\lambda^2}{2\pi} nf_2 \tag{4.4}$$

where r_0 is the classical electron radius, and n is the number of atoms per unit volume. A material of thickness t results in a total amplitude transmission $\tilde{T}$,

$$\tilde{T} = e^{-i(2\pi/\lambda)\tilde{n}} = e^{-i(2\pi/\lambda)t}T \tag{4.5}$$

where

$$T = e^{(-2\pi/\lambda)\beta t}e^{i(2\pi/\lambda)\delta t} \tag{4.6}$$

is the product of two terms that describe the absorption of the amplitude of the incident wave in the material and the phase shift relative to a wave propagating in vacuum. In the following only the amplitude transmission T is considered.

Consider a homogeneous object of material of type 2. An incident plane wave with complex amplitude A_0 is transmitted by the object, and the resulting amplitude is A_2. A small particle of material of type 1 which is inserted into material 2 causes an amplitude distribution behind the object which is given by the amplitude A_2 and the amplitude of the diffracted wave A_d. The amplitude A_d can be calculated from the amplitude A_1 directly behind the particle,

$$A_1 = A_0 e^{-(\mu 1/2)t} e^{i\eta_1 t} \tag{4.7}$$

and the amplitude A_2, in the absence of the particle,

$$A_2 = A_0 e^{-(\mu 2/2)t} e^{i\eta_2 t} \tag{4.8}$$

These three amplitudes are related by $A_1 = A_2 + A_d$.

The amplitudes A_0, A_1, A_2, and A_d and their corresponding intensities and photo numbers are referred to an area F in the object plane. The expressions in the image plane, A'_1, etc, are referred to the area V^2F, where V is the magnification. The plane wave with amplitude A_2 is focused onto a phase plate that has thickness t_p and an amplitude transmission

$$T_p = e^{-(\mu_p/2)t_p} e^{i\eta_p t_p} \tag{4.9}$$

In the image plane the amplitude is given by

$$A'_2 = A_2 T_p \tag{4.10}$$

The diffracted wave with amplitude A_d covers, in the back focal plane, an area much larger than that of the phase plate so that A'_1 can be written as

$$A'_1 = A'_2 + A'_d = A'_2 + A_d = A'_2 + (A_1 - A_2) \tag{4.11}$$

because $A_d = A_1 - A_2$. Thus,

$$A'_1 = A_0 \, (e^{-(\mu_2/2)t} e^{i\eta_2 t} e^{-(\mu_p/2)t_p} e^{i\eta_p t_p} + e^{-(\mu_1/2)t} e^{i\eta_1 t} - e^{-(\mu_2/2)t} e^{i\eta_2 t}) \tag{4.12}$$

The intensity of the image of the small particle of material 1 is thus given by

$$\begin{aligned} I'_1 = A'_1 A^{*\prime}_1 = A_0 A^*_0 \{ & e^{-\mu_2 t}(e^{-\mu_p t_p} + 1) + e^{-\mu_1 t} \\ & + 2e^{-(\mu_1/2)t} e^{-(\mu_2/2)t} e^{-(\mu_p/2)t_p} \cos[(\eta_1 - \eta_2)t - \eta_p t_p] \\ & - 2e^{-(\mu_1/2)t} e^{-(\mu_2/2)t} \cos[(\eta_1 - \eta_2)t] \\ & - 2e^{-\mu_2 t} e^{-(\mu_p/2)t_p} \cos(\eta_p t_p) \} \end{aligned} \tag{4.13}$$

and the intensity of the image of the surroundings is given by

$$I'_2 = A'_2 A^{*\prime}_2 = A_0 A^*_0 e^{-\mu_2 t} e^{-\mu_p t_p} \tag{4.14}$$

The contrast K in the image can then be calculated from

$$K = \frac{I_1' - I_2'}{I_1' + I_2'} \tag{4.15}$$

It is now necessary to distinguish between negative and positive phase contrast and negative and positive image contrast in the x-ray region. This distinction will be made using the same definitions as for the visible region of the spectrum, namely:

Positive phase contrast:	Acceleration of the phase of the direct radiation by the phase plate
Negative phase contrast:	Retardation of the phase of the direct radiation by the phase plate
Positive image contrast:	Image detail bright compared to surroundings
Negative image contrast:	Image detail dark compared to surroundings

Compared with visible light it should be noted that, because the refractive index n is less than unity in the x-ray region, a quarter-wave phase plate accelerates the phase of the radiation and produces positive phase contrast. Negative phase contrast can be obtained using a three-quarter wave phase plate. Another way in which X rays differ from visible light is that, in practice, no phase plate can be made without being accompanied by absorption. Image contrast is given by interference of the diffracted spectra with the central-order radiation (A_d' and A_2' of the previous calculation). Normally, the central-order radiation is of much higher intensity than that of the diffracted spectra. Thus, when the phase shift is applied to the central-order radiation, a contrast enhancement is obtained from absorption in the phase plate. For a given material, absorption for negative phase contrast is higher than for positive phase contrast because a $3\lambda/4$ shift requires a thicker phase plate.

It should also be possible to optimize image contrast through the use of different phase plate materials in order to combine the best possible phase shifts and absorption values. To obtain the best contrast for a given combination of an object and its surroundings, the fact that components of amplitude contrast are added to the phase contrast must also be taken into account. This amplitude contrast can also be either positive or negative, thus increasing or decreasing the total image contrast compared to the case of pure phase contrast. In consequence, to obtain optimum contrast for a given object, the choice of positive or negative phase contrast has to be made, depending on the components of amplitude and phase contrast that are expected. Furthermore, the choice of the phase plate material has to be made so that optimum contrast enhancement is obtained.

4.3. Radiation Dosage

For x-ray microscopy experiments, especially of biological specimens, it is not only the image contrast that is important. It is even more important to determine the dosage necessary to detect an object with a given signal-to-noise ratio.

The dosage is given by the ratio of the energy absorbed in the volume tF and the mass ρtF of the object, where ρ is the density. To determine the dosage the first step is to calculate N_0, the number of photons per area F, necessary to produce an image with a given signal-to-noise ratio. As described above, the requirement is to image a small particle with the parameters μ_1, η_1, t, and F—for example, protein—which generates N_1 detected photons, surrounded by material with the parameters μ_2, η_2, and t—for example, water—which generates N_2 detected photons. For these calculations, it is assumed that the detector in the image plane has a detective quantum efficiency of 100%, and that the zone plate has a diffraction efficiency of 100%.

The number of detected photons is given, in general, by $N_i' = kI_i'$, where k is a constant of proportionality. The signal-to-noise ratio (S/N) is given by

$$\frac{S}{N} = \frac{N_1 - N_2}{\sqrt{N_1 + N_2}} = \frac{kI_1' - kI_2'}{\sqrt{kI_1' + kI_2'}} \tag{4.16}$$

that is,

$$\left(\frac{S}{N}\right)^2 = k\frac{(I_1' - I_2')^2}{I_1' + I_2'} = \frac{N_2'}{I_2'}\,\frac{(I_1' - I_2')^2}{I_1' + I_2'} \tag{4.17}$$

However, because

$$N_2' = N_0 e^{-\mu_2 t} e^{-\mu_p t_p} \tag{4.18}$$

it follows that

$$N_0 = \left(\frac{S}{N}\right)^2 e^{\mu_2 t} e^{\mu_p t_p} \frac{I_2'(I_1' + I_2')}{(I_1' - I_2')^2} \tag{4.19}$$

The energy E absorbed by the small particle is given by

$$E = (N_0 - N_1)\frac{hc}{\lambda} = N_0(1 - e^{-\mu_1 t})\frac{hc}{\lambda} \tag{4.20}$$

and the dosage D is given by

$$D = \frac{E}{\rho t F} = \left(\frac{S}{N}\right)^2 \frac{hc}{\lambda \rho t F} e^{\mu_2 t} e^{\mu_p t_p} (1 - e^{-\mu_1 t}) \frac{I_2'(I_1' + I_2')}{(I_1' - I_2')^2} \qquad (4.21)$$

4.4. Results

Using the formalism described above, the image contrast and dosage have been calculated for a 50-nm protein structure and for a 50-nm nucleic acid structure in water surroundings. The calculations were made for a signal-to-noise ratio of 3. Amplitude contrast and dosage for amplitude contrast are obtained by setting $t_p = 0$. Figures 4.1–4.4 show results of the calculations. Figure 4.1 shows the total contrast for a protein structure in water based on negative phase contrast obtained with a nickel phase plate. Compared with the amplitude contrast (dashed line) the total contrast based on negative phase contrast is higher for nearly all wavelengths. Figure 4.3 shows the total contrast based on positive phase contrast for a structure of nucleic acid in water; this is also higher at nearly all wavelengths compared to amplitude contrast (dashed line). The phase plate is adjusted to $\lambda/4$ or $3\lambda/4$ shift as required for any wavelength.

It should be noted that full phase contrast ($\lambda/4$ or $3\lambda/4$ wavelength shift) in combination with amplitude contrast does not necessarily result in the best possible image contrast. Calculations in which the phase shift is a free parameter may lead to slightly improved data. Figure 4.2 shows the radiation dosage that

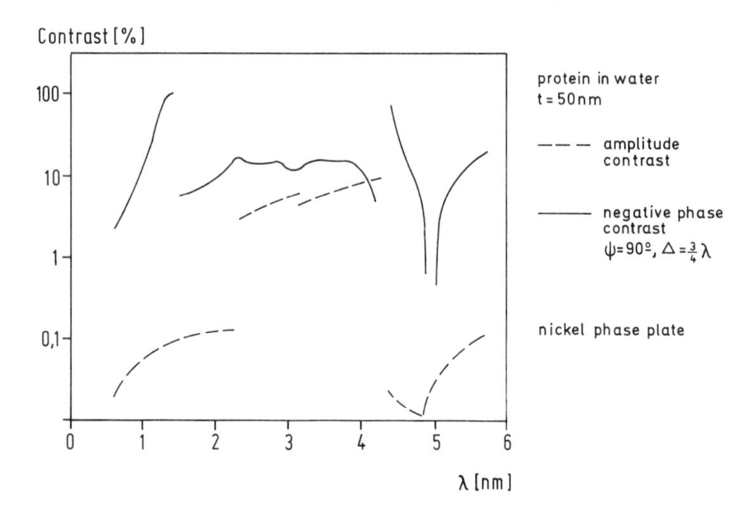

Figure 4.1. Amplitude contrast and total contrast based on negative phase contrast for a protein structure in water.

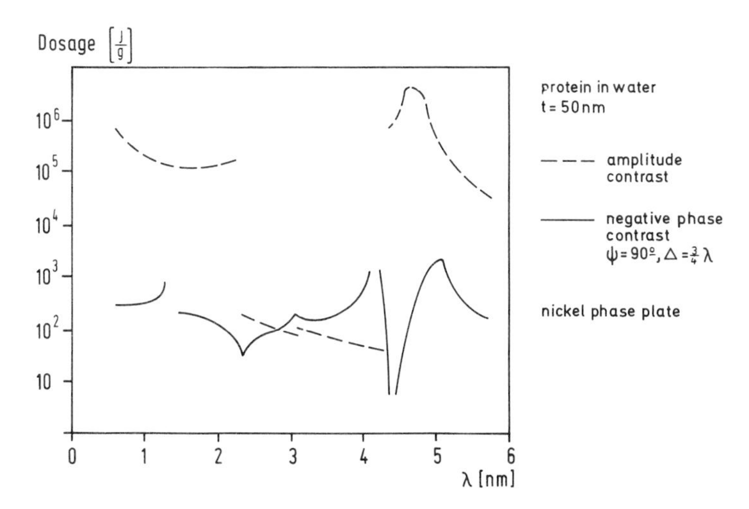

Figure 4.2. Dosage necessary to image a protein structure in water according to Figure 4.1 with a signal-to-noise ratio of 3.

has to be applied to a protein structure of 50 nm in water for negative contrast based on phase contrast compared to amplitude contrast (dashed line), and Figure 4.4 shows the same for a structure of nucleic acid in water. Figures 4.2 and 4.4 show that for most wavelengths the dosage in phase contrast is much lower than in amplitude contrast. It is especially interesting that the dosage at short wavelengths remains approximately constant. This indicates that phase contrast will

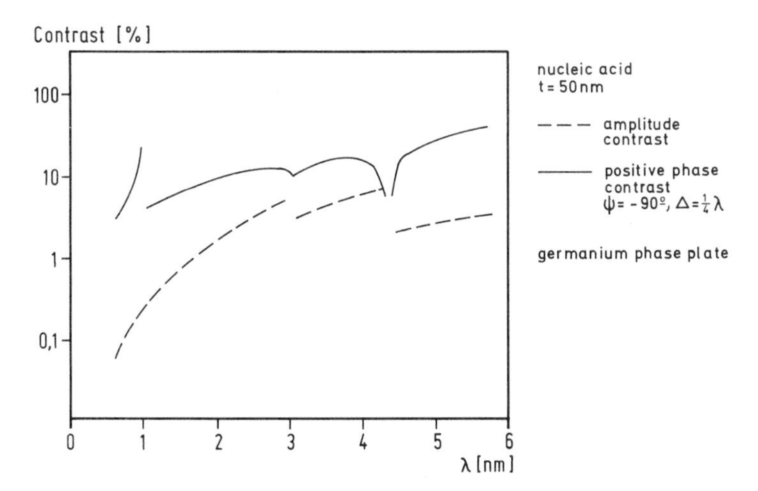

Figure 4.3. Amplitude contrast and total contrast based on positive phase contrast for a nucleic acid structure in water.

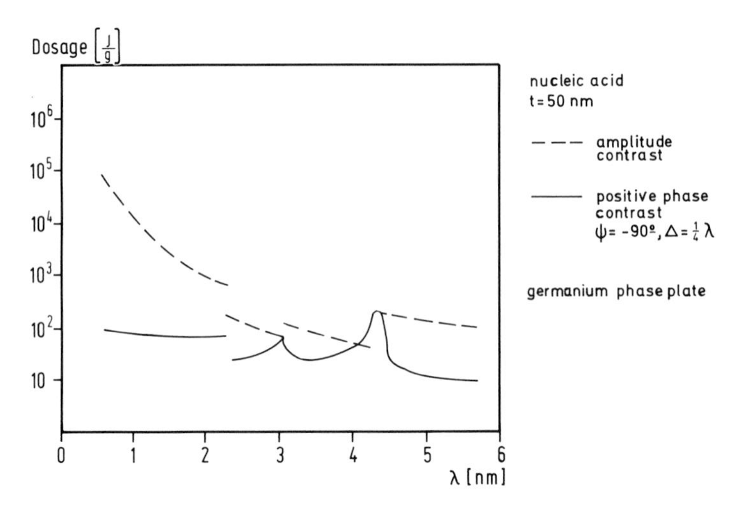

Figure 4.4. Dosage necessary to image a nucleic acid structure in water according to Figure 4.3 with a signal-to-noise ratio of 3.

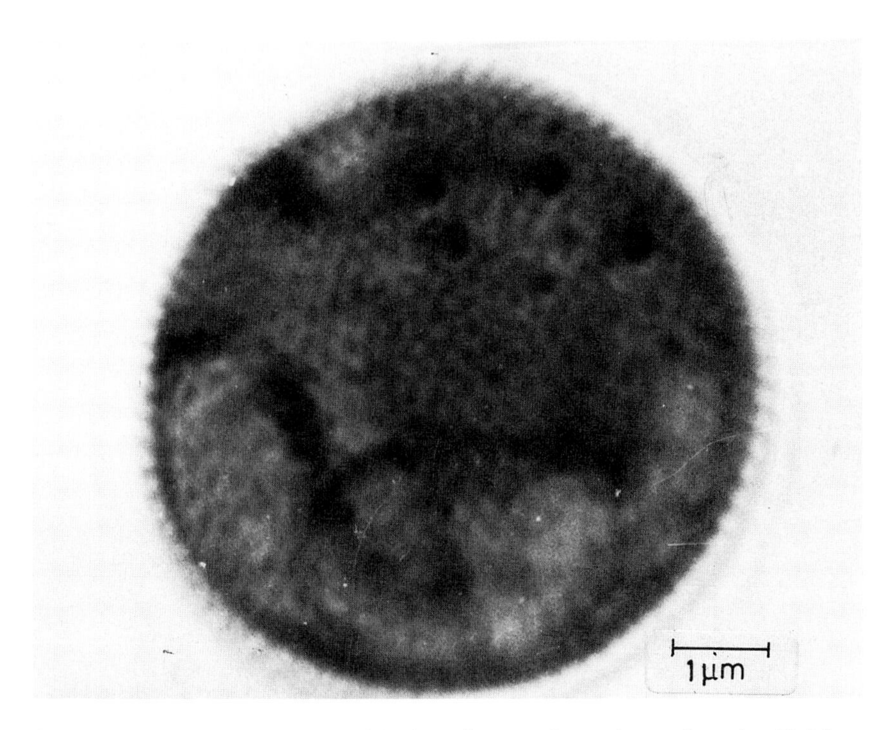

Figure 4.5. A phase-contrast image of the Australian moss *Dawsonia superba* made with 4.5 nm radiation. The x-ray magnification is 350×, and the exposure time was 40 s with an electron-beam current of 136 mA in the BESSY storage ring.

enable x-ray microscopy to be extended to wavelengths shorter than 2.2 nm and possibly even shorter than indicated in Figures 4.1–4.4, which were limited by the data available.[6]

Phase-contrast x-ray microscopy requires coherent illumination so that only the central-order radiation is shifted in phase. However, x-ray microscopes up to now have not allowed coherent illumination without further improvements to the x-ray optical system[4] so that, up to the present time, experiments have been made with partially coherent illumination. As an example, Figure 4.5 shows an image of the Australian moss *Dawsonia superba* made with 4.5 nm radiation. The x-ray magnification was 350×, and the exposure time was 40 s with an electron-beam current of 136 mA in the BESSY storage ring.

ACKNOWLEDGMENTS. The imaging of the spore of *Dawsonia superba* was performed in collaboration with V. Sarafis, Hawkesbury Agricultural College, Australia. The experimental work has been funded by the German Federal Minister of Research and Technology (BMFT) under contract number 05 320DAB.

References

1. G. Schmahl and D. Rudolph, "Proposal for a phase-contrast x-ray microscope," in: *X-ray Microscopy: Instrumentation and Biological Applications* (P. C. Cheng and G. J. Jan, eds.), pp. 231–238, Springer, Berlin (1987).
2. M. Hart and U. Bonse, "Interferometry with X rays, *Physics Today* **23**(8), 26–31 (1970).
3. U. Bonse and W. Graeff, "X-ray and neutron interferometry," in: *X-Ray Optics, Application to Solids* (H. J. Queisser, ed.), Topics in Applied Physics, Vol. 22, pp. 93–143, Springer, Berlin (1977).
4. G. Schmahl, D. Rudolph, and P. Guttmann, "Phase-contrast x-ray microscopy—experiments at the BESSY storage ring," in: *X-Ray Microscopy II* (D. Sayre, M. Howells, J. Kirz, and H. Rarback, eds), pp. 228–232, Springer, Berlin (1988).
5. D. Rudolph, B. Niemann, G. Schmahl, and O. Christ, "The Göttingen x-ray microscope and x-ray microscopy experiments at the BESSY storage ring," in: *X-Ray Microscopy* (G. Schmahl and D. Rudolph, eds.), Springer Series in Optical Sciences, Vol. 43, pp. 192–202, Springer, Berlin (1984).
6. B. L. Henke, P. Lee, T. J. Tanaka, R. L. Shimabukuru, B. K. Fujikawa, "The atomic scattering factor, $f_1 + if_2$, for 94 elements and for the 100–2000 eV photon energy range," in: *Low-Energy X-Ray Diagnostics* (D. T. Attwood and B. L. Henke, eds.), American Institute of Physics Conference Proceedings, Vol. 75, pp. 340–388, New York (1981).

5

Scanning X-Ray Microscopy

C. J. Buckley and H. Rarback

5.1. Introduction

Microscopy using soft X rays has considerable potential for examining a variety of specimens at a higher resolution than is possible with visible light. This is because the wavelengths of soft X rays (1–10 nm) are about two orders of magnitude shorter than visible wavelengths. The predominant contrast mechanism in x-ray microscopy is one of absorption, which strongly depends on both wavelength and atomic number. The resulting sensitivity to the elemental composition of a specimen is particularly advantageous for the examination of hydrated biological specimens that are a few micrometers thick. This is because the absorption of X rays by water in the wavelength range 2.4–4.4 nm is small compared to that by material containing carbon. Thus, wet specimens can be imaged at atmospheric pressure without the need for fixing, sectioning, or staining.

Another advantage of x-ray microscopy is that the amount of damage caused to specimens can be less when imaging with X rays than with electrons. This has been examined in detail,[1,2] and it appears that a resolution of 10 nm may be obtainable for biological specimens in their natural state (wet and unstained) when imaged by soft X rays, while the resolution obtainable with electron microscopy on these specimens is about an order of magnitude worse because of the structure-changing high doses necessary for high resolution imaging of natural specimens. These and other factors[3] provide considerable motivation for the use of soft X rays in microscopy.

C. J. Buckley and H. Rarback • NSLS, Brookhaven National Laboratory, Upton, New York 11794. *Present address for C. J. B.:* Department of Physics, King's College, University of London, London WC2R 2LS, United Kingdom.

A number of approaches can be taken to high-resolution soft x-ray imaging (see Chapters 4, 6, and 7). These are contact microscopy using resists,[4,5] holography,[6,7] diffraction,[8] and scanning and nonscanning x-ray microscopy (using soft x-ray focusing optics). X-ray microscopy using focusing optics is closest to conventional visible light microscopy. The nonscanning microscope uses a condenser/objective arrangement, while in scanning x-ray microscopy, X rays are brought to a small focus which the specimen is scanned over, and the transmitted flux is recorded by a detector such as a gas flow proportional counter.

Of these techniques, scanning x-ray microscopy imparts the least dose to the specimen at comparable signal-to-noise ratios. This is because photons transmitted by the specimen are efficiently detected in a proportional counter instead of being lost by inefficient optics or low quantum efficiency area detectors. Interpretation of images obtained from the scanning microscope is also by far the simplest of the techniques. The image is built up as an absorption map in a serial fashion with the aid of a computer, which makes image processing convenient (Figure 5.1). Images can also be taken on either side of an absorption edge and subtracted to obtain an elemental map.[9]

Up to the present, scanning x-ray microscopy has not had an impact on biological research. This has mainly been due to the lack of suitable x-ray sources and of high-resolution x-ray optics. During the last few years, however, considerable progress has been made in both these areas. It is predicted that in the near future scanning x-ray microscopy will be asked to tackle some interesting problems in cell and tissue culture biology. This chapter introduces the

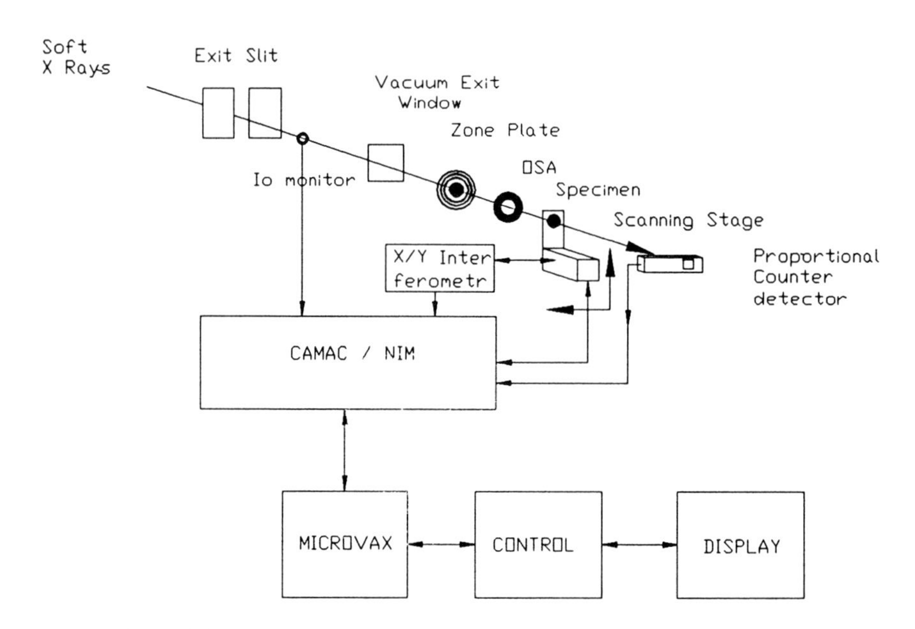

Figure 5.1. Schematic diagram of a scanning transmission soft x-ray microscope.

technology behind the current generation of scanning x-ray microscopes and shows recent results obtained with them.

5.2. X-Ray Optics

The resolution of the scanning x-ray microscope is determined by the x-ray probe size. All practical x-ray sources must be demagnified if submicrometer resolution is to be obtained. It is possible to use a pinhole to produce a small probe but, since all x-ray sources are divergent, this approach discards a large fraction of the useful radiation and requires that the specimens be placed within a micrometer or two of the pinhole if diffraction is not to inflate the spot size. Improved use of the radiation and much greater working distances can be obtained if a focusing device is placed between the source and the specimen. Unfortunately the refractive indices of all materials are close to unity at soft x-ray wavelengths, and refractive lenses of the type used in visible light microscopy are not applicable. This leaves reflection optics and diffraction optics.[10] The reflectivity of X rays is high only for small grazing angles, where hyperbolic and elliptical surfaces are suitable imaging elements.[11] To use mirrors at near normal angles of incidence, multilayer coatings may be employed. [12–14] These layered coatings reinforce reflections at boundaries in much the same way as reinforcements are obtained by Bragg reflection. Multilayers can be deposited on figured surfaces and used to focus soft X rays.

Both of the above methods have been used to produce optical elements for microscopes[15–17] but to date considerably higher spatial resolution has been achieved in microscopes that use Fresnel zone plates to focus the soft X rays.[18–20] A Fresnel zone plate is a circular diffraction grating comprised of concentric opaque and transparent zones (Figure 5.2a). When the zone plate is illuminated by monochromatic radiation the radiation is dispersed into a series of diffraction

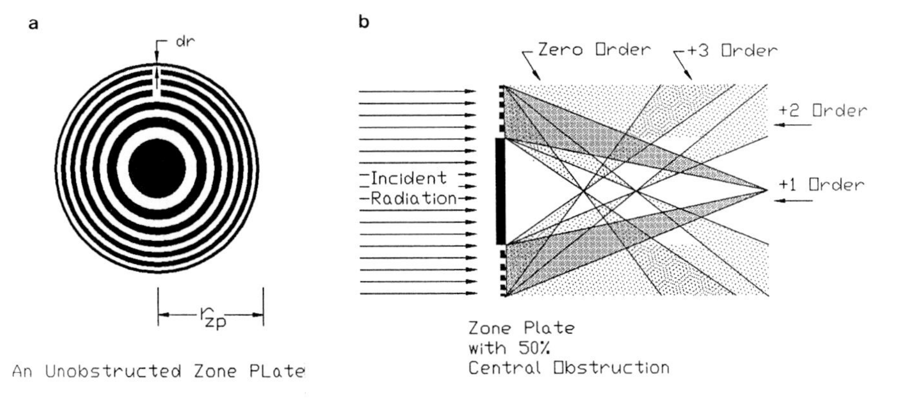

Figure 5.2. (a) A Fresnel zone plate. (b) Diffraction orders from a zone plate.

orders on the zone plate axis (Figure 5.2b). The form of the intensity distributions of these orders on axis is very similar to that produced by a convex refractive lens making the zone plate a viable focusing element for use in scanning and nonscanning x-ray microscopes. The resolving power of an x-ray microscope that uses zone plates is dependent primarily on the width of the finest zone and on the choice of diffraction order used for imaging. To a good approximation the full width at half maximum (FWHM) of the probe produced by a coherently illuminated zone plate is given by

$$\text{FWHM} = \frac{dr}{m} \tag{5.1}$$

where dr is the finest zone width, and m is the diffraction order. A perfectly formed zone plate will produce positive and negative diffraction orders that have the relative intensities

$$I_0 = \left(\frac{1}{2}\right)^2, \; I_m = \left(\frac{1}{m\pi}\right)^2 \qquad m = \pm 1, \pm 3, \ldots \tag{5.2}$$

where I_0 represents the undiffracted zero order, and I_m represents the intensities of the diffracted radiation. Negative values of m represent virtual orders, and it is the real positive orders that are useful in scanning x-ray microscopy. Of these, the positive first order contains the highest intensity (about 10% of the incident radiation) and gives rise to the longest focal length f_1,

$$f_m = \frac{2r_{zp}dr}{m\lambda} \tag{5.3}$$

and is therefore the most practical for the scanning microscope. In equation (5.3) r_{zp} is the radius of the zone plate and λ is the wavelength of the illuminating radiation.

5.2.1. Status of Zone Plate Optics

During the past few years increasing effort has been applied to making Fresnel zone plates to provide focusing elements for soft x-ray microscopes. These zone plates have been made by holography (using UV light) and by electron-beam lithography. The zone plates that have been made using holography have finest zone widths of 55 nm,[21] the limiting factor being the UV wavelengths used.

The smallest zone widths that have been fabricated to date have been made by electron-beam lithography. Electron beams can be focused to produce probes that are only a few nanometers in diameter. Thus, in principle it should be possible to define zone plate patterns that have finest zone widths of this order. There are a number of technical difficulties, however, in defining an accurate zone plate pattern in a substrate which will result in a zone plate useful for soft X rays. Rapid progress has been made in this area in the last 2 years, and useful zone plates now being made by electron-beam lithography have diameters in the 50–100 μm region and finest zones in the range 30–50 nm.[22–24]

It is likely that zone plates that have even finer zone widths will be fabricated in the near future, as processing techniques and electron-beam writing algorithms improve. Also, the advent of high-brightness x-ray sources may facilitate the production of zone plates made by x-ray holography—which could result in zone plates with zone widths of less than 10 nm.

5.2.2. Source Requirements: Coherence and Brilliance

For a zone plate to be able to produce a probe that has the dimensions of the finest zone width, it must be illuminated by X rays that have sufficient spatial and temporal coherence. Ideally, the zone plate would be illuminated by X rays from a point source, and the focal spot would have the form of the point spread function of the zone plate (a modified Airy pattern). If the source has a width of S and the zone plate has a focal length f_1, however, then the zone plate produces an image of the source that is the convolution of the geometric image and the point spread function. The first-order point spread function has a FWHM equal to the finest zone width (dr), and the width (s) of the geometric image is Sd/D, where D and d are the respective distances of source and image from the zone plate (Figure 5.3). All practical sources have dimensions that are large compared to the point spread function of the zone plate and therefore must be demagnified, in which case $d \ll D$ and $d \approx f_1$. Hence,

$$s = S\frac{f_1}{D} \tag{5.4}$$

and, from equation (5.3),

$$s = S\frac{2r_{zp}dr}{\lambda D}$$

or

$$s = \frac{S\,dr\Theta_{zp}}{\lambda} \tag{5.5}$$

where Θ_{zp} is the angle subtended by the zone plate at the source (see Figure 5.3). Clearly, the smaller S or Θ_{zp} the smaller will be the geometric image of the

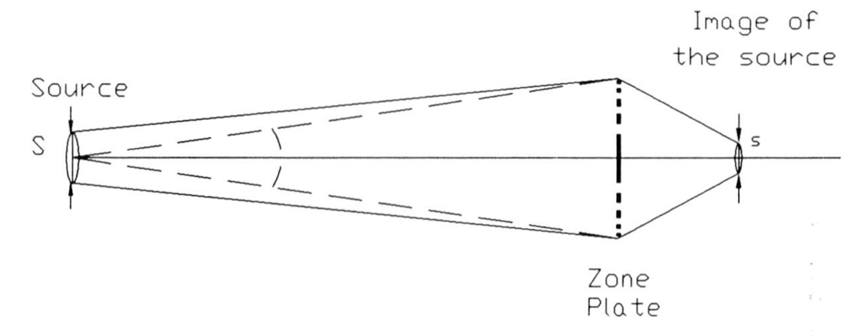

Figure 5.3. Demagnification of an x-ray source by a zone plate.

source. Little is gained from demagnifying the source to dimensions smaller than the finest zone width because then the width of the point spread function dominates. Further, the smaller the geometric image, the smaller the flux contained within the probe. A good compromise between the size of the probe and the flux contained within it is achieved when the width of the geometric image of the source is equal to that of the finest zone ($s = dr$) when the probe should have a FWHM of 1.1 dr.[25] This then leads to the statement that, for the zone plate to perform with near diffraction-limited resolution, the spatial coherence of the source must be such that

$$S\Theta_{zp} = \lambda \tag{5.6}$$

The zone plate is a highly chromatic lens [equation (5.3)], and thus the source must have a sufficiently low bandpass that the width of the probe is not enlarged significantly. Ray constructs lead to a spectral resolving power $\lambda/\Delta\lambda$ requirement of

$$\frac{\lambda}{\Delta\lambda} \geq \frac{r_{zp}}{2\,dr} \tag{5.7}$$

where $\delta\lambda$ is the spread of wavelengths about λ, while recent work[26] suggests that the modulation transfer function is not significantly altered when the spectral resolving power is a factor of two less than this. Provided the requirements on spatial and temporal coherence are met (and the unwanted orders are removed by an aperture and a central stop equal to 40% of the zone plate radius), the FWHM of the probe should be equal to 1.1 dr for a well-made zone plate.

For the microscope to be useful it must be capable of producing images with an acceptable signal-to-noise ratio in a reasonable time. Consider an image of a specimen that has 20% contrast and 256^2 pixels. For the signal-to-noise level to be acceptable (e.g., $\geq 5:1$) about 1000 transmitted photons per pixel are required on average. Thus, for the whole image, a total of about 7×10^7 photons are required. To produce such an image in, say, 30 minutes, about 10^5 photons s^{-1} are required at the zone plate focus if the specimen is 50% transmissive on average. Because of the rather low efficiencies of practical amplitude zone plates (typically 1–5% in the first order), and monochromators (typically 5%), the x-ray source must supply a minimum of 10^8 spatially and temporally coherent photons each second. The brilliance of a source is a measure of the coherent flux and is measured in units of photons per second per solid angle per source area in a given spectral bandwidth. With the current generation of x-ray optics the source brilliance required is 4×10^{12} photons s^{-1} $mrad^{-2}$ mm^{-2} per 0.1% bandpass at a wavelength of about 3.5 nm. Until recently very few sources could deliver soft X rays with this brilliance. The following section introduces the current and forthcoming generation of sources which make practical scanning x-ray microscopy possible; for a fuller description see Chapter 3.

5.3. X-Ray Sources

A traditional x-ray source is the rotating anode electron bombardment source, which has been used widely in, for example, x-ray diffraction. Unfortunately, this type of source has a soft x-ray brilliance of only 10^6–10^7 photons s^{-1} $mrad^{-2}$ mm^{-2} per 0.1% bandpass. However, much brighter sources are becoming available. These are plasma sources and synchrotron sources. The former can be produced, for example, by pulsed laser bombardment of a suitable target and by focused gas discharges. Laser–plasma sources are under development, and have demonstrated time-averaged brilliances of 10^{12}–10^{13} photons s^{-1} $mrad^{-2}$ mm^{-2} per 0.1% bandpass. Such plasma sources have the drawback that they produce debris that have to be removed from the x-ray beam—for example, by means of relay optics. There is some question of the usefulness of laser–plasma sources for scanning x-ray microscopes, though a feasibility study by Michette et al.[27] indicates that further improvement in the repetition rate of laser–plasma sources could lead to a viable one-shot-per-pixel pulsed source. X-ray lasers are another form of plasma source[28] but, while these have good temporal coherence, their spatial coherence is poor, and they do not have sufficiently high average brilliances to be used for scanning x-ray microscopy at present. Plasma focus sources[29] are also under development, but these are insufficiently stable and have too low a repetition rate to be useful at present.

Synchrotrons are currently the brightest sources of soft X rays. A highly collimated x-ray beam is emitted from the relativistic electrons as their path is changed by bending magnets that keep the electrons circulating in the ring. The actual brilliance of a synchrotron source depends on electron-beam parameters such as energy and size. Bending-magnet sources typically have a soft x-ray brilliance of 10^{12}–10^{14} photons s^{-1} $mrad^{-2}$ mm^{-2} per 0.1% bandpass and have been used for scanning x-ray microscopy.[3,30] A recent innovation in synchrotron technology has been the use of multiperiod magnets (known as undulators) to produce even brighter sources of soft X rays. An undulator can be placed in a straight section of the storage ring where the electrons wiggle as they traverse the alternating magnetic field, causing emission of a highly collimated x-ray beam in the forward direction. Recently synchrotron/undulator sources have been used for scanning x-ray microscopes at the Brookhaven NSLS X17T beamline and at the Daresbury SRS 5U beamline. In both cases the monochromator optics initially used to relay the X rays to the zone plates were poorly matched and caused a sizeable reduction in the coherent flux. However, new optics have been designed for both sources to make use of the high brilliances (10^{14}–10^{15} photons s^{-1} $mrad^{-2}$ mm^{-2} per 0.1% bandpass at the SRS 5U and 10^{16} photons s^{-1} $mrad^{-2}$ mm^{-2} per 0.1% bandpass at NSLS X1). The following is a description of the scanning microscopes that have been used with synchrotron radiation and early results from them.

5.4. Scanning X-Ray Microscopes at Synchrotron Sources

5.4.1. The King's College Microscope at Daresbury

The high-resolution contamination-written zone plates made at King's College London give the potential for sub-50 nm resolution, while the 5U undulator beamline at Daresbury provides a soft x-ray source of sufficient brilliance to give acceptable imaging times. The 5U beamline has a 10-period undulator and the x-ray flux is spectrally tuned by means of an SX700-type monochromator.[31] The exit slit of this monochromator serves as a low-bandpass source of X rays which is demagnified by a zone plate 1 m downstream of it. The unwanted orders produced by the zone plate are removed by an aperture mounted on a three-dimensional stage. The specimen is located at the first-order focus of the zone plate and is scanned by a piezoelectric stage whose position can be monitored to an accuracy of about 10 nm by induction transducers. This fine stage has a range of 40 μm and is mounted on a coarse stage that has a range of a few millimeters. The detector is a gas flow proportional counter mounted behind the specimen (Figure 5.4). Initial alignment of the specimen and zone plate optics is done with a travelling light microscope in conjunction with a Michelson interferometer which, by forming white light fringes on the specimen, can be used to locate specimens at the zone plate focus without the need for focal scans.[32] Stage scanning, data acquisition, and display are controlled by a microcomputer.

The SX700 in use on the 5U beamline is not well suited to scanning x-ray microscopy and does not deliver the full potential of the source flux to the zone plate. This has made it impractical to use the low-efficiency carbon zone plates.

Figure 5.4. The KCL/Daresbury STXM showing (a) the zone plate mount, (b) the positioning stage for the order-selecting aperture, (c) the specimen mount, (d) the fine scanning stage, (e) the coarse scanning stage, and (f) the detector (rotated out of position).

Instead, gold replicas of the carbon zone plates (made by x-ray lithography[22]) have been used. This combination has been supplying an average of 25,000 photons s^{-1} into a diffraction-limited spot. Imaging times are therefore in the region of 30 minutes. This group will soon improve their beamline to use a spherical grating monochromator, which should reduce imaging times substantially.

First Results. The images shown in Figures 5.5 and 5.6 are of rabbit muscle. These were test specimens and were not hydrated. The wavelength used was 3.2 nm. Figure 5.5 shows unstained rabbit muscle fiber section (0.5 μm thick), fixed in glutaraldehyde and postfixed in osmium tetroxide. The image is composed of 100 × 100 pixels and took about 15 minutes. The image in Figure 5.6 is of two unstained myofibrils, fixed in glutaraldehyde with no postfixing. The image is also 100 × 100 pixels and took about 25 minutes to obtain. These

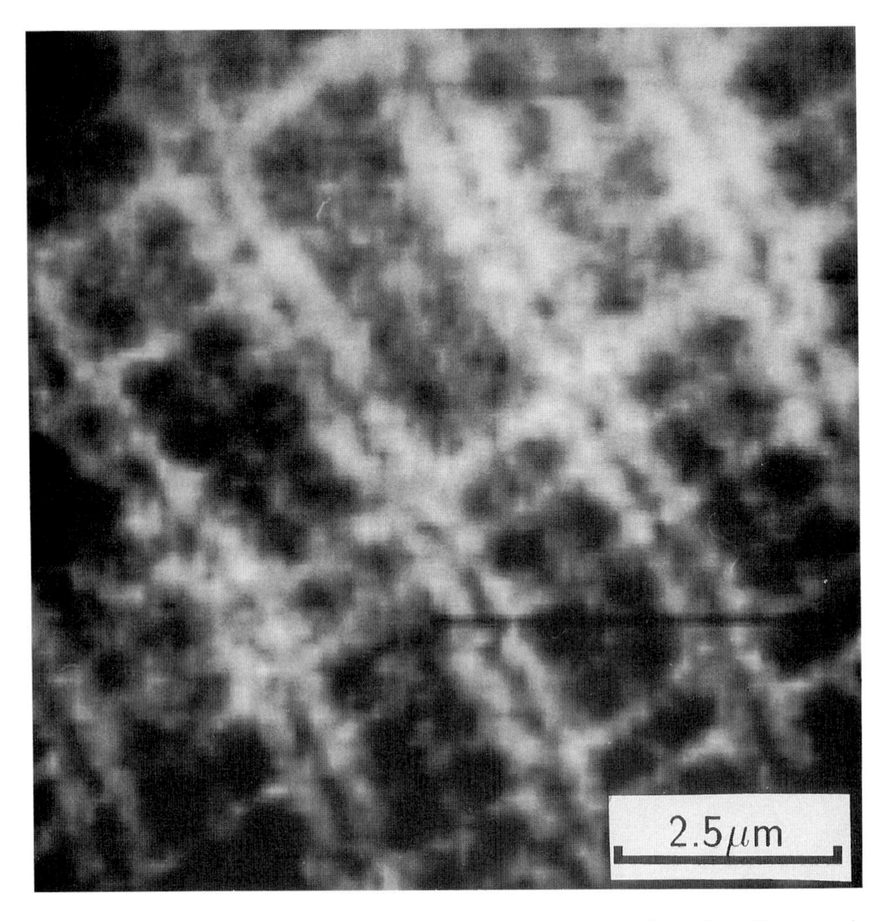

Figure 5.5. Unstained 0.5-μm thick glutaraldehyde-fixed rabbit muscle section. The image is composed of 100^2 pixels.

test specimens were imaged to investigate their contrast. A future study will compare stained and unstained muscle to help understand the contrast obtained with stained muscle specimens in electron micrographs.

5.4.2. The Stony Brook/NSLS Scanning Microscope at Brookhaven

The Stony Brook/NSLS group has been active in scanning x-ray microscopy (using synchrotron radiation) since 1980. Most of the group's work has come from experiments using bending magnet radiation from the NSLS 750-MeV storage ring where zone plates made by IBM/LBL[3] and KCL[32] were employed. Use of this source resulted in a coherent photon flux at the zone plate

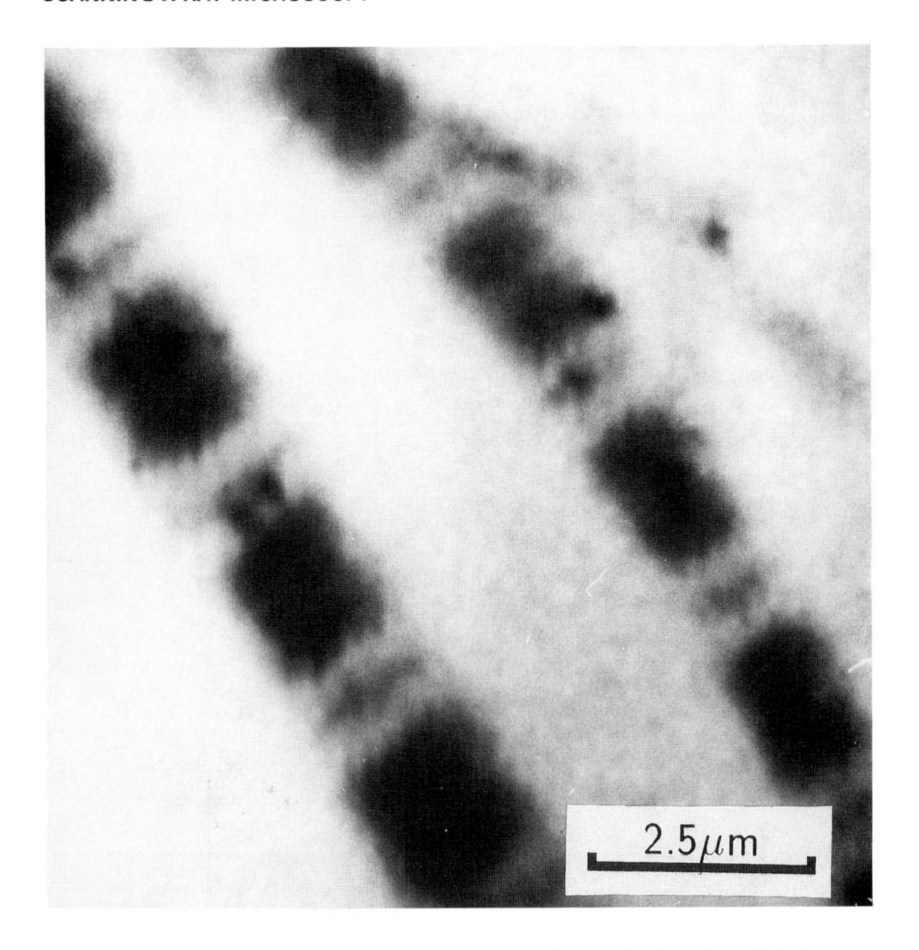

Figure 5.6. Two unstained myofibrils. The myofibrils were fixed in glutaraldehyde.

focus of an average of 10^4 photons per second. This level of flux restricted imaging times to about an hour at a resolution of about 150 nm. More recently the group has utilized radiation from a 10-period undulator (X17T) on the NSLS 2.5-GeV storage ring in a brief experiment before the source was shut down for an extensive upgrade. The beamline optics were not optimal for maximum flux use, but an average of 3×10^5 photons s^{-1} was obtained in a 75 nm spot.(19)

As with previous experiments the group has employed a grazing-incidence toroidal grating monochromator and a contamination barrier to relay coherent photons from the undulator source to the zone plate. The microscope was mounted on a massive granite bench supported on pneumatic mounts to reduce vibrations, with coupling to the monochromator by flexible bellows. The current version of the Stony Brook/NSLS microscope uses a silicon nitride vacuum window, a piezoelectric driven fine stage, and a gas flow proportional counter in much the same configuration as the KCL/Daresbury microscope. However, this microscope employs a He–Ne single-pass laser interferometer[34] to keep track of the stage position. The minimum step count is set by this interferometer, and at present is 30 nm. Focusing of the microscope is achieved by moving the zone plate at right angles to the raster scan, thereby maintaining the interferometer in a fixed geometry. Figure 5.7 shows a view of the microscope. Control of the stage, data acquisition and real-time image display are handled by a minicomputer via a CAMAC crate.

Results. In the latest series of experiments, wet and dry unstained biological specimens were imaged in the scanning microscope. For example, a study of wet zymogen granules showed evidence that the granules have internal structure.[35,36] Demyelinated axons from medullated shrimps have also been studied. This specimen is of interest because it has the fastest reported conduction velocity for a nerve (about 200 m^{-1} at 20°C). One unusual feature of the axon is that it is surrounded by a thick layer of longitudinally disposed microtubules.[37] These are thought to provide mechanical strength and play a role in axonal transport. It is hoped that the axons can be imaged in their natural state but with specific elements attached to molecules that are thought to be transported in the microtubule system. Images could be taken at x-ray wavelengths on either side of the absorption edge of the attached element. Figure 5.8 is an image of an unstained demyelinated axon from medullated shrimp which was dried on a microscope grid. The specimen was imaged as part of a feasibility study. The signal-to-noise ratio in this image is poor, but it is likely that with the expected improvement in source brightness elemental imaging of wet shrimp axon will be possible. The source has now been updated, and studies on these and other specimens continues.

5.4.3. The Göttingen Scanning Microscope at BESSY

The Göttingen scanning microscope[30] differs from those described above in several details. Principally, it uses radiation from a synchrotron bending magnet at BESSY in Berlin, it uses a zone plate monochromator, the specimen is mounted in vacuum rather than at atmospheric pressure, and the scanning mechanism is based on lever reduction. To date, only test specimens such as other zone plates and diatoms have been imaged.

5.5. Summary and Future Work

Scanning x-ray microscopy is just reaching the point where useful research should become possible. For biology, scanning microscopes with zone plate

Figure 5.7. The Stony Brook/NSLS STXM showing (A) the silicon nitride window vessel, (B) the zone plate/order-selecting mount, (C) the specimen mount, (D) the detector, (E) part of the interferometer system, (F) the piezoelectric scanning stage, and (G) the alignment microscope.

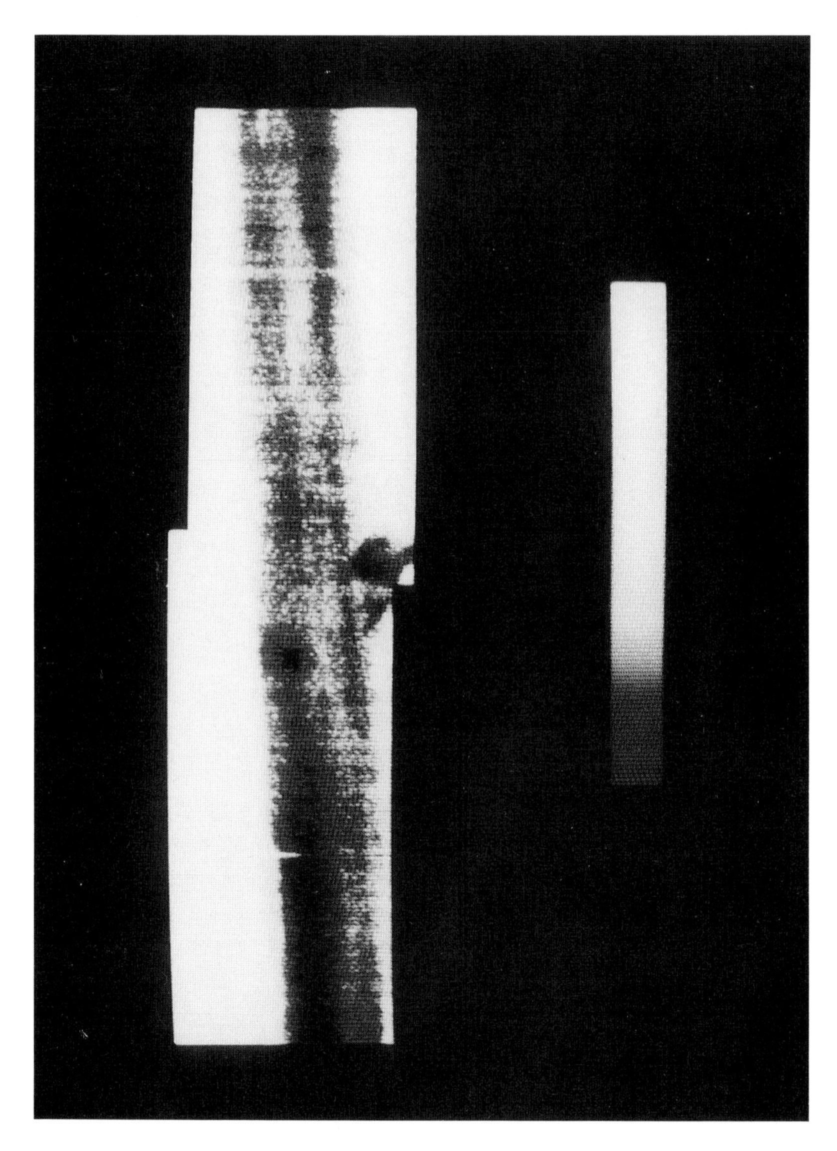

Figure 5.8. Scanning x-ray micrograph of demyelinated axon of medullated shrimp nerve. The wavelength used was 3.2 nm. The transverse dimension of the structure is approximately 4.5 μm.

optics will soon be at the stage where wet specimens will be imaged with a resolution of 50 nm in about a minute, and further progress is expected. Scanning x-ray microscopy should also have applications in the materials sciences. There are several groups developing scanning photoemission microscopy (which has a considerably higher signal-to-noise ratio than electron-beam induced Auger electron spectroscopy) at a spatial resolution of 50–1200 nm. More than one channel of spectral information can be obtained at this high resolution.

The joint developments of high-resolution soft x-ray optics and high-brightness sources have brought scanning x-ray microscopy to the level where it should be possible to obtain important biological information. The imaging of wet specimens a few micrometers thick at resolutions in the range 10–50 nm should be achieved. Scanning photoemission microscopy should also make significant contributions to surface science and materials science.

ACKNOWLEDGMENTS. The authors would like to acknowledge the following people who have been involved in the realization of the scanning x-ray microscopes and preparation of specimens giving rise to the images shown here. King's College London: J. T. Beswtherick, M. T. Browne, R. E. Burge, R. C. Cave, P. Charalambous, G. Foster, A. R. Hare, C. P. B. Hills, J. M. Kenney, A. G. Michette, D. Morris, G. R. Morrison, and T. Taguchi. Daresbury Laboratory: P. J. Duke, A. MacDowell, C. Mythen, H. A. Padmore, and J. B. West. Stony Brook (SUNY): H. Ade, S. -F. Fan, N. Iskander, C. Jacobsen, J. Kirz, and I. McNulty. NSLS (BNL): D. Shu. LBL and IBM (Yorktown Heights): Y. Vladimirsky and D. P. Kern.

References

1. D. Sayre, J. Kirz, R. Feder, D. M. Kim, and E. Spiller, "Transmission microscopy of unmodified biological materials: Comparative radiation dosages with electrons and ultrasoft x-ray photons," *Ultramicroscopy* **2,** 337–341 (1977).
2. J. Kirz, "Energy deposition by X rays and Electrons," in: *Examining the Submicron World* (R. Feder, J. W. McGowan, and D. M. Shinozaki, eds.), pp. 159–167, Plenum, New York.
3. J. Kirz and H. Rarback, "Soft x-ray microscopes," *Rev. Sci. Instrum.* **56,** 1–13 (1985).
4. P. C. Cheng, D. M. Shinozaki, and K. H. Tan, "Recent advances in contact imaging of biological materials," in: *X-Ray Microscopy: Instrumentation and Biological Applications* (P. C. Cheng and G. J. Jan, eds.), pp. 65–104, Springer, Berlin (1987).
5. D. M. Shinozaki, "High-resolution image storage in polymers," in: *X-Ray Microscopy II,* Springer Series in Optical Sciences, Vol. 56, pp. 118–123, Springer, Berlin, (1988).
6. D. Joyeux, S. Lowenthal, F. Polack, and A. Bernstein, "X-ray microscopy by holography at LURE," in: *X-Ray Microscopy II,* Springer Series in Optical Sciences, Vol. 56, pp. 246–252, Springer, Berlin (1988).
7. C. Jacobsen, "X-Ray Holographic Microscopy of biological Specimens Using an Undulator," Ph.D. Thesis, State University of New York at Stony Brook (1988).
8. D. Sayre, "Diffraction-imaging possibilities with soft X rays," in: *X-Ray Microscopy: Instrumentation and Biological Applications* (P. C. Cheng and G. J. Jan, eds.), pp. 213–223, Springer, Berlin (1987).
9. J. M. Kenney, C. Jacobsen, J. Kirz, H. Rarback, F. Cinotti, W. Thomlinson, R. Rosser, and G. Schidlovsky, "Absorption microanalysis with a soft x-ray microscope," *J. Microscopy* **138,** 321–328 (1985).
10. A. Franks, "X-ray optics," *Sci. Prog. (Oxford)* **64,** 371–422 (1977).

11. H. Wolter, "Mirror systems with grazing incidence as image-forming optics for X rays," *Ann. Phys. 6th Ser.* **10,** 94–114 (1952).
12. J. Dumond and J. P. Youtz, "An x-ray method of determining rates of diffusion in the solid state," *J. Appl. Phys.* **11,** 357–365 (1940).
13. T. W. Barbee, "Multilayers for x-ray optics," in: *Applications of Thin-Film Multilayered Structures to Figured X-Ray Optics, Proc. SPIE* **563,** 2–28 (1985).
14. E. Spiller, "Experience with the *in situ* monitor system for the fabrication of x-ray mirrors," in: *Applications of Thin-Film Multilayered Structures to Figured X-Ray Optics, Proc. SPIE* **563,** 367–375 (1985).
15. A. Franks and B. Gale, "Grazing-incidence optics for x-ray microscopy," in: *X-Ray Microscopy* (G. Schmahl and D. Rudolph, eds.), Springer Series in Optical Sciences, Vol. 43, pp. 129–138, Springer, Berlin (1984).
16. E. Spiller, "A scanning soft x-ray microscope using normal incidence mirrors," in: *X-Ray Microscopy* (G. Schmahl and D. Rudolph, eds.), Springer Series in Optical Sciences, Vol. 43, pp. 226–231, Springer, Berlin, (1984).
17. J. Underwood and T. W. Barbee, "Soft x-ray microscopy with a normal incidence mirror," *Nature* **294,** 429–431 (1981).
18. G. Schmahl, D. Rudolph, B. Niemann, and W. Meyer-Ilse, "X-ray microscopy with synchrotron radiation," *J. Microsc. Spectrosc. Electron.* **11,** 389–396 (1986).
19. H. M. Rarback, D. Shu, S. C. Feng, H. Ade, J. Kirz, I. McNulty, D. P. Kern, T. H. P. Chang, Y. Vladimirsky, N. Iskander, D. Attwood, K. McQuaid, and S. S. Rothman, "Scanning x-ray microscope with 75-nm resolution," *Rev. Sci. Inst.* **59,** 52–59 (1988).
20. G. R. Morrison, M. T. Browne, C. J. Buckley, R. E. Burge, R. C. Cave, P. Charalambous, P. J. Duke, A. R. Hare, C. P. B. Hills, J. M. Kenney, A. G. Michette, K. Ogawa, A. M. Rogoyski, and T. Taguchi, "Early experience with the King's College–Daresbury x-ray microscope," in: *X-Ray Microscopy II,* Springer Series in Optical Sciences, Vol. 56, pp. 201–208, Springer, Berlin (1988).
21. G. Schmahl, D. Rudolph, P. Guttmann, and O. Christ, "Zone plates for x-ray microscopy," in: *X-Ray Microscopy* (G. Schmahl and D. Rudolph, eds.), Springer Series in Optical Sciences, Vol. 43, pp. 63–74, Springer, Berlin (1984).
22. C. J. Buckley, M. T. Browne, R. E. Burge, P. Charalambous, K. Ogawa, and T. Taguchi, "Zone plates for scanning x-ray microscopy: Contamination writing and efficiency enhancement," in: *X-Ray Microscopy II,* Springer Series in Optical Sciences, Vol. 56, pp. 88–94, Springer, Berlin (1988).
23. V. Bögli, P. Unger, and H. Beneking, "Microzone plate fabrication by 100-keV electron beam lithography," in: *X-Ray Microscopy II,* Springer Series in Optical Sciences, Vol. 56, pp. 80–87, Springer, Berlin (1988).
24. Y. Vladimirsky, D. Kern, T. H. P. Chang, D. Attwood, H. Ade, J. Kirz, I. McNulty, H. Rarback, and D. Shu, "High-resolution Fresnel zone plates for soft X rays," *J. Vac. Sci. Technol.* **B6** (1988).
25. C. J. Buckley, "The Fabrication of Gold Zone Plates and Their Use in Scanning X-Ray Microscopy," Ph.D. Thesis, University of London, p. 121 (1987).
26. J. Thieme, "Theoretical investigations of imaging properties of zone plates using diffraction theory," in: *X-Ray Microscopy II,* Springer Series in Optical Sciences, Vol. 56, pp. 70–79, Springer, Berlin (1988).
27. A. G. Michette, R. E. Burge, and A. M. Rogoyski, "The potential of laser plasma sources in scanning x-ray microscopy," in: *X-Ray Microscopy II,* Springer Series in Optical Sciences, Vol. 56, pp. 59–62, Springer, Berlin (1988).
28. C. H. Skinner, D. E. Kim, A. Wouters, D. Voorhees, and S. Suckewer, "X-ray laser sources for microscopy," in: *X-Ray Microscopy II,* Springer Series in Optical Sciences, Vol. 56, pp. 36–42, Springer, Berlin (1988).

29. W. Neff, J. Eberle, R. Holz, F. Richter, and R. Lebert, "A plasma focus as a radiation source for a laboratory x-ray microscope," in: *X-Ray Microscopy II*, Springer Series in Optical Sciences, Vol. 56, pp. 22–29, Springer, Berlin (1988).
30. B. Niemann, P. Guttmann, R. Hilkenbach, J. Thieme, and W. Meyer-Ilse, "The Göttingen scanning x-ray microscope," in: *X-Ray Microscopy II*, Springer Series in Optical Sciences, Vol. 56, pp. 209–215. Springer, Berlin (1988).
31. H. Petersen and H. Baumgartel, "BESSY SX/700: A monochromator system covering the spectral range 3 eV $< h\nu <$ 1000 eV," *Nucl. Instrum. Methods* **208,** 315–318 (1983).
32. C. J. Buckley, "The Fabrication of Gold Zone Plates and Their Use in Scanning X-Ray Microscopy," Ph.D. Thesis, University of London, p. 162 (1987).
33. C. Jacobsen, J. Kirz, I. McNulty, R. Rosser, C. Buckley, R. E. Burge, M. T. Browne, R. Cave, P. Charalambous, P. J. Duke, J. M. Kenney, A. G. Michette, G. R. Morrison, F. Cinotti, H. Rarback, and J. Pine, "Scanning soft x-ray microscope at the national synchrotron light source," in: *Short-Wavelength Coherent Radiation: Generation and Applications* (D. T. Attwood and J. Bokor, eds.), AIP Conference Proceedings, Vol. 147, pp. 57–63, (1986).
34. D. Shu, D. P. Siddons, and H. Rarback, "Two-dimensional laser interferometric encoder for the soft x-ray microscopy at the NSLS," *Nucl. Instrum. Methods Phys. Res.* **A266,** 313–317 (1988).
35. N. Iskander, "Scanning transmission x-ray microscopy of unaltered biological specimens," internal report, Department of Physics, University of California at Berkeley, May 1987.
36. S. S. Rothman, N. Iskander, K. McQuaid, D. T. Attwood, J. H. Grendell, Y. Vladimirsky, D. Kern, H. Ade, J. Kirz, I. McNulty, H. Rarback, and D. Shu, "The biology of the cell and the high-Resolution x-ray microscope", in: *X-Ray Microscopy II*, Springer Series in Optical Sciences, Vol. 56, pp. 372–377. Springer, Berlin (1988).
37. K. Hama, "The fine structure of the Schwann cell sheath of the nerve fiber in the shrimp (*Penaeus Japonica*), *J. Cell Biol.* **31,** 624–632 (1966).

6

X-Ray Microradiography and Shadow Projection X-Ray Microscopy

P. C. Cheng, S. P. Newberry, H. G. Kim, and I. S. Hwang

6.1. Introduction

The fundamental advantages of soft X rays over electrons in the examination of fine structures have been described in detail by various authors.[1–5] Theoretically, x-ray microscopy provides higher resolution than light microscopy, higher penetration ability than electron microscopy, and, most importantly, x-ray microscopy promises the potential for imaging hydrated specimens. Therefore, x-ray microscopy could occupy a niche in biological research in three-dimensional imaging of samples in the resolution range of the electron microscopy but of substantially greater thickness. This would greatly simplify the observation and interpretation of three-dimensional ultrastructures of living specimens beyond the resolution limit of the light microscope.

It has long been proposed and recently demonstrated that x-ray microscopy

P. C. Cheng • Advanced Microscopy Laboratory, Department of Anatomical Sciences, School of Medicine and Biomedical Sciences/School of Engineering and Applied Sciences, State University of New York at Buffalo, Buffalo, New York 14214. ***I. S. Hwang*** • Department of Electrical and Computer Engineering, State University of New York at Buffalo, Buffalo, New York 14260. ***S. P. Newberry*** • CBI Labs, Schenectady, New York 12306. ***H. G. Kim*** • Laboratory for Laser Energetics, University of Rochester, Rochester, New York 12306.

can be used to reveal the elemental composition of a specimen.[6–8] But x-ray microscopy is incapable of distinguishing many molecules such as proteins, which have very similar elemental compositions and packing density, but which have very different structures and functions. Furthermore, many biologically important elements, such as calcium and magnesium, are present in living cells at concentrations too low to be detected by x-ray absorption microscopy. Recent advances in cell biology have included the development of various antibody probes for tagging specific structural proteins, membrane receptors, and organelles. The use of colloidal gold-labelled antibodies as x-ray dense probes with molecular specificity could become a powerful technique for x-ray imaging, and the use of Golgi (silver) stain on neurons for x-ray microscopy has been reported by several workers.[9] With the use of specific stains, x-ray microscopy would gain a competitive edge over light microscopy in resolution and electron microscopy in thick-specimen handling capability. It is important, therefore, not to restrict the application of x-ray microscopy to unstained specimens alone.

Two of the older techniques of x-ray microscopy, namely, microradiography and shadow projection microscopy, will be discussed in this chapter. X-ray microscopy involving focusing elements has been discussed in Chapters 4 and 5 of this book, and elsewhere.[10–12] The methods of contact imaging, and some of the recent results in a variety of applications, have been well documented.[5,13–15] Shadow projection microscopy, in its wavelength range (0.1–0.2 nm) and resolution domain (comparable to light microscopy), is the most convenient form of x-ray microscopy. Past work has shown that shadow projection x-ray microscopy could be a very useful tool in a variety of applications including studies of processed foods, rubber compounds, alloys, microcircuit boards, and anatomical structures of plants and animals.

6.2. X-Ray Sources

Since the early part of the century, one of the major stumbling blocks in the development of x-ray microscopy has been the lack of high-intensity sources. The only x-ray source available has been the electron impact source. In the past decade, however, there has been a rapid development of intense x-ray sources using synchrotron radiation[16,17] and high-temperature plasmas. See Chapter 3 for a fuller coverage.

A synchrotron storage ring provides electromagnetic radiation in a continuous spectrum, and therefore a monochromator or filters must be employed to obtain narrow-band x-rays of a wavelength suitable for x-ray imaging. Figure 6.1 shows the Canadian Synchrotron Radiation Facility (CSRF) beamline at the 1 GeV storage rings of the University of Wisconsin. This is the beamline on which the x-ray contact imaging described in this chapter has been conducted. The beamline

Figure 6.1. The Canadian Synchrotron Radiation Facility beamline used for contact imaging.

consists of a Mark IV grasshopper monochromator, a differential pumping station, and a specimen chamber.

It is well known that hot plasmas of small dimensions may be produced by focusing a high-intensity laser beam (up to 150 J at a wavelength of 1054 nm) in a 1 ns pulse onto a target. Extremely hot plasmas with electron temperatures of approximately 1 keV (10^7 K) can be achieved.[18] From such high-temperature plasmas efficient production of X rays with energies below 8 keV may be obtained. The x-ray production can be greatly enhanced by using a shorter laser wavelength such as that obtained from frequency-tripled Nd : glass laser pulses. The x-ray conversion may approach 30% of the laser energy by the use of such shorter wavelength lasers.[18] Figure 6.2 demonstrates an order of magnitude increase in x-ray conversion efficiency into the $1s^2$–$1s2p$ resonance lines of various elements by the use of laser pulses with λ = 350 nm as compared to λ =

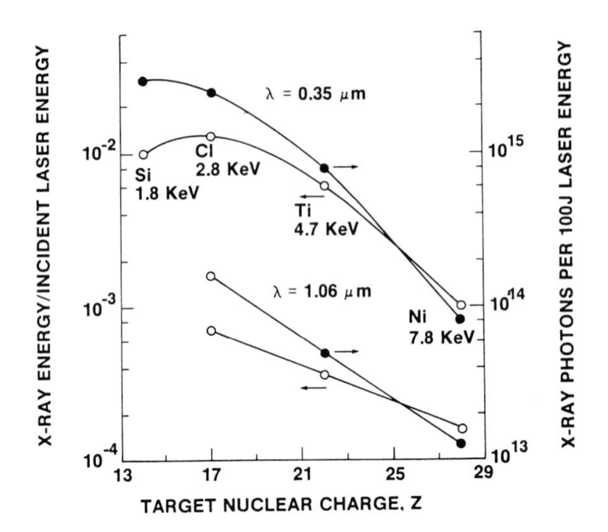

Figure 6.2. Conversion of laser energy into x-ray energy for targets of various elements. An order of magnitude increase in the conversion efficiency is observed when a frequency-tripled laser (λ = 0.35 μm) is used as compared with the fundamental frequency (λ = 1.06 μm).

1060 nm. The x-ray spectrum consists of a broad continuum of energies around 1 keV with strong line components with energies of several kiloelectronvolts (if the constituent atoms of the target have lines in this energy range). Figure 6.3 shows the spectral output of a molybdenum plasma formed by a 350-nm laser pulse of 1 ns duration. The broad continuum in Figure 6.3 is particularly useful for contact x-ray microscopy of biological specimens, the difference in the absorption coefficients of water and protein at an x-ray wavelength of 2.5 nm being about ten-fold. Figure 6.4 shows the basic source set-up and the spectral output of a molybdenum plasma (Figure 6.3). For x-ray contact microradiography, a suitable filter should be used to isolate X rays from the rest of the radiation produced, which is unwanted UV and visible radiation. Otherwise, a contact image will be produced primarily by UV or visible light instead of x-rays.

The spectral and temporal characteristics of the laser pulse are reproduced by the radiation from the plasma. For short pulse length lasers this is the most useful property, along with the very small source size, of the laser plasma x-ray source for x-ray microscopy of living biological specimens.

A very small x-ray source (<1 μm diameter), which is accessible for a close approach by the specimen, is needed to achieve high resolution for shadow projection x-ray microscopy. A focused electron beam impacting on a thin-film transmission target is commonly used to generate X rays. Both electromagnetic lenses and electrostatic lenses have been used successfully to focus the electron beam. Some investigators have used the focused electron beam in a scanning electron microscope to generate the required small x-ray source size.[19]

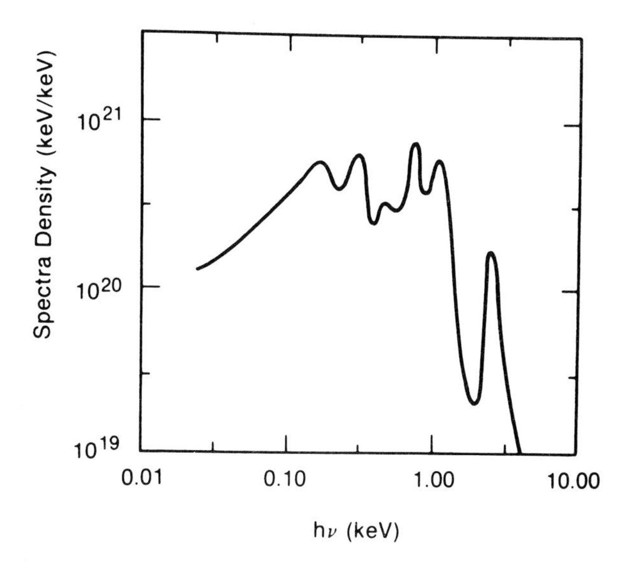

Figure 6.3. X-ray spectrum emitted by a molybdenum target irradiated by a 1 ns, $\lambda = 0.35\ \mu m$ laser pulse of 50 J. (From Ref. 42.)

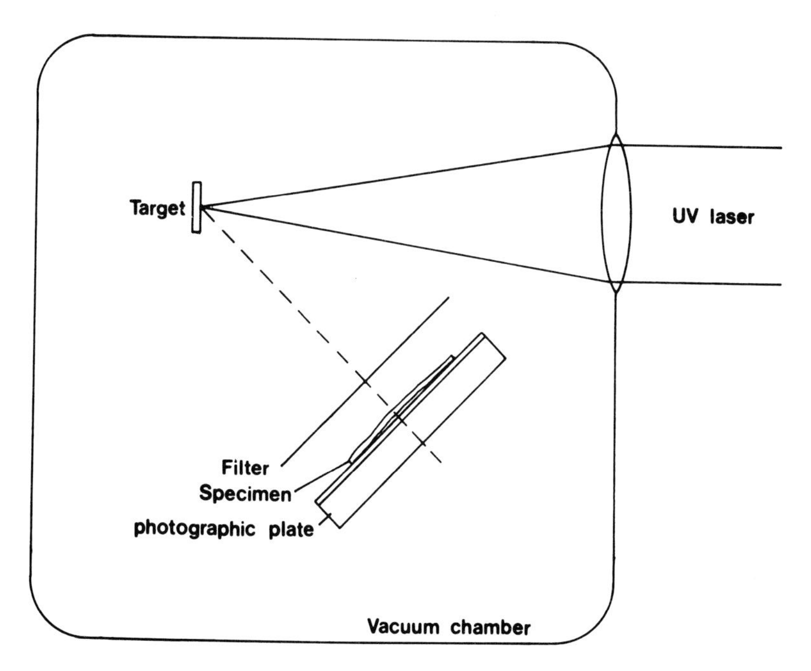

Figure 6.4. Experimental configuration for x-ray microradiography using a laser-produced plasma source.

6.3. X-Ray Detectors

In both x-ray contact imaging and shadow projection x-ray microscopy the x-ray images are recorded by using photochemical detectors. Silver halide emulsion is used as the photochemical detector in conventional x-ray microradiography where the resolution is limited by the silver grain size and distribution (the graininess of the film). Agfa holographic plate and Kodak High Resolution glass plates are commonly used for x-ray contact microradiography. Some workers have searched for photochemical detectors with higher resolution—for example, a Saran polymer film.[20] Modern x-ray contact imaging uses high molecular weight polymers (photoresists) as the photochemical detector in order to achieve higher resolution.[1] The commonly used positive photoresists are polymethylmethacrylate (PMMA), polymethylmethacrylate–methacrylic acid copolymer Co(PMMA– MAA), and polybutylsulfone (PBS). Operationally speaking, silver halide emulsion has much higher sensitivity ($\times 10^4$) than photoresist,[21] but silver emulsion only has a practical pixel size down to 0.2 μm, while PMMA can achieve pixel resolution in the range of tens of nanometers.[22] In terms of practical exposure times, ordinary thermionic electron-impact x-ray sources give useful results, using silver halide emulsion, in the resolution range of the optical microscope. A large increase in x-ray intensity, however, is needed to allow access to the higher-resolution region made possible for contact work by use of PMMA as the detector. In summary, the choice of silver halide emulsion or photoresist is dependent on the intensity of the x-ray source used, the x-ray wavelength, and the desired resolution.

In shadow projection work the detector plane intensity can only be lowered by the shadow magnification, so while projection equipment may be used as a contact source, with some practical advantages, its resolution is limited by the necessity to use silver halide detectors. Early on one of us (SPN) demonstrated, but reported only at local meetings, that a charged Xerographic plate is also an attractive detector for projection work. Polaroid type 52 or P/N 55 Land film is best suited for survey work in projection. However, fine grain glass plates, such as Kodak Lantern Slide (medium contrast), should be used for serious image recording.

Recent advances in shadow projection microscopes include the use of low-light TV cameras (SIT), image intensifiers, CCD devices, and microchannel plates as the detectors.[23,24] In contact imaging, photocathodes have been used as the x-ray detectors in an x-ray photoelectron microscope.[25]

6.4. X-Ray Microradiography

X-ray contact microradiography was pioneered by Goby in the study of plant tissue and other biological materials.[26,27] Later, he also developed a method of obtaining 3-D image pairs of thick specimens.[28] Work on the improvement of the technology continued for many decades.[29–31] Contact microradiography became very popular during the 1940s and 1950s,[32] and many fine results were published during this period.[6,7] However, during the same period, electron microscopy gradually emerged as a powerful tool for biological sciences. This is one of the reasons why x-ray contact microradiography slowly lost its popularity. Not until the late 1970s did it reemerge[1] with the use of high molecular weight photoresist technology developed by the microelectronics industry. The technique was reintroduced as "x-ray contact microscopy."

In x-ray contact microradiography, the specimen is placed in close contact with an x-ray-sensitive material (e.g., silver halide emulsion or polymer resist) and the sandwich is irradiated by X rays. Figure 6.5 shows typical set-ups for

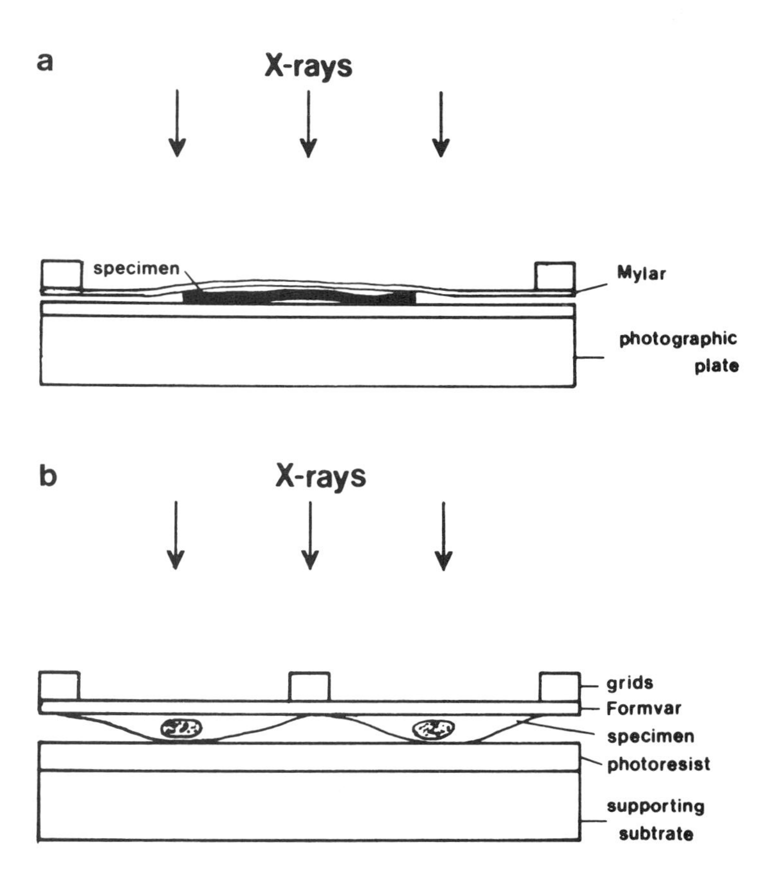

Figure 6.5. Diagrammatic representation of contact x-ray microradiography set-ups (a) using silver emulsion as the detector and (b) using photoresist as the detector.

contact microradiography. After exposure, the x-ray sensitive material is developed, in a photographic developer for silver emulsion and in an appropriate solvent for polymer resist. A suitable developer for PMMA is a mixture of isopropanol and methyl isobutyl ketone in various ratios (e.g., 1 : 1). For a positive resist such as PMMA, the more heavily irradiated areas of the resist have lower molecular weights and dissolve more readily, and thus the resultant image is a map of the relative x-ray opacity of the different regions of the specimen. The topography of the developed resist surface can be examined in a light microscope (phase contrast or differential interference contrast), a scanning electron microscope (secondary electron mode or low-loss electron mode), or a transmission electron microscope. It is at this stage that the image is magnified. The information in the image is essentially contained in the topography of the exposed and developed resist surface, and the resolution of the method depends on the properties of the resist, the irradiation and development conditions, and the method of examination of the developed resist.

The three commonly used polymer resists, PMMA, Co(PMMA–MAA) and PBS can all be dissolved in solvents such as chlorobenzene, trichloroethane, and chloroform. To prepare a resist film, a drop of the solution, with typical concentration in the range 5–10% (w/v), is placed on the center of a Si_3N_4 window wafer, a silicon wafer, or a glass wafer (glass cover slip) and spun at high speed (1000–5000 rpm) for about 60 s. A standard resist spinner as used in microelectronics fabrication is needed to produce a uniform thin film of resist on the surface of the substrate. The thickness of the resist can be controlled by changing the viscosity (i.e., the concentration) of the polymer solution or by varying the speed of the spinner. After spinning, the resist-coated wafer is baked in an oven to remove residue solvent. Typical thicknesses of resist used for microradiography range from 0.4 to 2.0 μm.

When silver emulsion is used, the resultant miniature contact image can be magnified using either a macrophotography set-up (Figure 6.6) or a conventional light microscope. A typical application of this type of contact microradiography is demonstrated by studies of the development of silica cells in the leaf of *Zea mays* (var. Golden Beauty) (Figure 6.7). This image was obtained by using a laser-produced plasma x-ray source. A frequency-tripled Nd : glass laser beam (GDL at the Laboratory for Laser Energetics, University of Rochester) was focused onto a molybdenum thin-film target to produce a high-temperature plasma, which gave an integrated x-ray emission of 32 J. A 200-nm-thick evaporated aluminium layer on a 2-μm-thick mylar film was used as a filter to block unwanted visible and UV radiation from the source. Exposure was made at a distance of 25 cm from the source. The leaf was fixed in Cheng's fixative,[(33)] dehydrated in acetone, and critical point dried. Note the high x-ray absorption of the clover-leaf-shaped silica cells. The silica cells are arranged in files and are closely associated with vascular bundles (Figures 6.7a and 6.7b). The bases of epidermal hairs also show high x-ray absorbance. In older leaves, silica deposi-

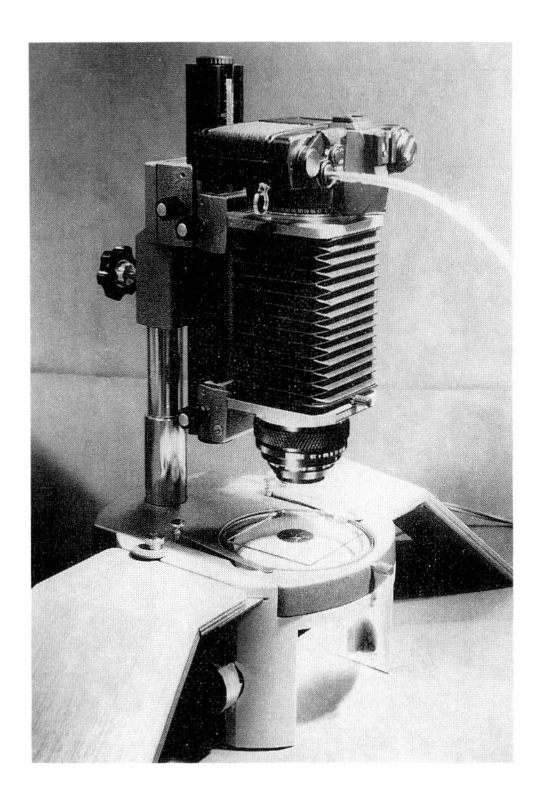

Figure 6.6. A typical macrophotography set-up used to magnify contact microradiographs recorded on silver emulsion. A mirror housing between the lens and object can be used to control the image contrast.

tion is no longer restricted to the clover-leaf-shaped silica cells, but also appears in selective epidermal cells and hairs along the edges of the leaf (Figure 6.7c and 6.7d). In contrast to conventional x-ray microanalysis, which is restricted to the very surface layer of the specimen, x-ray contact microradiography can detect silica deposits deep inside the leaf tissue. X-ray contact imaging could be one of the most convenient ways to study the distribution of silica deposition in higher plants in general. The nature of the contact imaging technique allows a large leaf area to be imaged, thus providing sufficient sample size for a proper study of the development and distribution of silica cells.

X-ray contact microradiography has also been used in the study of thick and dense specimens in materials science—for example, electromigration in simple microdevices.[34] One of the authors (HGK) used x-ray contact microradiography to inspect Parylene coatings on the outer surfaces of inertial fusion targets. Figure 6.8 shows an x-ray contact image (taken on photographic plate) of an inertial fusion target (glass balloon) with an outer Parylene coating. The target

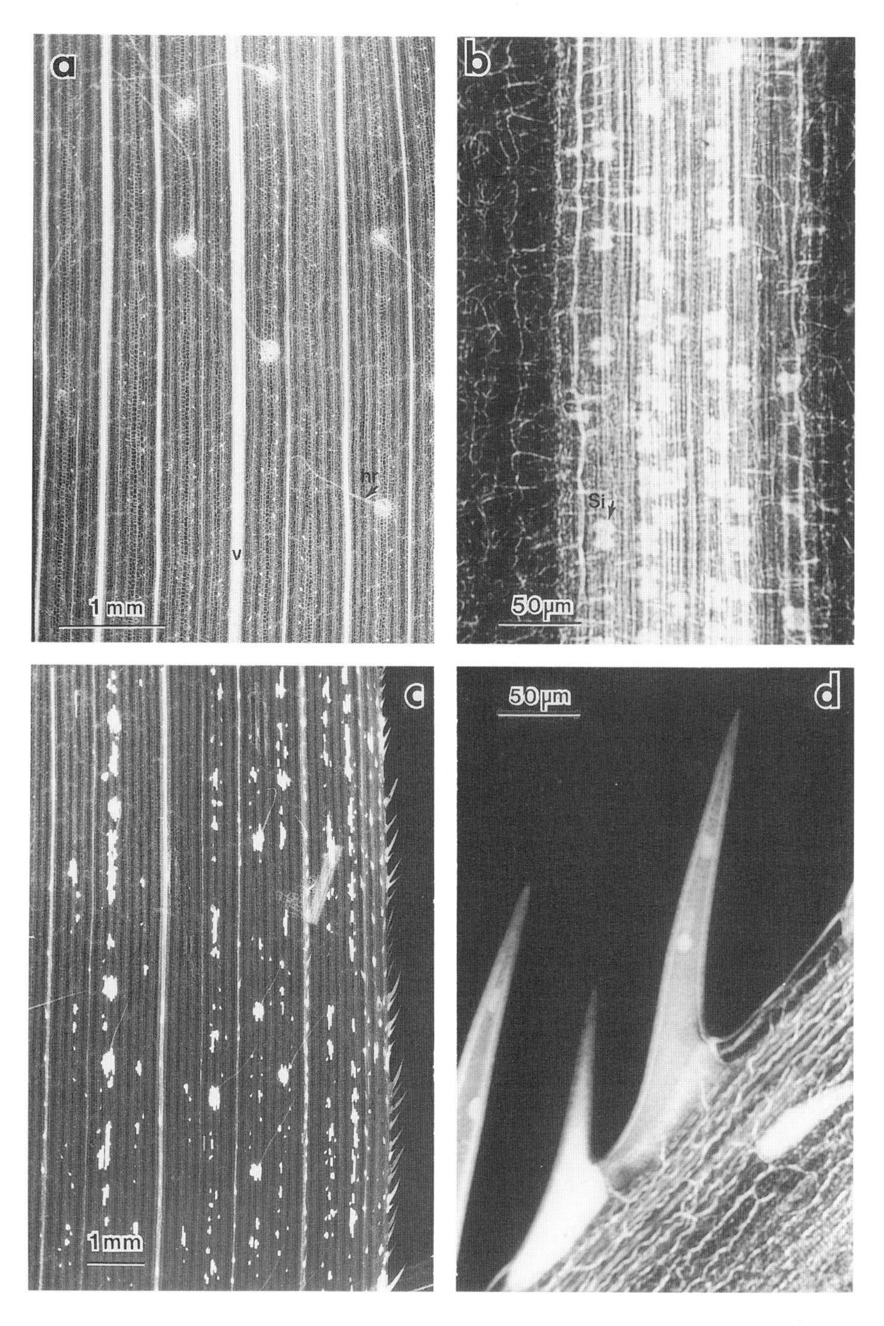

Figure 6.7. Silica cell distribution in the leaf blade of *Zea mays* L. The images are negative, that is, brighter parts correspond to higher x-ray absorbance: (a) Low magnification view of corn leaf shows vascular bundles and epidermal hairs (hr). (b) High magnification view of the vascular bundle showing the epidermally located silica cells (Si). (c) In an older leaf, silica deposition occurs in epidermal cells and hairs along the edge of the leaf blade. (d) High-magnification view of hairs showing silica deposition.

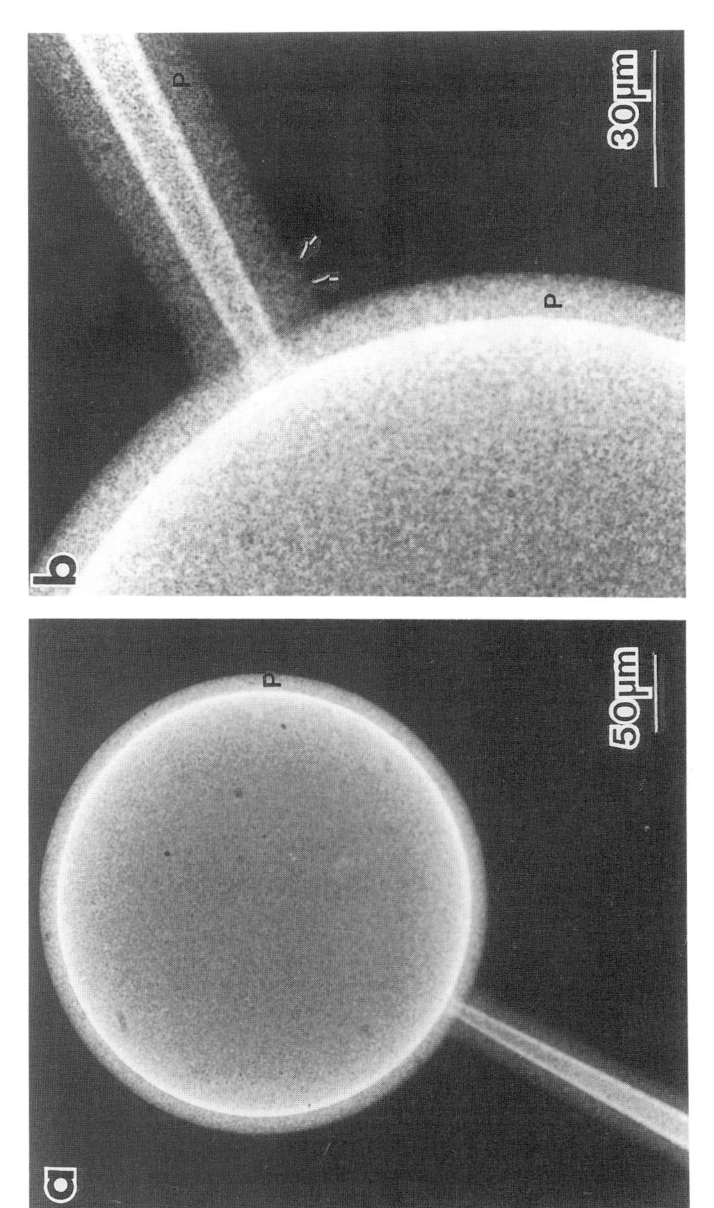

Figure 6.8. X-ray contact image of an inertial fusion target (glass sphere) overcoated with a layer of Parylene (poly-paraxylyene) (P). The inertial fusion target was mounted on a glass stick prior to the Parylene coating. Small irregularities (arrowed) are clearly visible.

was supported on the tip of a glass micropipette prior to coating. Small irregularities on the target are clearly visible.

X-ray contact imaging using photoresist provides significantly higher resolution than when silver halide emulsion is used. When a photoresist is used as the detector, the contact microradiograph is recorded on the developed resist surface as topographic map of the relative x-ray opacity of the specimen. The resultant x-ray image can be magnified using either an epi-illuminated interference phase contrast microscope or a transmitted-phase contrast microscope (if the supporting substrate is transparent). Figure 6.9 shows an x-ray contact image of mouse 3T3 fibroblast magnified by using a transmitted light phase contrast microscope. In this example the x-ray resist was supported on a transparent Si_3N_4 window. However, a glass wafer (such as microscope cover slip) is generally used as the supporting substrate.

Clearly the use of a light microscope limits the resolution of the resultant image. Scanning electron microscopy (SEM) can be used to obtain magnified images with higher resolution. Generally, the resist surface is coated with a thin layer of gold and observed in the secondary electron (SE) mode. Recent work has demonstrated that, for the study of small surface features on a photoresist, it is

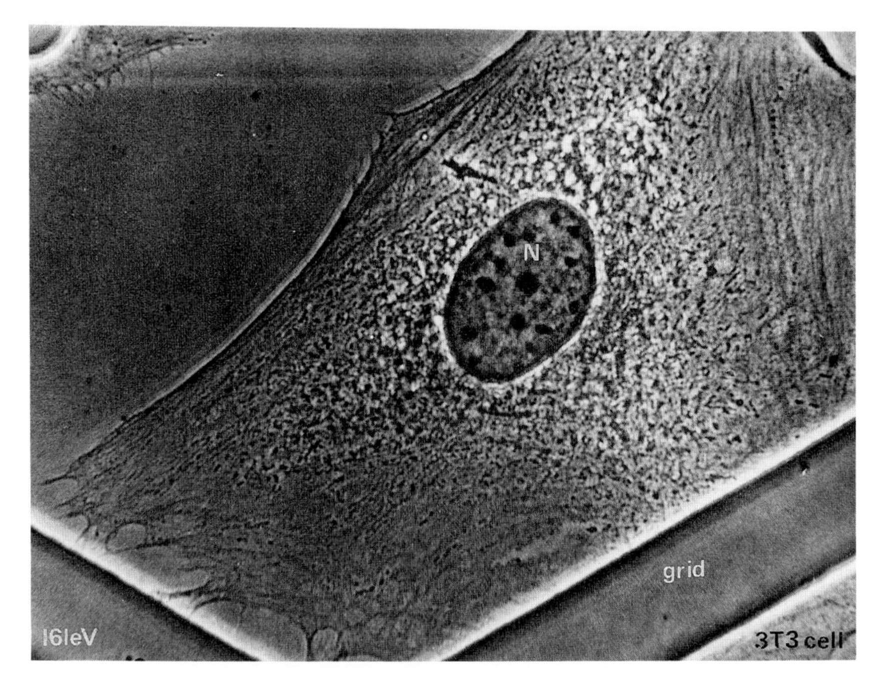

Figure 6.9. X-ray contact image of fibroblasts magnified by using a phase-contrast microscope. The image was taken on PMMA resist with 161-eV X rays.

better to operate the scanning electron microscope in the low-loss electron (LLE) mode.(35,36) This mode requires no surface coating on the resist, and provides high detectability for small surface features. Figure 6.10 shows a contact image of a human fibroblast magnified using the SE mode (Figure 6.10a) and the LLE mode (Figure 6.10b). In this example, the PMMA has been coated with 10 nm of Au/Pd. The LLE imaging mode was operated at $V_0 = 10$ kV and $V_{loss} = 200$ V in a Cambridge S-250 Mk III scanning electron microscope equipped with a LaB_6 single-crystal cathode and a LLE detector.(37)

One drawback of using SEM as the magnifying tool for x-ray contact imaging is the difficulty in interpreting the resultant image, which is a surface topographical view of the resist. The use of transmission electron microscopy (TEM) can circumvent this problem by converting the three-dimensional surface topographical profile to the two-dimensional density map that represents the x-ray absorbance of the specimen. TEM observation requires the resist to be relatively thin (typically 0.5 μm) and supported by a very thin light-element window. Commonly used window materials are carbon film, Si_3N_4, and SiB in the thickness range 50–100 nm.(15,38) Figure 6.11 shows a typical image of a fibroblast obtained on Si_3N_4-supported PMMA and magnified by TEM.

Photoresists are highly sensitive to electron-beam irradiation and can be damaged easily.(14,15,39) On electron-beam irradiation, the resist can undergo significant mass loss, wrinkling, and even bubbling. The potential artifacts introduced by the electron-beam damage should not be overlooked. By using a surface replica technique the problem of electron-beam damage to the resist can be bypassed while maintaining the high resolution of electron microscopy. A freestanding surface replica of the resist can be easily obtained by evaporating Au/Pd and C onto it, and by dissolving away the resist with suitable solvents.(15) The replica technique provides the highest resolution available in x-ray contact microradiography. Figure 6.12 shows a comparison between an x-ray image magnified by direct TEM viewing and by using the replica technique. The resolution of the contact imaging technique is limited by surface roughness caused by the chemical development.(40)

X-ray microradiography, while technically simple and inexpensive, has certain drawbacks. Since it is a two-step process, image formation on the photoresist followed by magnification of the miniature contact print, it is difficult to use in the study of dynamic biological processes. Using a high-intensity pulsed x-ray source a "frozen frame" of a dynamic process could be obtained; only plasma x-ray sources, such as the *z*-pinch and laser generated x-ray sources, can meet this requirement. However, modern biological research generally requires careful follow-ups and multiple recordings of a specimen in real time. For example, the study of cell secretion requires multiple observations of a complex and dynamic process. X-ray microradiography may have difficulties meeting these requirements.

In the past few years, there has been increased activity in x-ray contact

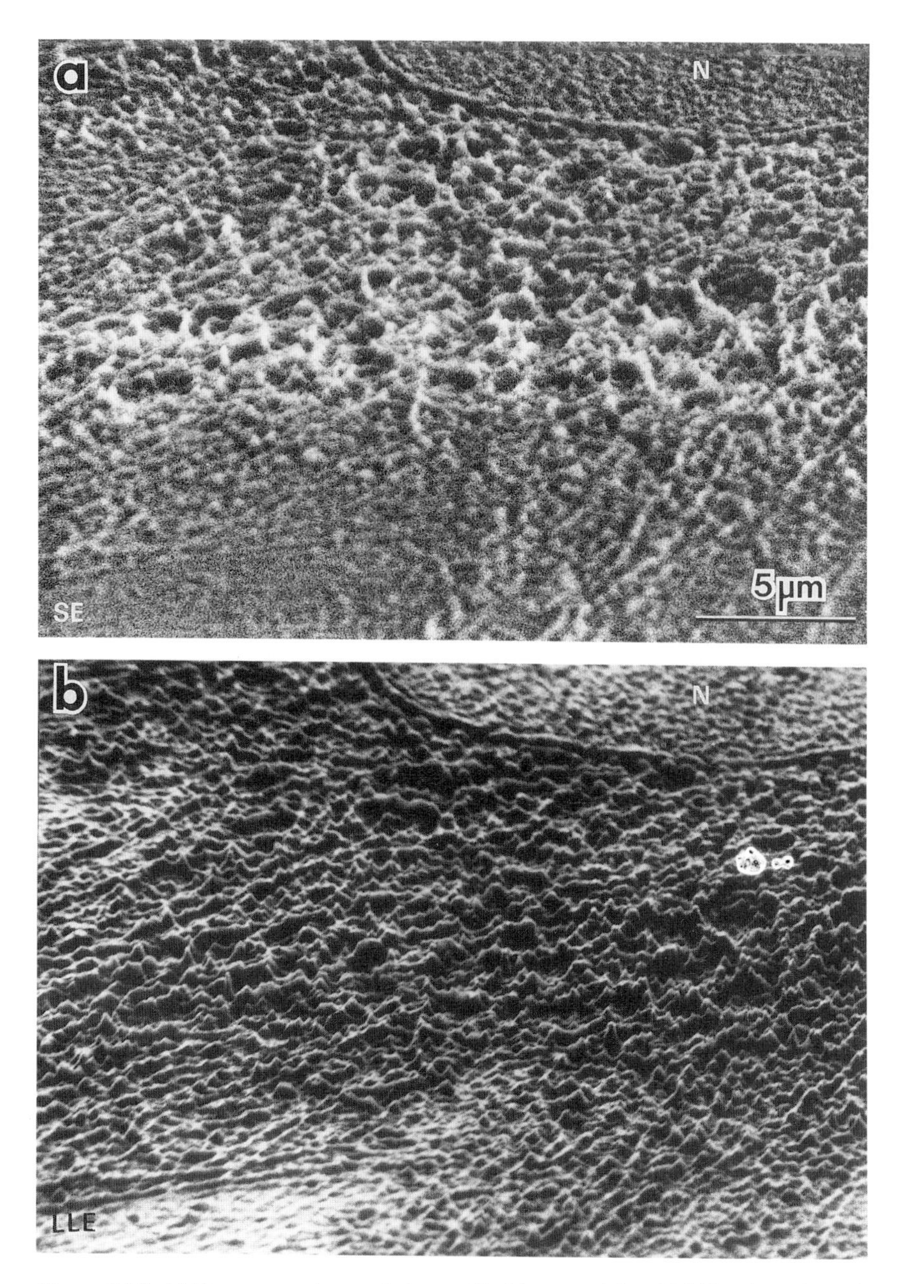

Figure 6.10. (a) X-ray contact image of a human fibroblast recorded on PMMA resist and magnified by a SEM operated in the secondary electron mode. (b)The LLE image shows the topography more clearly. (From Refs. 35 and 36.)

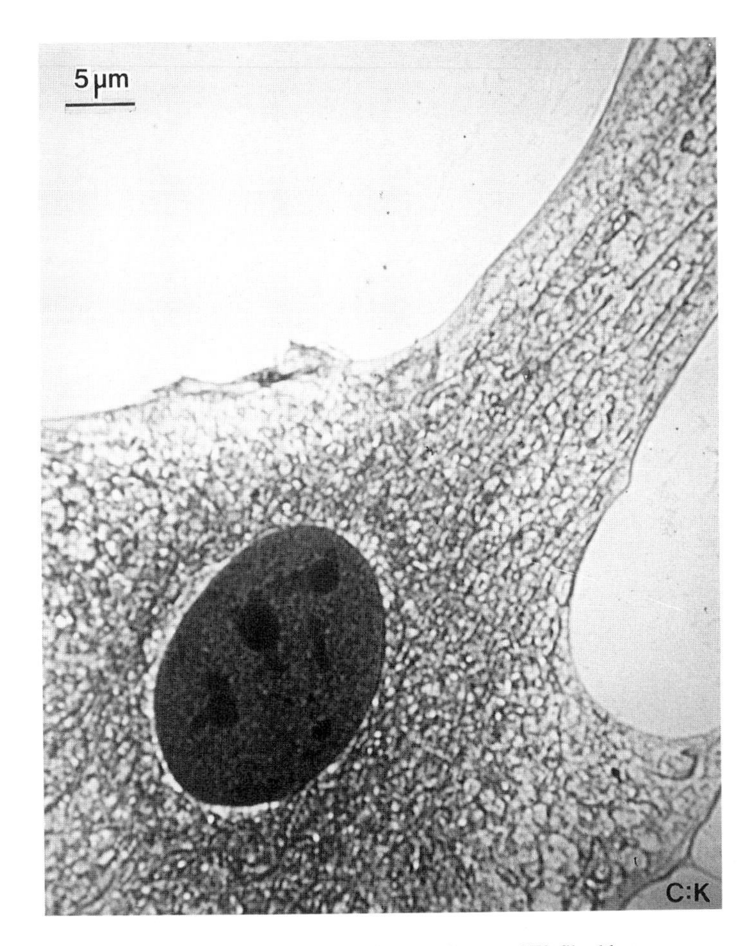

Figure 6.11. TEM-magnified image of mouse 3T3 fibroblast.

imaging using both *z*-pinch and laser-generated x-ray sources. A single x-ray pulse from a high-peak-power laser-produced plasma has been shown to be sufficient to form images in x-ray lithography and contact microradiography. In submicron x-ray lithography studies using PBS or poly(glycidyl-methacrylate-ethyl acrylate) resists, relief images were formed with incident x-ray fluxes about an order of magnitude smaller than normally required. Although the reason for this is not clear at the moment, it may be due to the transient character of the x-ray exposure.[41] High resolutions are possible with such sources. For example, an x-ray source size of 100 μm and an exposure at 10 cm from the source gives a penumbral blurring of 10 nm, assuming a separation of 10 μm between the mask and the resist. An example of a single-pulse exposure is given in Figure 6.13. The characteristics of the laser were $\lambda = 350$ nm, pulse length $= 1$ ns, and

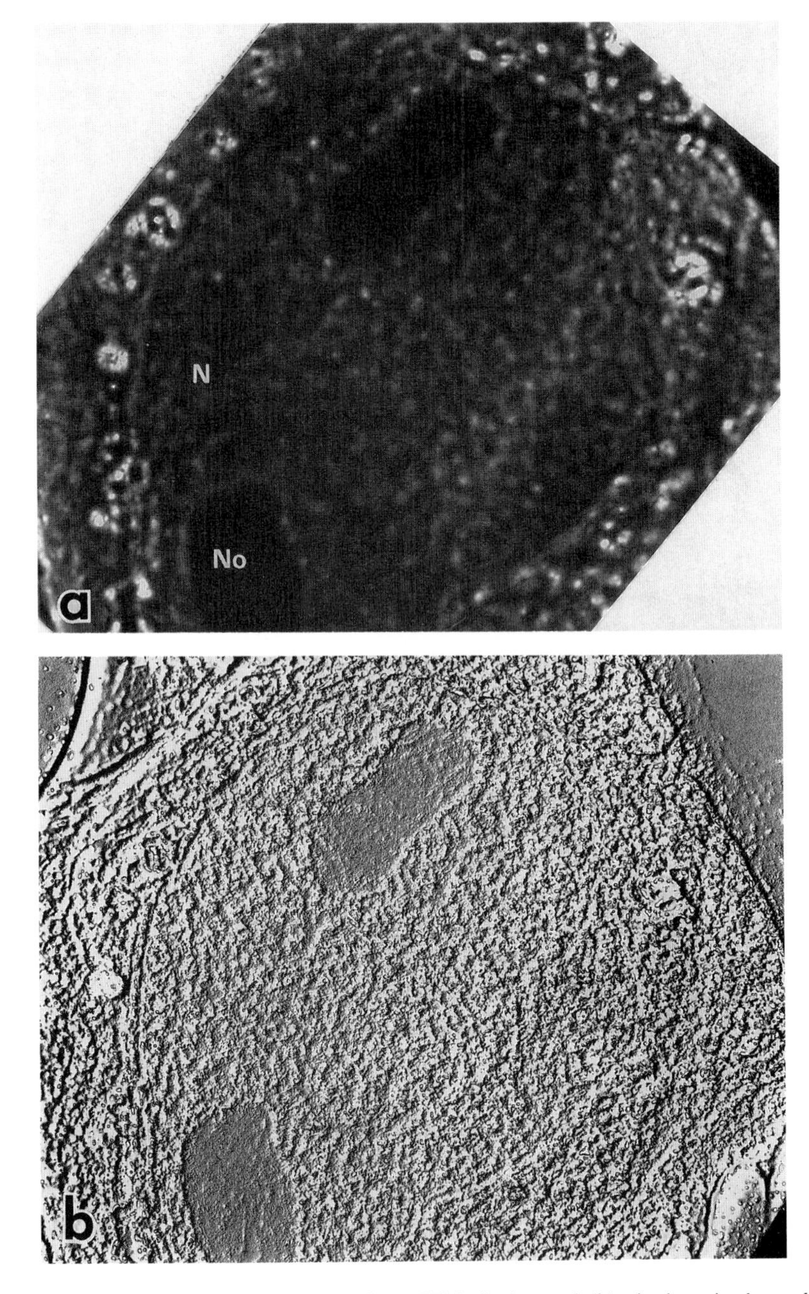

Figure 6.12. Comparison between (a) direct TEM viewing and (b) viewing via the replica technique.

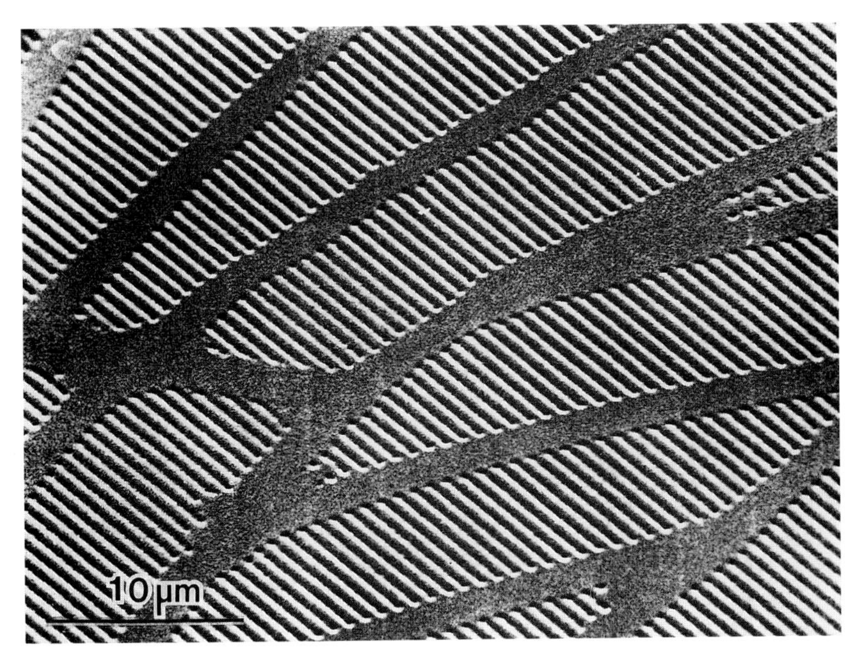

Figure 6.13. Submicron x-ray lithography using a UV laser-produced plasma as a source. Characteristics of the laser were $\lambda = 0.35$ μm, pulse duration 1 ns, and energy 35 J. PBS resist was used as the detector.

energy = 35 J. The resist was PBS. A gold grating of 0.45 μm linewidth, which was used as a mask, is clearly resolved.

For biologists intending to use pulsed x-ray sources for their research, it is important to note that both the z-pinch and laser-produced plasma sources generate not only X rays but also high-intensity UV, visible, and IR radiation. The photoresists used in x-ray contact imaging are also highly sensitive to UV radiation as well as subject to thermal damage caused by IR radiation. It is therefore essential to use suitable filters or monochromating devices to remove the UV and IR components. Recent results indicate that, with a 25-nm-thick Al filter in place, PMMA is not sensitive enough to form an image using a single x-ray pulse obtained by focusing 356 nm, 40 J laser radiation to a 100 μm spot on a molybdenum target. Heavy exposure, however, was observed when irradiation was made without the filter. Therefore, it is clear that a contact image dominated by UV is obtained if the exposure is made without a suitable UV-blocking filter. When PMMA was exposed to unfiltered radiation from a laser-generated plasma source, a significant increase in surface roughness was obtained even prior to chemical development. This is believed to be due to UV ablation of the resist. No ablation was observed when a 25-nm-thick Al filter was placed between the source and the resist.[42]

Three-Dimensional Imaging by X-Ray Microradiography

To obtain three-dimensional information about a cell or tissue, the general practice is to serial section the specimen and subsequently reconstruct the structure by using wax models or computer graphics. Occasionally, if the cell is sufficiently thin (e.g., cultured fibroblasts), stereo image pairs can be obtained with a TEM. Due to the high penetration power of X rays (especially for shorter-

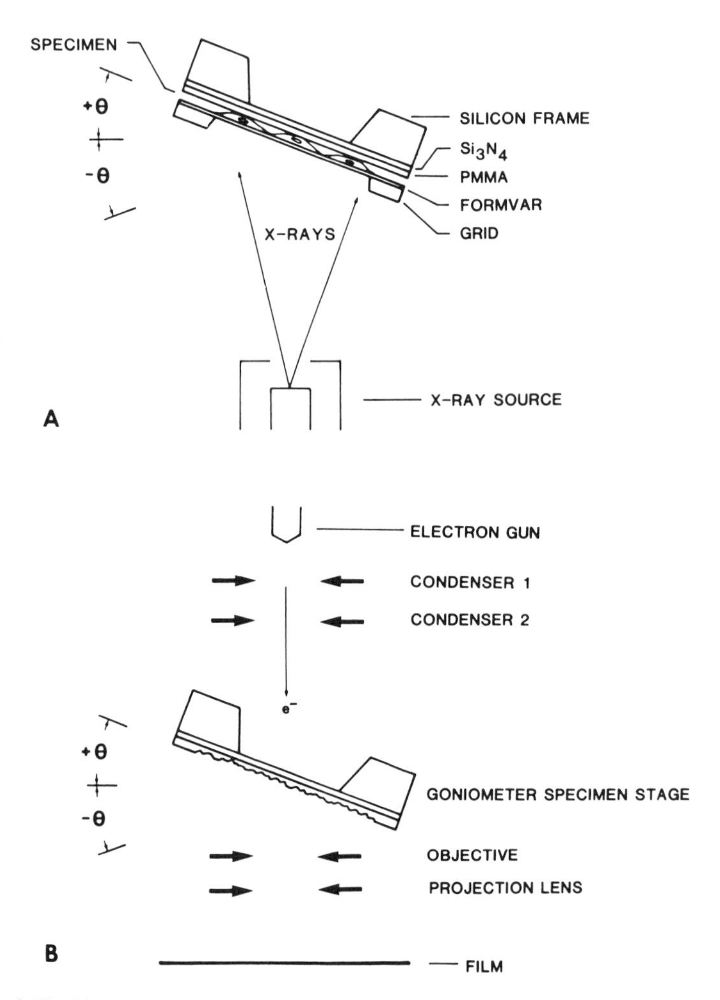

Figure 6.14. Diagrammatic representation of a three-dimensional imaging set-up. (A) The resist-coated window is placed in intimate contact with a cell-bearing grid. Exposures are made while the specimen-window complex is tilted at specific angles (e.g., ±20°). (B) TEM viewing of the contact images is done with the resists tilted at the same angles.

wavelength radiation), x-ray microscopy could be the ideal tool for the study of thick and dense specimens. Three-dimensional image pairs obtained from contact microradiography using photographic plates have been shown in many early papers. Recently, a technique to obtain image pairs from contact images formed on a photoresist has been demonstrated.[14,15]

For conventional transmission electron microscopy, stereo image pairs can be obtained by projecting the same specimen at different tilt angles. A similar technique can be used in x-ray contact imaging to obtain image pairs. However, some modification of the imaging procedure is required. A specimen is placed in intimate contact with a Si_3N_4-supported resist and the exposure is made with the x-ray source positioned at a certain angle (e.g., 20°). Then, the specimen is reloaded with a new resist and exposed at a different angle (e.g., −20°). It is essential to estimate the final magnification and specimen thickness, because the optimum results are dependent on a proper tilt angle; the optimum tilt angle can be determined from the analysis published by Hudson and Markin.[43] After exposure, the two resists are developed identically and viewed under a TEM equipped with a goniometer stage. To facilitate TEM operation, a specially made window holder is used that allows window rotation for alignment against the tilt axis.[15] The TEM projection images of the x-ray contact images are obtained with the photoresists positioned at the proper tilt angles corresponding to the angles at which the exposures were made (Figure 6.14). Stereo image pairs can then be viewed with the aid of a stereo map viewer. Figure 6.15 shows a three-dimensional image pair of a cultured fibroblast.

6.5. Shadow Projection Microscopy

The shadow projection method has the distinction of being the only form of x-ray microscopy that has become commercially established. During a period of almost 15 years several versions were manufactured and sold. It is a very convenient form of microscopy especially suited to industrial problems. The appropriate wavelength X rays for industrial subjects is in the range 0.1–0.2 nm where the specimen may be in normal atmospheric air. The reasons for its abandonment and the prognosis for its revival with improved technical competence are outlined in previous publications.[44,45]

Conceptually the microscope is quite simple; it is the use of shadow enlargement from a point source of X rays. Realization, however, is complex as may be seen in the schematic diagram of Figure 6.16. An electron microscope type column is required to produce a small and bright electron spot, which in turn creates an x-ray spot of sufficient brightness and small enough diameter to provide a useful magnification (several hundred diameters). Unlike a normal x-ray tube, the x-ray target is on a thin window in the vacuum wall. The X rays are transmitted through the target rather than emitted from the surface of electron

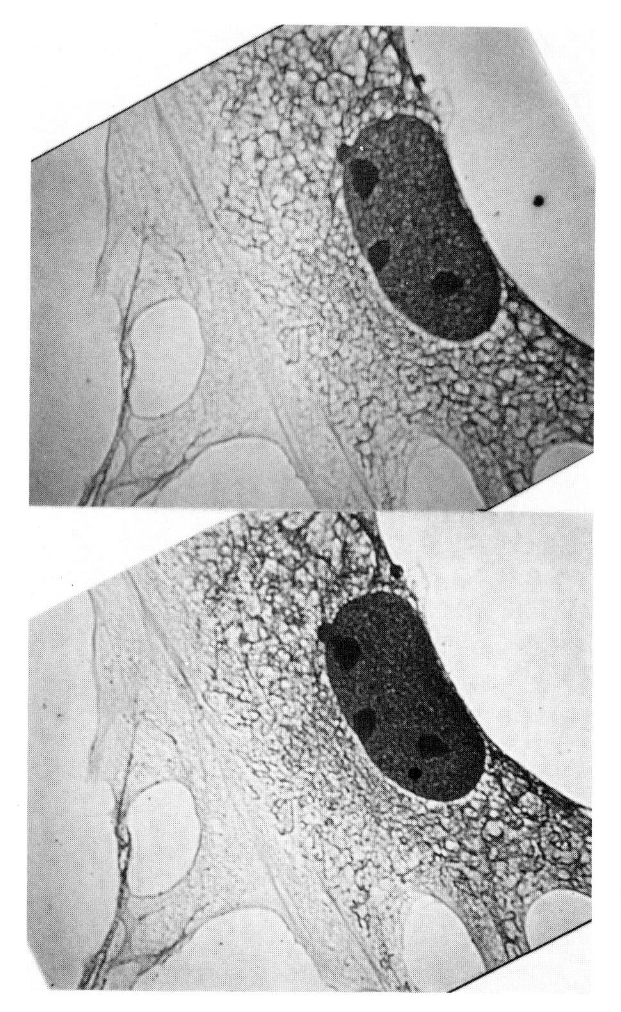

Figure 6.15. Three-dimensional image of a mouse 3T3 fibroblast. The cell was grown on the surface of a Formvar film, fixed in 1% glutaraldehyde and critical-point dried. The contact images were taken at $\pm 20°$ by a stationary target x-ray source with a carbon target.

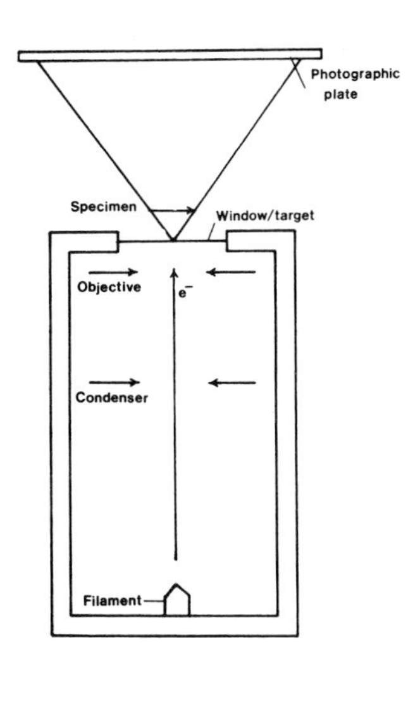

Figure 6.16. (a) Diagrammatic representation of a shadow projection microscope. (b) The shadow projection microscope built by General Electric in the 1950s.

impingement on a massive solid target as in classical x-ray sources. The classical source does not need to be near the specimen irradiation. Here, for magnification up to several hundred diameters the specimen-to-source distance must be 0.1 mm or less for a useful microscope. Fresnel diffraction is also minimized by placing the specimen close to the source.(32) At least one design has placed a solid target close to the specimen by use of a needle-shaped anode close to an x-ray window,(46) but the transmission target as used by all the commercial microscopes can be shown to be the best choice. The design of the transmission target window is a key part of the technology and is discussed more fully below.

To the microscope designer, the electron beam region below the target/window structure (Figure 6.16a) is of major concern. To the microscopist the target/window structure and the x-ray path beyond it are the most important regions, the major factors being the x-ray wavelength, the x-ray intensity, the means of viewing, the means of recording, specimen mounting, and specimen manipulation. The x-ray wavelength is dependent on the tube voltage, which is variable in a range up to 40 kV or more, and on the target material. The x-ray spectrum can be modified by suitable thin filters placed on either side of the specimen. The x-ray intensity has been, in practice, limited by the electron-beam intensity possible in a well-defined spot of diameter 1 μm or less, and not, as might be expected, by melting of the target. Thermal calculations and experience have shown that, because the ratio of area to circumference goes down linearly with spot diameter, at spot sizes of ≈1 μm thermionic electron sources could not melt refractory metal targets. The intensity limit has a major impact on the design and use of the microscope. Both x-ray production by electron impact and light production by fluorescent screens have conversion efficiencies of only a few percent. The product of these two efficiencies results in a visual image that is very dim. At useful magnifications the fluorescent screen image can only be viewed in total darkness and following dark adaptation for up to 30 min for critical focusing. The low x-ray intensity also makes photographic recording very difficult, but fast Polaroid film can be used for both survey work and noncritical images. The ability to expose and process a Polaroid image in one or two minutes makes the shadow projection microscope a practical instrument. Better images are obtained using lantern slide emulsion (e.g., Kodak Lantern Plate, medium contrast) or other wet-processed plates rather than instant film, but with 2–20 min exposures. This range of exposure for good-quality images is set by user tolerances because a clean microscope column is stable for an hour during long exposures.[47] An arbitrary limitation of exposure time to a few minutes and the use of available recording media determine the basic parameters of the microscope design. A series of engineering compromises were independently but uniformly arrived at by the manufacturers. These are a camera distance (source-to-film distance) usually 50 mm and a source-to-specimen distance down to 50 μm. This close approach, in turn, dictates that the target/window is on the outer surface of the vacuum wall with a large coplanar area surrounding it to give adequate space for a specimen holder and manipulator. The design is completed by making the imaging space light tight and totally shielded against x-ray leakage with adequate safety interlocks. A vacuum camera is useful for softer X rays (≥0.8 nm). A complete microscope is shown in Figure 6.16b. This unit (formerly manufactured by General Electric) is the type currently used by two of the authors (PCC and SPN) in their respective laboratories. The major parts of the microscope are identified in the schematic (Figure 6.16a) and follow the same geometric order in the instrument.

Specimen sizes and thicknesses can have large ranges depending on the application. The large electron-accelerating voltage range available easily accommodates samples from the whole periodic table and thicknesses from a few micrometers up to a centimeter or more. Since the most important use of the microscope is the display of three-dimensional structure, the surface finish of the specimen is generally not important. At low voltages (corresponding to long wavelengths) specimen mounting presents a problem. There is no equivalent in the x-ray region to glass transparency in the optical region of the spectrum. A mounting substrate must be thin and of low molecular weight yet hold biological or plastic subjects in proper alignment and be free of vibration or electrostatic forces from x-ray induced ionization. Refractive index variations are small in the x-ray region and thus, unlike in optical microscopy, voids in the specimen are generally welcomed. These may be filled with absorption-contrast material, but it is not necessary to suppress multiple reflections as is done by a clearing agent in the optical case. It is difficult to remove the embedding material from a section without inducing structural drift or disintegration. The use of the polyethylene glycol (PEG) sectioning technique[48] provides a method whereby the embedding matrix can be removed without difficulty after the section has been cut. Heavy metal staining with silver, for example, to stain neurons can be quite useful. With general specimens, preparation and mounting is less difficult for x-ray microscopy than for optical microscopy.

Figure 6.17 shows an x-ray projection image of a terminal influorescence of Northern teosinte (*Zea mays,* subsp. mexicana, race Norbogam). The influorescence was fixed in Cheng's fixative,[33] postfixed in 1% OsO_4, dehydrated in ethanol and critical-point dried. The x-ray image was taken on Kodak Lantern Slide (medium contrast) using a tungsten target and an acceleration voltage of 20

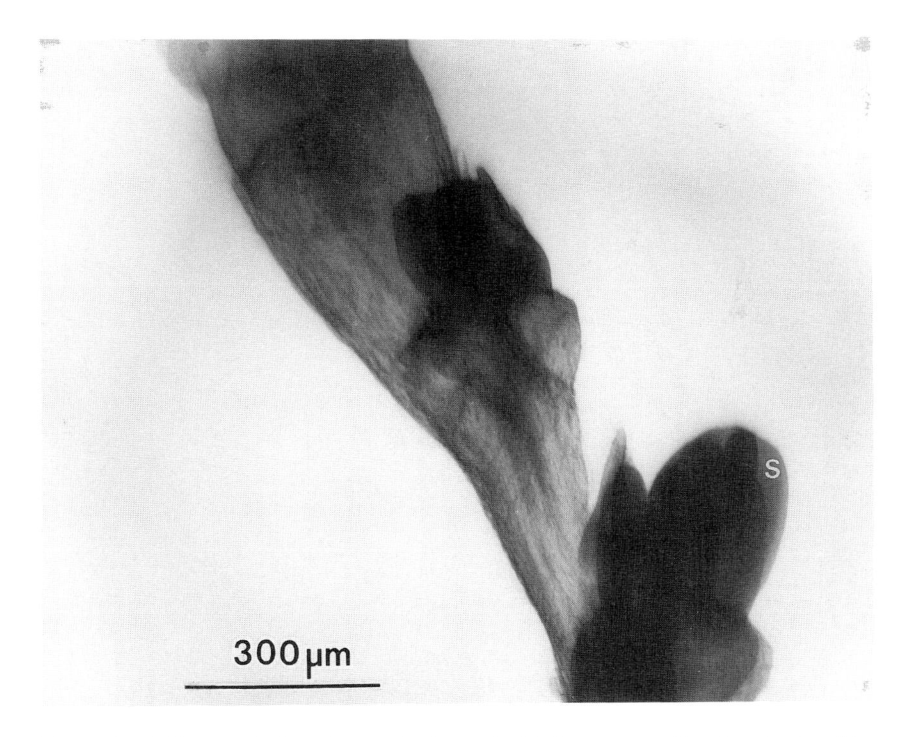

Figure 6.17. X-ray projection micrograph of a terminal influorescence of Northern teosinte (*Zea mays,* subsp. mexicana, race Norbogam). The x-ray image was taken on Kodak Lantern Slide (medium contrast) using a tungsten target (250 nm on a 20 μm beryllium window) at an acceleration voltage of 20 kV. Note the developing stamens (S).

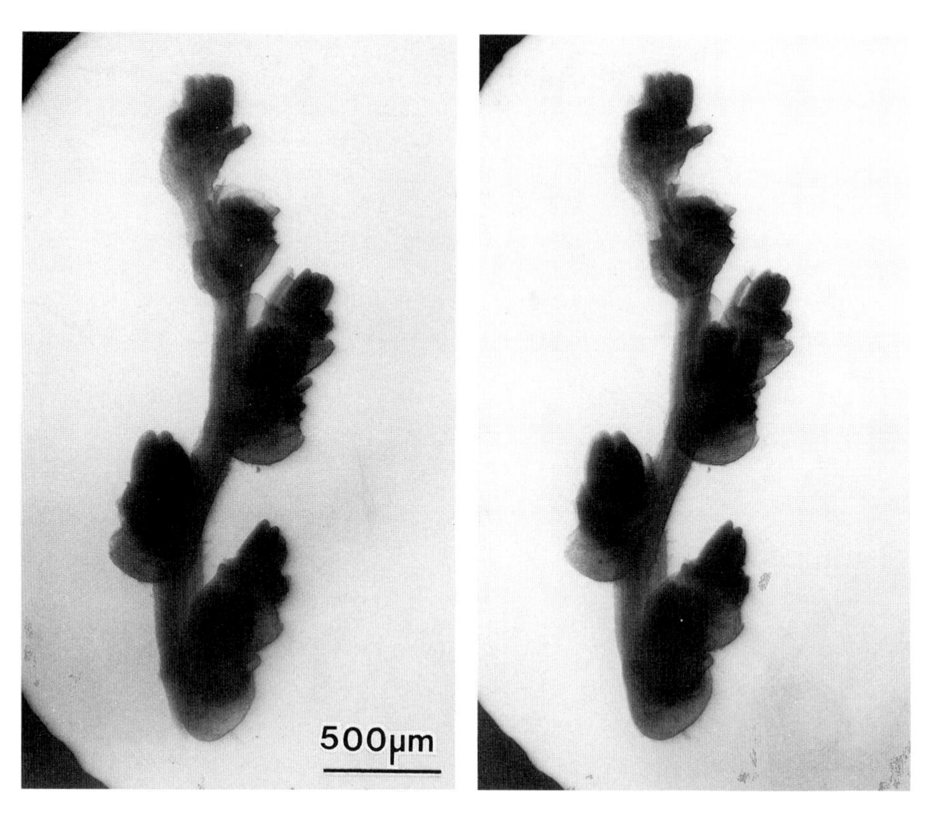

Figure 6.18. X-ray stereo image pair of a terminal influorescence of Northern teosinte. The x-ray images were taken on Kodak Lantern Slide (medium contrast) by using a General Electric shadow projection x-ray microscope operated at 20 kV with a tungsten target (250 nm on 20 μm beryllium).

kV. Figure 6.18 is a stereo image pair of a terminal influorescence of Northern teosinte. Figure 6.19 shows an x-ray projection image of carbon particles.

6.5.1. Thin-Film Targets

Conventionally, the target material is made of 20 μm sheet beryllium coated with a suitable target layer such as 0.25 μm of tungsten. The size of the x-ray spot (typically $\approx$1 μm) is determined by the size of the focused electron beam and the thickness of the target material. Electron scattering in the target layer and in the beryllium support results in beam broadening normal to the electron beam, which increases with increasing film thickness as demonstrated in Figure 6.20 by Monte Carlo calculations for 20-kV electrons. This broadening limits the resolution in shadow projection x-ray microscopy due to increasing of the x-ray spot size. Therefore, the modern approach in achieving a high-resolution projection

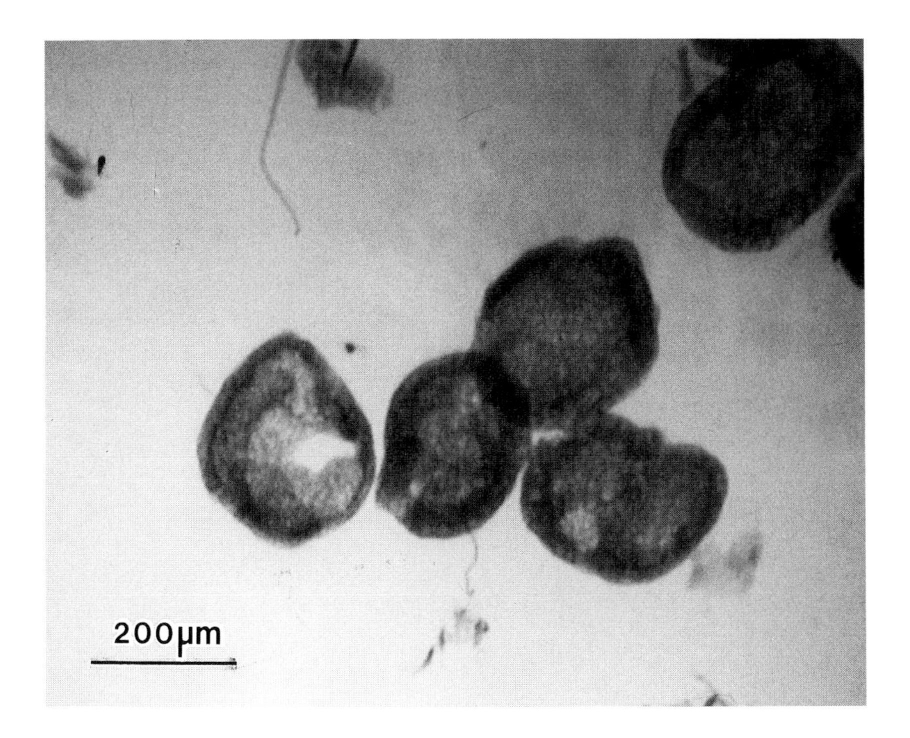

Figure 6.19. X-ray shadow projection micrograph of carbon particles obtained from combustion of viscous oil. The x-ray shadow projection microscope was operated at 10 kV with a tungsten target. The projection image was recorded on Kodak Lantern Slide (medium contrast). (Samples provided by D. F. Dryer and Dr. I. V. Heilweill, Mechanical Engineering and Aerospace Systems, Princeton University.)

image is to use thin-film targets. Small-area (200 μm × 200 μm) silicon nitride thin films (100 nm) have been used in synchrotron radiation research as window materials separating 1 atm pressure difference. Coating of a material onto the silicon nitride window produces a suitable target for shadow projection x-ray microscopy provided that a filter is added to capture the electrons which penetrate the target window into the image space. Figure 6.20 shows that spot broadening is less in thin targets than in thick targets.

Recently, Kolarik and Svoboda[49] attempted to get around the need for an ultra-thin window by using an aluminum filter to remove the Be K alpha radiation while passing the copper radiation. This helps a little but will not solve the problem. The scattered electrons in the Be produce enough white radiation which will pass through the filter and obscure the small source in the copper or other hard target. Filter can remove the Be K alpha and associated wavelength (e.g., B or C filter) might give a low contrast high resolution image, but the technique is no help for the usual energy range.

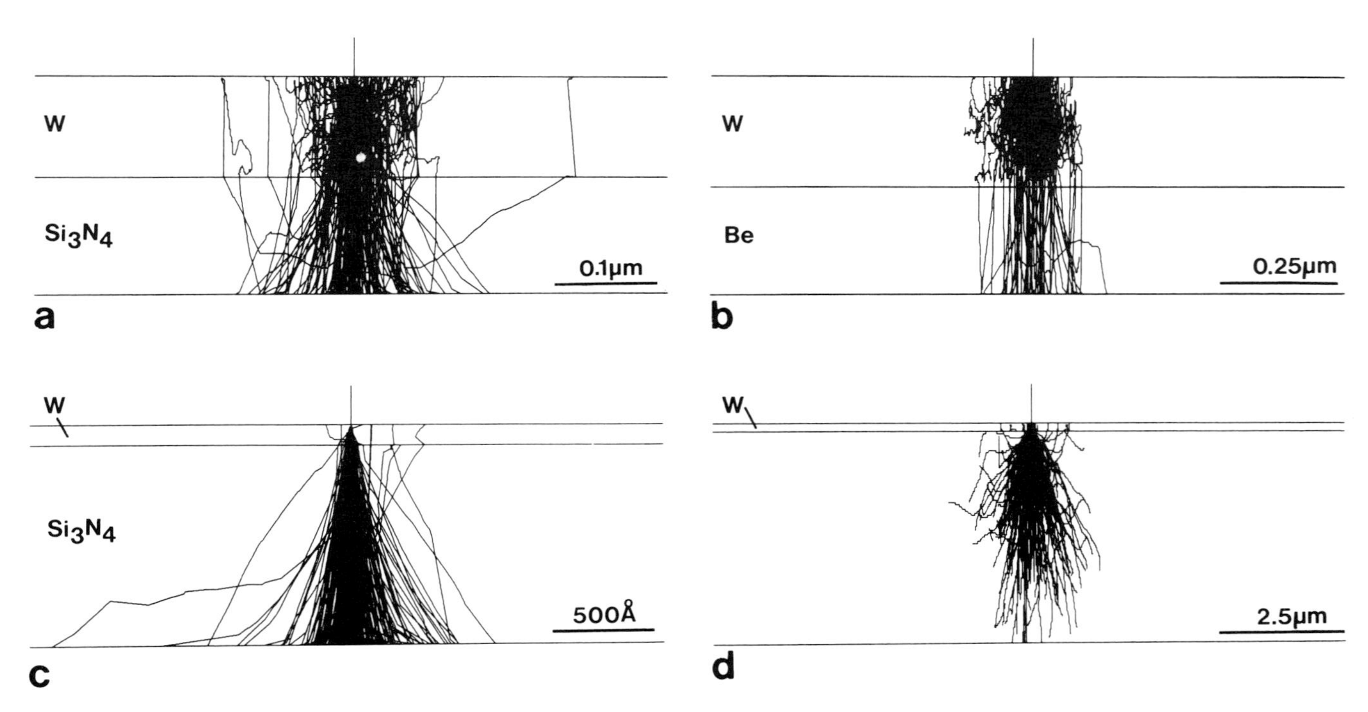

Figure 6.20. Monte Carlo calculation of trajectory of 250 electrons of energy 20 kV in thick and thin targets: (a) Electron trajectory in a conventional target/window (250 nm W on 20 μm Be); (b) electron trajectory in a thinner target (50 nm on 20 μm Be); (c) electron trajectory in a thin-film target (250 nm W on 120 nm Si_3N_4); (d) electron trajectory in a thin-film target (25 nm W on 120 nm Si_3N_4). [Original software was developed and provided by Dr. D. Joy and modified by one of the authors (ISH).]

6.5.2. Real-Time Imaging and Image Processing

The low-intensity projected image can be captured from a fluorescence screen by means of a low-light TV camera, a microchannel plate, or an image intensifier coupled with either a conventional TV camera or a CCD camera.[23,24] Integration of the noisy pictures from the low-light camera is needed to

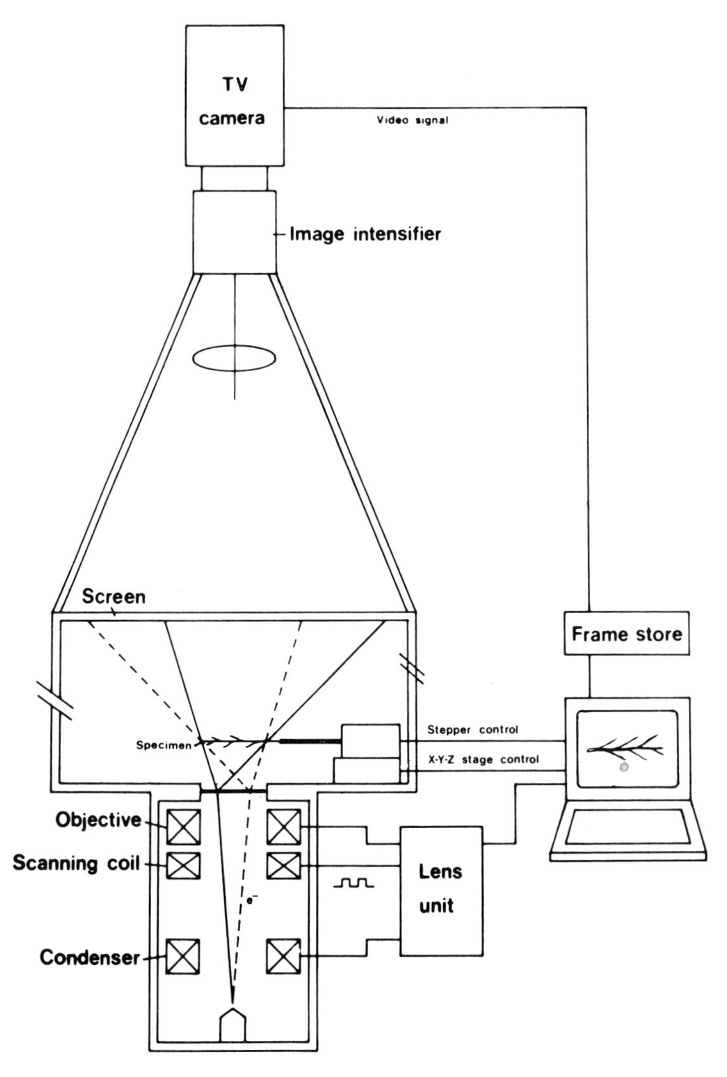

Figure 6.21. A proposed shadow projection x-ray microscope equipped with an image intensifier CCD system and a computer-based image processor.

improve the S/N ratio of the resulting image. Recently, the use of an image intensifier coupled with a TV camera to obtain real-time images from a shadow projection x-ray microscope has been demonstrated.[23] For three-dimensional imaging, the electron beam could be electronically deflected to form two x-ray sources. The two projected images can be grabbed by a low-light TV camera in synchrony with the deflecting signals (Figure 6.21). The two images can be viewed on a monitor in the form of a red–green image pair or using liquid crystal shutter glasses synchronized with the alternating display of the two video images. Implementation of microtomography on the shadow projection x-ray microscope is now a distinct possibility. As is generally known, x-ray tomography is a recent but well-established technique in the medical field with a resolution of about 2 mm. The proposed microscopic adaptation would consist of a low-light TV camera or image intensifier for grabbing the low-intensity fluorescence image formed on a fine-grain phosphor screen. Multiple projection images with different orientations about an axis could be obtained for the reconstruction of the final tomograph.

6.6. Conclusions

X-ray contact microradiography, with its historical development and recent refinement, should become a useful tool in solving specific problems in biological sciences. As has been pointed out, one of the potential applications is the study of mineralization of large area of tissues. The advantage of x-ray microradiography, besides being technically simple and inexpensive, is that the technique offers very high resolution images if the contact image is obtained on a photoresist, and viewed and magnified properly. Being a two-step imaging process, however, contact imaging followed by magnification, x-ray contact microradiography could not be used in studies of dynamic processes that require real-time imaging capability. The use of polymer resists as the photochemical detectors in contact imaging provides significant improvement in resolution over silver halide emulsion. This by no means infers, however, that the silver halide emulsion is outdated. Photographic plates are ideal for the applications in the ≈ 1 μm resolution range.

To date the shadow projection method has been more successful in industrial problems than in biological ones, because contrast and specimen preparation difficulties were not frequently encountered on industrial samples. It is now possible, after 30 years of technological progress, to make a much more convenient and useful x-ray microscope incorporating TV image capture, image processing, ultraclean vacuum, solid-state power supplies, silicon nitride vacuum windows, multilayer x-ray optical elements, and brighter electron sources. It is not inconceivable to consider small bench-top sealed tubes with field-emission sources, one for each target material, as the basic equipment of the projection and contact methods within the next decade. In applications where tunable x-ray capability is essential, then synchrotron sources will probably dominate, and focusing microscopes will probably be the preferred technology. Wherever possible, improvements to the small portable microscope should employ parts and materials from consumer products or other low-cost sources if the microscope is to serve the average laboratory or institution. The combination of TV capture with real-time image processing can simplify column adjustment and dispense with most photography. It would be especially helpful for time-lapse recording and microtomography. A clean vacuum would provide the necessary environment for high-brightness, long-life electron sources and would eliminate the need for frequent adjustment of the optics. Solid-state electronics permit substantial size and weight reductions. Silicon nitride windows and multilayer x-ray optical components can provide contrast improvement and better filtering.

ACKNOWLEDGMENTS. This work was partially supported by the U.S. Department of Energy under agreement DE-AS08-88DP10782 and DE-FC08-85DP40200 and by the Laser Fusion Feasibility Project at the LLE, University of Rochester, which has the following sponsors: Empire State Electric Energy Research Co., New York State Energy Research an Development Authority, Ontario Hydro, and the University of Rochester. Such support does not imply endorsement of the content by any of the above parties. Partial support was also given by grant BRSG SO RR 07066 awarded by the Biomedical Research Support Grant Program, Division of Research Resources, National Institutes of Health, Bethesda, Maryland. We are grateful for the wonderful technical assistance of Mr. M. D. Wittman at the LLE and many helpful suggestions from Dr. Victor K-H. Chen. Special thanks go to the Canadian Synchrotron Radiation Facility at the University of Wisconsin for providing valuable beamtime for the x-ray contact imaging work. The work on x-ray shadow projection microscopy was conducted at the laboratory owned by S. P. Newberry.

References

1. E. Spiller, R. Feder, J. Topalian, D. Eastman, W. Gudat, and D. Sayre, "X-ray microscopy of biological objects with carbon K_α and with synchrotron radiation," *Science* **191,** 1172–1174 (1976).
2. D. Sayre, J. Kirz, R. Feder, D. M. Kim, and E. Spiller, "Potential operating region for ultrasoft x-ray microscopy of biological objects," *Science* **196,** 1339 (1977).
3. R. Feder, E. Spiller, J. Topalian, A. N. Broers, W. Gudat, B. Panessa, J. A. Zadunaisky, and J. Sedat, "High-resolution soft x-ray microscopy," *Science* **197,** 259 (1977).
4. E. Spiller and R. Feder, "The optics of long-wavelength X rays," *Sci. Am.* **239-5,** 70–78 (1978).
5. J. W. McGowan, B. Borwein, J. A. Medeiros, T. Beveridge, J. D. Brown, E. Spiller, R. Feder, J. Topalian, and W. Gudat, "High-resolution microchemical analysis using soft x-ray lithographic techniques," *J. Cell Biol.* **80,** 732–735 (1979).
6. A. Engström, "Quantitative micro- and histochemical elementary analysis by Roentgen absorption spectrography," *Acad. Radiol. Scand., Suppl.* **63,** 1–106 (1946).

7. A. Engström, "Quantitative microchemical and histochemical analysis of elements by X rays," *Nature* **158,** 664–665 (1946).
8. F. Cinotti, M. C. Voisin, C. Jacobsen, J. M. Kenney, J. Kirz, I. McNulty, H. Rarback, R. Rosser, and D. Shu, "Studies of calcium distribution in bone by scanning x-ray microscopy," in: *X-Ray Microscopy: Instrumentation and Biological Applications* (P. C. Cheng and G. J. Jan, eds.), pp. 311–327 Springer, Berlin (1987).
9. R. L. Saunders, "Biological applications of projection x-ray microscopy," in: *5th International Congress on X-Ray Optics and Microanalysis, Gubingen University,* pp. 550–560, Springer, Berlin (1969).
10. B. Niemann, "Current status of the Göttingen scanning x-ray microscope: Experiments at the BESSY storage ring," in: *X-Ray Microscopy: Instrumentation and Biological Applications* (P. C. Cheng and G. J. Jan, eds.), pp. 39–52, Springer, Berlin (1987).
11. W. Meyer-Ilse, G. Nyakatura, P. Guttmann, B. Niemann, D. Rudolph, G. Schmahl, and P. C. Cheng, "Status of x-ray microscopy experiments at the BESSY Laboratory," in: *X-Ray Microscopy: Instrumentation and Biological Applications* (P. C. Cheng and G. J. Jan, eds.), pp. 34–38, Springer, Berlin (1987).
12. G. Schmahl and D. Rudolph, "Proposal for a phase contrast x-ray microscope," in: *X-Ray Microscopy: Instrumentation and Biological Applications* (P. C. Cheng and G. J. Jan, eds.), pp. 231–238, Springer, Berlin (1987).
13. P. C. Cheng, K. H. Tan, J. W. McGowan, R. Feder, H. B. Peng, and D. M. Shinozaki, "Soft x-ray contact microscopy and microchemical analysis of biological specimens," in: *X-Ray Microscopy, Springer Series in Optical Sciences* (G. Schmahl and D. Rudolph, eds.), Vol. 43, pp. 285–293, Springer, Berlin (1984).
14. P. C. Cheng, J. W. McGowan, K. H. Tan, R. Feder, and D. M. Shinozaki, "Ultrasoft x-ray contact microscopy: A new tool for plant and animal cytology," in: *Examining the Submicron World* (R. Feder, J. W. McGowan, and D. M. Shinozaki, eds.), pp. 299–350, Plenum, New York (1986).
15. P. C. Cheng, D. M. Shinozaki, and K. H. Tan, "Recent advances in contact imaging of biological materials," in: *X-Ray Microscopy: Instrumentation and Biological Applications* (P. C. Cheng and G. J. Jan, eds.), pp. 65–104, Springer, Berlin (1987).
16. H. Winick and S. Doniach, *Synchrotron Radiation Research,* Plenum, New York, (1982).
17. E. Koch, D. E. Eastman, and Y. Farge, "Synchrotron radiation: A powerful tool in science," in: *Handbook on Synchrotron Radiation* (D. E. Eastman and Y. Farge, eds.), Vol. 1, pp. 1–63, North-Holland, Amsterdam (1983).
18. B. Yaakobi, P. Bourke, Y. Conturie, J. Delettrez, J. M. Forsyth, R. D. Frankel, L. M. Goldman, R. L. McCrory, W. Seka, and J. M. Soures, "High x-ray conversion efficiency with target irradiation by a frequency-tripled Nd : glass laser," *Opt. Comm.* **38,** 196–200, (1981).
19. N. Takahashi, S. Takahashi, M. Kagayama, and K. Yada, "Three-dimensional visualization of Golgi-stained neurons by a projection x-ray microscope converted from a scanning electron microscope," *Tohoku J. Exp. Med.* **1141,** 249–256 (1983).
20. H. H. Pattee, "High-resolution radiosensitive materials for microradiography at long wavelengths," in: *X-Ray Microscopy and Microanalysis* (A. Engstrom, V. Cosslett, and H. H. Pattee, eds.), pp. 61–65, Elsevier, Amsterdam (1960).
21. D. J. Nagel, "Comparison of x-ray sources for exposure of photoresists," *Ann. N.Y. Acad. Sci.* **342,** 235–251 (1980).
22. D. C. Flanders, "Replication of 170 Ålines and spaces in PMMA using x-ray lithography," *Appl. Phys. Lett.* **36,** 93–96 (1980).
23. S. P. Newberry, "Image capture in the projection shadow x-ray microscope," in: *X-Ray Microscopy II,* Springer Series in Optical Sciences, Vol. 56, pp. 306–309, Springer, Berlin (1988).
24. R. L. Davis, N. A. Flores, and K. T. Evans, "Development and assessment of an image intensifier for real-time x-ray microscopy," *Br. J. Radiol.* **59,** 273–276 (1986).

25. F. Polack and S. Lowenthal, "Photoelectron x-ray microscopy: recent developments," in: *X-Ray Microscopy, Springer Series in Optical Sciences* (G. Schmahl and D. Rudolph, eds.), Vol. 43, pp. 251–260, Springer, Berlin (1984).
26. P. Goby, "Une application nouvelle des rayons X, La microradiographie," *Compt. Rend. Acad. Sci. Paris* **156,** 686–688 (1913).
27. P. Goby, "A new application of Roentgen rays, microradiography," *J. Roy. Mic. Soc.*, August, 373–375 (1913).
28. P. Goby, "La microradiographie stereoscopique en relief et en pseudo-relief. La stereomicroradiographie," *Compt. Rend. Acad. Sci. Paris* **180,** 735–737 (1925).
29. A. Dauvillier, "Sur un tube a rayons X de longueur d'onde effective enable a 8 units Ångstrom," *Compt. Rend. Acad. Sci.* **185,** 1460–1462 (1927).
30. A. Dauvillier, "Realisation de la microradiographie integrale," *Compt. Rend. Acad. Sci.* **190,** 1278–1289 (1930).
31. P. Lamarque, "Technique de l'historadiographie," *Comptes Rendus de l'Association des Anatomistes* **31,** 197–206 (1936).
32. V. E. Cosslett and W. C. Nixon, *X-Ray Microscopy,* University Press, Cambridge (1960).
33. P. C. Cheng, D. B. Walden, and R. I. Greyson, "Improved plant microtechniques for TEM, SEM, and LM specimen preparation," *Nat. Sci. Council Monthly. Rep. of China* **7,** 1000–1007 (1980).
34. D. Sayre, "Imaging properties of the soft x-ray photon," in: *X-Ray Microscopy: Instrumentation and Biological Applications* (P. C. Cheng and G. J. Jan, eds.), pp. 13–31, Springer, Berlin (1987).
35. O. C. Wells and P. C. Cheng, "Examination of uncoated photoresist by the low-loss electron method in the scanning electron microscope," *J. Appl. Phys.* **62,** 4872–4877 (1987).
36. O. C. Wells and P. C. Cheng, "Examination of soft x-ray contact images by the low-loss electron method in the scanning electron microscope," in: *X-Ray Microscopy II,* Springer Series in Optical Sciences, Vol. 56, pp. 316–318, Springer, Berlin (1988).
37. O. C. Wells, "Low-loss image for surface scanning electron microscope," *Appl. Phys. Lett.* **19,** 232–235 (1971).
38. J. Pawlak, P. C. Cheng, and D. M. Shinozaki, "A simple procedure for the fabrication of Si_3N_4 windows," in: *X-Ray Microscopy: Instrumentation and Biological Applications* (P. C. Cheng and G. J. Jan, eds.), pp. 336–345, Springer, Berlin (1987).
39. D. M. Shinozaki and B. W. Robertson, "The examination of topographic images in resist surfaces," in: *X-Ray Microscopy: Instrumentation and Biological Applications* (P. C. Cheng and G. J. Jan, eds.), pp. 105–125, Springer, Berlin (1987).
40. D. M. Shinozaki, P. C. Cheng, and R. Feder, "Soft x-ray induced surface roughness in PMMA," in: *Proceedings of the XI International Congress on Electron Microscopy* (S. Maruse, T. Imura and T. Suzuki, eds), pp. 1763–1764, Japanese Society of Electron Microscopy, Kyoto (1986).
41. B. Yaakobi, H. Kim, J. M. Soures, H. W. Deckman, and J. Dunsmuir, "Submicron x-ray lithography using laser-produced plasma as a source," *Appl. Phys. Lett.* **43,** 686–688 (1983).
42. P. C. Cheng, H. G. Kim, and M. D. Wittman, "Microradiography with laser-produced plasma sources: surface roughness on PMMA resist," *X Rays from Laser Plasmas, Proc. SPIE* **831,** 217–223 (1988).
43. B. Hudson and M. J. Markin, "The optimum tilt angle for electron stereo-microscopy," *J. Phys. E: Scientific Instruments* **3,** 311 (1970).
44. S. P. Newberry, "The shadow projection type of x-ray microscope," in: *X-Ray Microscopy: Instrumentation and biological Applications* (P. C. Cheng and G. J. Jan, eds.), pp. 126–141, Springer, Berlin (1987).
45. S. P. Newberry, "History of x-ray microscopy," in: *X-Ray Microscopy: Instrumentation and Biological Applications* (P. C. Cheng and G. J. Jan, eds.), pp. 346–360, Springer, Berlin (1987).

46. B. M. Rovinsky, V. J. Lutsau, and A. T. Avdeyenak, in: *X-Ray Microscopy and X-Ray Microradiography* (V. E. Cosslett, A. Engstrom, and H. H. Pattee, eds.), pp. 269–277, Academic, New York (1957).
47. S. P. Newberry and S. E. Summers, "The General Electric shadow x-ray microscope," *Proc. Int. Conf. Electron Microscopy,* 305–307 (1954).
48. J. J. Wolosewick, "The application of polyethylene glycol (PEG) to electron microscopy," *J. Cell Biol.* **86,** 675–681 (1980).
49. V. Kolarik and V. Svoboda, "An x-ray projection microscope with field emission gun," *J. of Mic.* **156**(2), 247–251 (1989).

7

Progress and Prospects in Soft X-Ray Holographic Microscopy

M. R. Howells, C. Jacobsen, J. Kirz, K. McQuaid, and S. S. Rothman

7.1. Introduction

The majority of x-ray imaging experiments currently use the contact technique or x-ray analogues of the optical microscope, either in direct imaging or scanning mode, as described in Chapters 4–6. It is also possible, however, to obtain sample information by exploitation of the diffracted field, as is done in crystallography. To do this a method for determining and using the phases of the diffracted wave is needed. In the soft x-ray region, holography is one way to provide such a method. Other ways have been proposed by Sayre.[1]

It was in 1948 that Gabor[(2)] first pointed out that it is possible to record both the amplitude *and* phase of a wave using an intensity detector provided that a suitably coherent phase reference wave is available to beat against the signal wave. He further showed how to use the recording as a diffracting structure to make a reconstructed image of the object emitting the original signal wave. X-ray holography seeks to use the information carried by the X-rays diffracted by a sample and to record and interpret it using the method of Gabor. The use of such optical methods is possible in the soft x-ray spectral region and there is a long history, beginning in the early 1950s, of attempts to make x-ray holograms using

M. R. Howells and C. Jacobsen • Center for X-Ray Optics, Lawrence Berkeley Laboratory, Berkeley, California 94720. ***J. Kirz*** • Department of Physics, State University of New York at Stony Brook, Stony Brook, New York 11794. ***K. McQuaid and S. S. Rothman*** • Schools of Medicine and Dentistry, University of California–San Francisco, San Francisco, California 94143.

x-ray tube sources and photographic film detectors. The hope was that the hologram could be recorded with X-rays and reconstructed with visible light, thereby achieving a three-dimensional "microscope" with resolution superior to the light microscope and without the need to fabricate a suitable lens.

These hopes were never realized. Holograms of reconstructable quality were obtained in relatively few cases(3,4) and none gave images with resolution higher than the light microscope. As a result, x-ray holography became a dormant field by the mid-1970s.

This general failure can be understood in terms of the poor coherence properties of the x-ray tube sources and the low resolution of the photographic film detectors that were used. Such an understanding gives insight into what is needed for successful x-ray holography, and forms the basis for the belief that current technological advances are opening the way to a productive future for x-ray holography. As explained below, high-coherence undulator sources of soft X-rays, high-resolution resist detectors, and digital image processing systems for making the reconstruction provide new capabilities for x-ray holography in the resolution regime between that of the optical and electron microscopes. Such approaches share the advantages of other soft x-ray imaging methods,(5) viz. applicability to samples in water in an atmospheric-pressure air environment, sufficient penetration to image unsectioned cells, contrast (without stains) based on x-ray absorption edges, and freedom from many of the background noise processes that afflict charged particle probes.

In what follows some of the latest developments in x-ray holography experiments are reported, and the limits of performance of the approaches currently in use are discussed. Also, some suggestions about where the techniques can (and cannot) go in the future are made.

7.2. The New Technologies

One of the main goals of all soft x-ray imaging techniques is to improve on the resolution of the light microscope, and so it is important to consider the ways in which this may be achieved in holography. In Gabor (in-line) holography (Figure 7.1), the resolution is limited by the detector resolution.(6) This means that a photoresist should be used as the high-resolution alternative to photographic film. Such a strategy was first used by Bjorklund et al.(7) and, for polymethylmethacrylate (PMMA) resist, extends the detector limit to a value variously estimated(8,9) to be in the range 5–20 nm. Of course, resist is much slower than film but this is a necessary concomitant for such a substantial improvement in resolution. The real measure of the wastefulness of a detector, the detective quantum efficiency, is reportedly not much different for resist(8) than for film(10) in the soft x-ray region.

For the large jump in resolution involved in changing to resist it is obviously

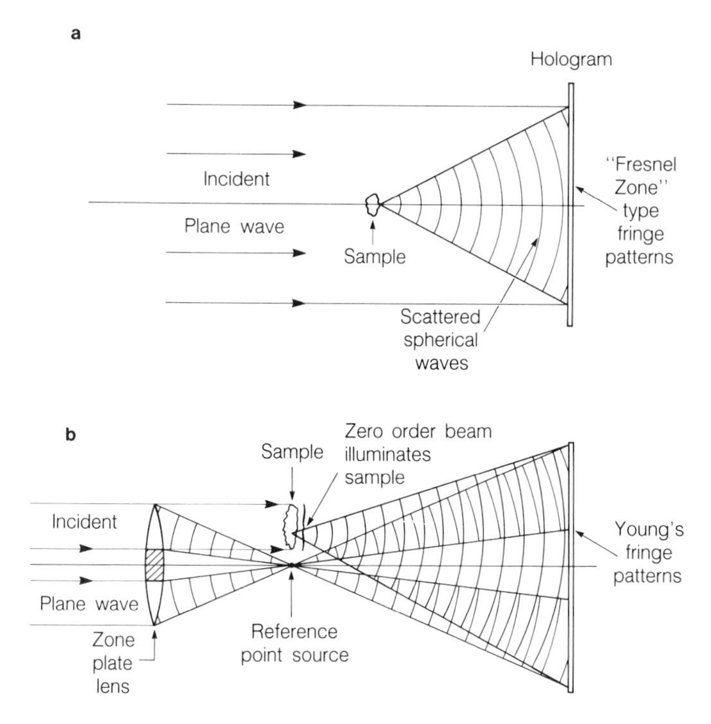

Figure 7.1. Layout of the Gabor (in-line) (a) and Fourier transform (b) holographic geometries.

necessary to provide many more illuminating photons and correspondingly higher radiation dose to the sample. Some attention has been given[11–13] to the scaling law of the needed photon fluence (and therefore the dose) as a function of resolution. It is argued[11–13] that the dose scales as the sixth power of the transverse resolution for a three-dimensional experiment, which means one where the sample thickness is significantly greater than the depth resolution. Otherwise, the experiment is two dimensional and the scaling law becomes a square law. The most popular way to provide an adequate flux of temporally and spatially coherent soft X-rays of wavelength 2.3–4.4. nm (which is the most appropriate range for making holograms of biological samples) is to use an undulator on an electron storage ring. A number of such devices are becoming available at the present time.[14] When operated on a storage ring that has a suitably low electron-beam emittance (the product of the spatial and angular widths) undulators provide near diffraction-limited soft x-ray beams with many laserlike properties.

As will be shown in the next section in more detail, the use of an undulator source and a resist detector makes it possible to record holograms with much finer fringe detail than the earlier workers achieved. Such holograms are now being made in France,[15] Japan,[16] and the United States.[17,18] The French group are based at the Institut d'Optique and LURE and have used an undulator on the ACO storage ring to make phase holograms of diatoms and other objects on preexposed resist. They have chosen to develop an analogue approach to reconstruction that uses sophisticated optical correction methods to deal with aberrations. The procedure is optimized with regard to speed, convenience, and signal-to-noise rather than resolution. Further improvements can be anticipated in the performance of this scheme when the program transfers to the new Super ACO storage ring. The group in Japan is using an undulator on the Photon Factory storage ring in a natural continuation of the earlier work which achieved the most successful holograms of the 1970s. The experiments so far reported are in an early stage, but recent improvements to the storage ring have made it a highly suitable source for the continuation of this widely admired holography program.

A further step forward in holographic imaging can be achieved by using numerical processing as a means of reconstructing the final image.[19] The availability and power of hardware capable of doing this is increasing at a rapid rate at the present time. Such an approach provides one way to avoid the resolution limitations normally involved in visible light reconstruction. It also gives much more flexibility in finding the focus; eliminating nonlinearities in the record-develop-read-out-digitize sequence; and in dealing with the twin-image problem inherent in the in-line holographic method. Furthermore, the end result in a computer reconstruction provides both the amplitude and phase of the image signal and this has potentially important applications. An example of the use of these digital methods is given in the next section.

7.3. X-Ray Holographic Experiments at the NSLS

The present authors have been implementing a program of x-ray holography at the National Synchrotron Light Source (NSLS) at Brookhaven National Laboratory for some time, and these experiments have recently begun to use an undulator source with resist as the recording medium.[17,18] The optical system that has been used to record Gabor x-ray holograms is shown in Figure 7.2. The NSLS X-17T mini-undulator beamline[20] was used to provide a spatially and temporally coherent beam of 2.5-nm X-rays. The temporal coherence (about 1 μm coherence length) was achieved by means of a monochromator (spectral filtering) with spatial coherence being given by a pinhole (spatial filtering). The coherent flux was about 10^8 photons s^{-1} which allowed the recording of a stack of holograms at 400 μm spacing (Figure 7.3) in about 1 hr.[31] The recordings were made on 200 nm thick layers of resist coated onto 120-nm-thick silicon nitride windows supported on silicon frames. The resist[7] was either PMMA or a

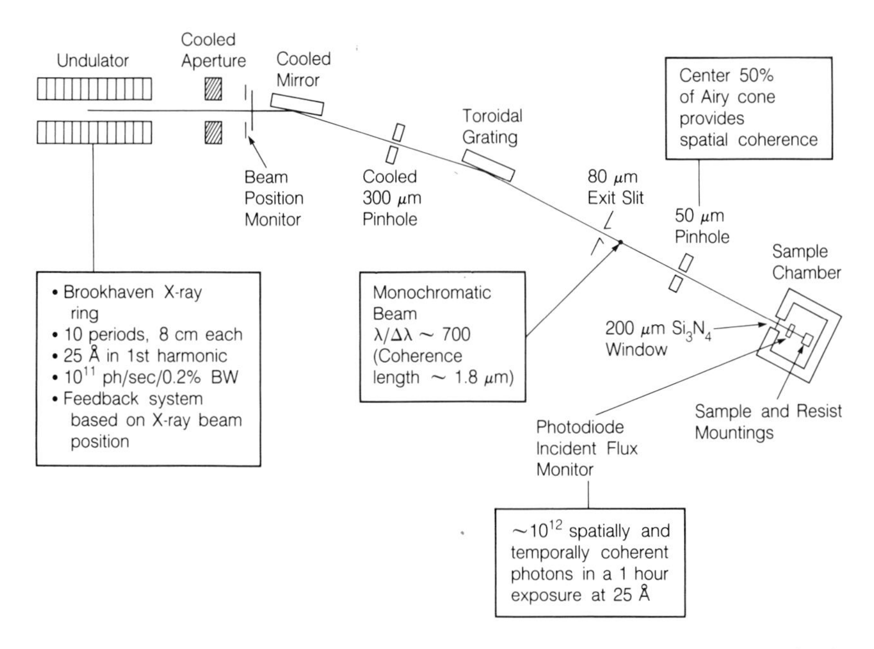

Figure 7.2. Optical layout and parameters for the recording of Gabor x-ray holograms using the X17T indulator beamline at the National Synchrotron Light Source at Brookhaven National Laboratory.

copolymer of 80% methyl methacrylate and 20% methacrylic acid (MMA–MAA). These are positive resists, which means that they are etched more quickly by solvents in regions of radiation exposure. After exposure the resists were developed by immersion in a solvent—methylisobutyl ketone diluted with iso-propanol. The interference fringes given by coherent superposition of the (roughly spherical) wave scattered by the sample and the plan incident wave were then formed on the resist surface as a relief pattern (Figure 7.4). To give good contrast, the resist was shadowed with gold–palladium at glancing incidence and imaged in a transmission electron microscope (TEM), which produced a photographic negative. The final step of the experiment was to digitize the negative with a microdensitometer to produce the numerical data that forms the starting point for the analysis.

The analysis takes advantage of the fact that all holograms can give an aberration-free reconstruction if they are illuminated with the original reference wave.(21) This would not be a useful thing to do in the present case because the reconstructed image would not be magnified. The same process can be mimicked in a computer, however, and the result displayed to get any required magnification. This procedure also has the other advantages mentioned earlier.

The calculation(17,18) consists of numerically simulating the propagation of

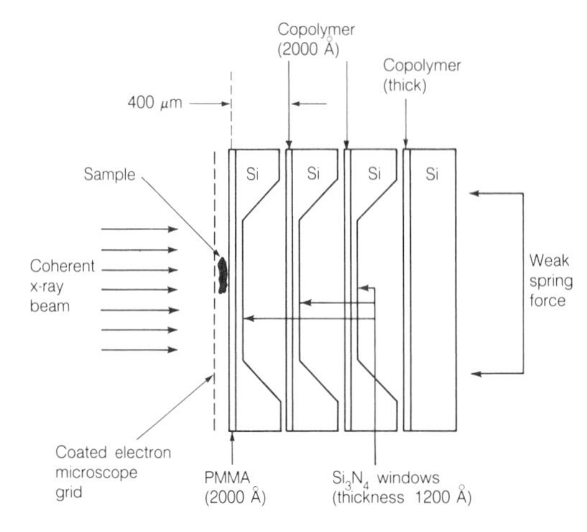

Figure 7.3. Arrangement for simultaneously recording several holograms and a contact micrograph of a sample.

the reference wave from the hologram to the real image in the Fresnel Approximation. This involves taking the Fresnel Transform of the data that are considered to represent a thin, amplitude hologram. The algorithm for doing this involves multiplying the data entries by a quadratic phase factor and taking the Fast Fourier Transform. On a MicroVAX II computer this takes about 5 min. The process of focusing takes place at this stage of the procedure and involves adjusting the propagation distance. An example of a reconstructed image is shown in Figure 7.5. The samples were mounted on electron microscope grids and, by reconstructing the image of the edge of one of the grid bars, the system resolution, defined as the distance from 25% to 75% of the step height, has been determined to be around 40 nm.

These experiments were carried out with the dual purpose of developing the technique and studying the process of secretion, particularly as revealed by structural details of secretion granules. The granules used in the experiments were obtained from the pancreatic acinar cells of fasted rats, and are known as zymogen granules. They were fixed in 1.5% glutaraldehyde in 150 m*M* sucrose, but were unsectioned and unstained. Efforts to understand the data in terms of the morphology of the granules are just beginning. For the moment the use of x-ray holography in making interesting images of biological samples has been shown, and resolution less than 100 nm has been demonstrated in such an experiment.

7.4. Three-Dimensional Imaging

One of the main goals of holographic imaging is to provide three-dimensional information. The diffraction limits to this are the same as in imaging with lenses and are related to the numerical aperture (NA); the transverse resolution is

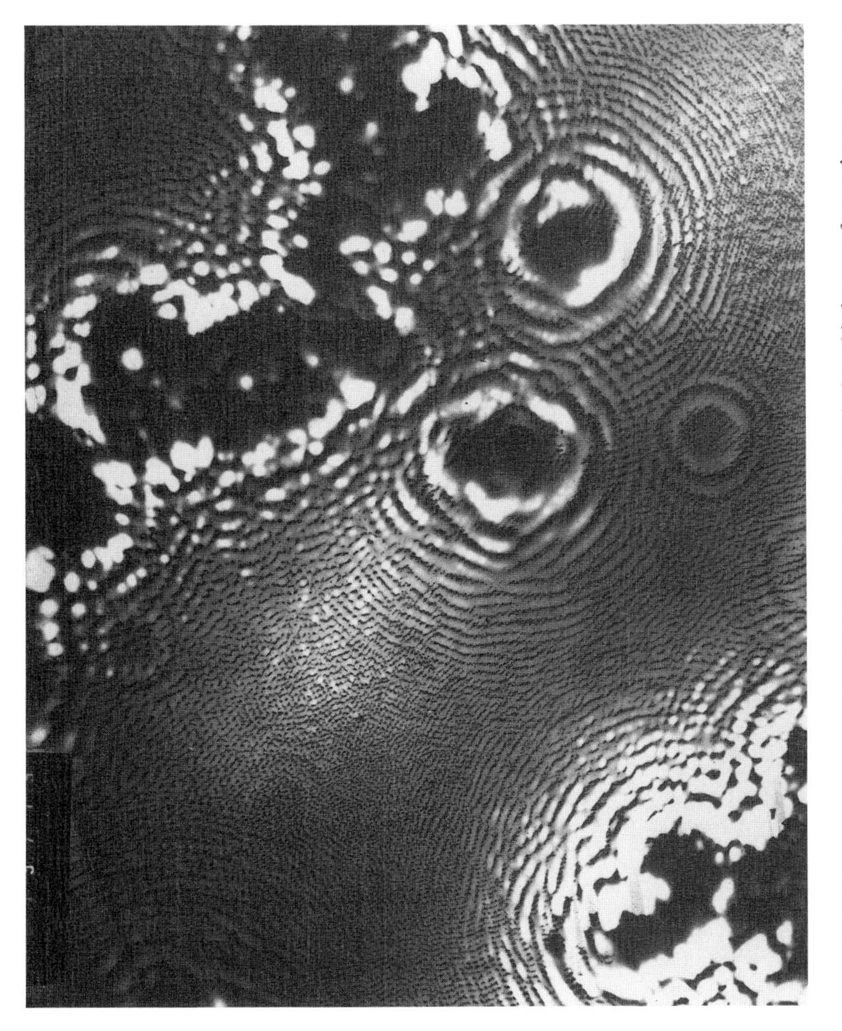

Figure 7.4. Electron micrograph of part of the glancing-incidence shadowed hologram of several zymogen granules, recorded on copolymer. The portion shown is 19 μm × 15 μm.

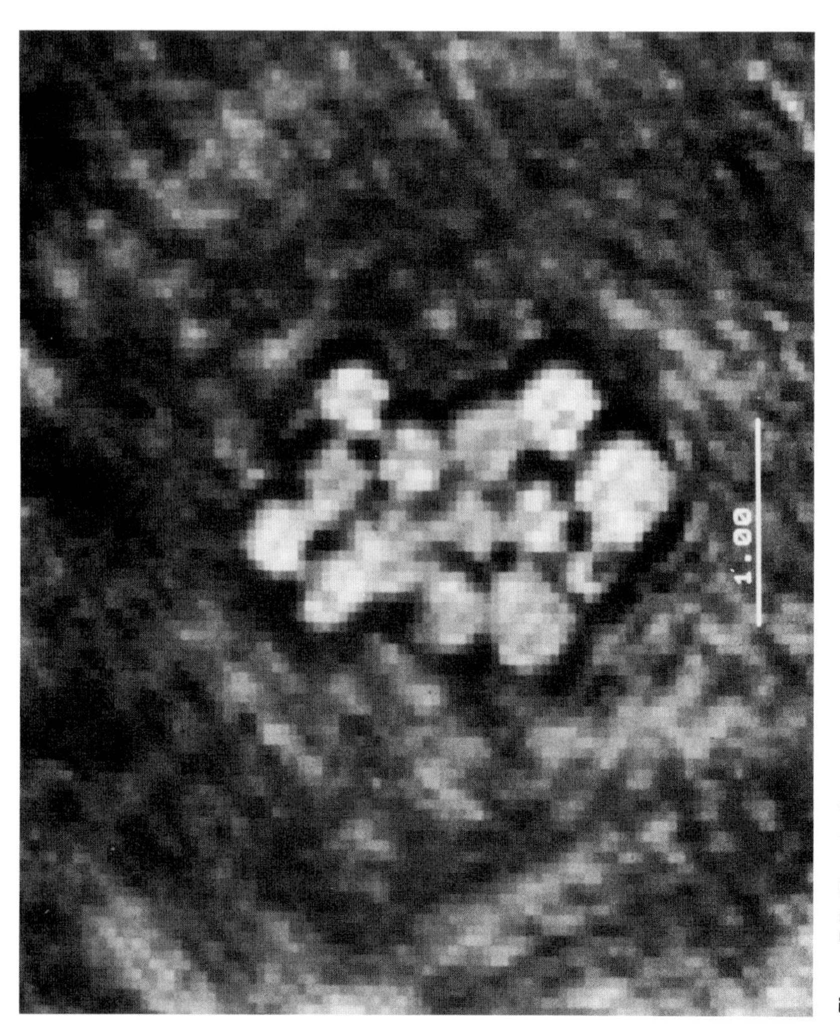

Figure 7.5. Numerically reconstructed image from the hologram shown in Figure 7.4. The diffraction structure near the center of the hologram is seen to be due to a clump of granules that are resolved in the image. The pixel size is 49 nm.

$\lambda/(2NA)$ and the longitudinal resolution is $\lambda/(NA)^2$. The NA in holography is not normally defined by the size of a physical aperture such as the edge of a lens but is determined in a complex way by the transfer function of the resist as a function of frequency (aperture angle), the power spectrum of the sample, and the coherent x-ray exposure. These parameters determine the roll-off of the signal-to-noise ratio at high frequencies, and hence the NA and resolution.

One conclusion that can be drawn from this is that the transverse and longitudinal resolutions are related through the numerical aperture, and that for experiments at low transverse resolution (low NA) the longitudinal resolution may be much larger than the sample thickness leading to an essentially two-dimensional image. To achieve three-dimensional imaging the numerical aperture, and thus both types of resolution, has to be improved. For example, the Brookhaven experiments reported above had a numerical aperture of about 1/40. If this could be improved to about 1/8, then a transverse resolution of 10 nm and a longitudinal one of 160 nm would be achieved, which would be quite useful. However, the dose required scales as the sixth power of the transverse resolution, so progressing toward 10 nm resolution would certainly imply that extraordinarily high doses would be needed. For samples composed mainly of biological material it would be hard to imagine making more than a few exposures (say two for a stereo-pair) if these limits were approached.

Another possibility may be exemplified as follows. Suppose that an exposure has been made at some given resolution and that a way exists to improve that resolution by a factor of two for a hologram from a single view direction. This would give four times better depth resolution and 64 times more dose. An alternative strategy, however, using the same dose, would be to forego the factor of two resolution improvement and make 63 more exposures at the same resolution with different illumination directions. This would provide greatly superior 3-D mapping and would lead to a form of tomography that is usually called diffraction tomography[22,23] because the wavefield emerging from the sample spreads out and propagates according to the laws of diffraction. This is contrary to the conditions normally encountered in computerized axial tomography (CAT) scanning, which is based on the laws of geometrical optics.

The two choices discussed above represent opportunities to seek either the best resolution or the best three-dimensionality. It is arguable that resolution is not the only criterion for useful imaging and that the advantage of being able to measure the full three-dimensional density map, or other type of information such as chemical, may sometimes justify some compromise of the resolution.

The choice between the two approaches is partly an instrumental one. If the available source was an x-ray laser, then the intrinsic characteristics of the source (large coherence length and short pulse length) would allow large doses to be used because the damage processes are slower than the pulse. This would favor pushing the resolution limit with single-shot imaging. With a synchrotron source the diffraction tomography option would be somewhat more compelling.

The other issue is not entirely a scientific one. It has to do with how people process visual information and draw inferences from it. It is not easy to know what type of three-dimensional information a biologist needs to have in order to advance understanding of a sample. We are accustomed to processing information about the surfaces bounding opaque objects because this is what perception based on visible light usually gives us. In the world of X-rays, however, nothing is fully opaque and we could be faced with genuine three-dimensional density distributions, perhaps containing both phase and amplitude information. Such a distribution will often be difficult to view without technical aids such as the ability to cut sections and remove parts, and will be highly unfamiliar to the observer. We are left with a software problem and a perception question. Is it necessary to map every tree in order to appreciate the essential nature of a forest or would a suitably three-dimensional picture from one viewpoint be enough?

A circumstance that makes it even harder to weigh the above two approaches against each other is our relative lack of experience in viewing single holograms of partially opaque objects even in visible light. Nevertheless, when looking at a reconstructed image of a good hologram of normal (fully opaque) objects, the feeling certainly exists of having more information about the scene than the Ewald sphere/information theory arguments would indicate. Perhaps, as so often is the case in choices of method, each will prove to have their own particular merit. At the present time it seems that both deserve to be pursued.

In view of the unfamiliar nature of diffraction tomography in both the soft x-ray and the optical microscopy communities, some of the basic ideas are reviewed in the next section, and some references where further information can be found are provided.

7.5. Diffraction Tomography

A diffraction tomography experiment could consist of illuminating the sample with monochromatic plane waves and measuring both the amplitude and phase of the diffracted field at some plane perpendicular to the illuminating direction and downstream of the sample. The same experiment would then be repeated many times for different illumination directions. Starting from an understanding of CAT scanning it might be supposed that, by allowing the wave to spread and diffract, the information content would be compromised severely. It was shown by Wolf in 1969,[24] however, that the information may in fact be preserved and can be extracted by taking the complex two-dimensional Fourier transform of the measured wavefield. Provided that the measured wavefield is the same as the Born approximation to the wavefield, this determines certain values of the three-dimensional Fourier transform of the complex refractive index distribution of the scattering object at Fourier frequencies lying on the Ewald sphere in frequency space. This is known as the generalized projection slice theorem.

By making further measurements at different illumination directions it is possible to fill in more data points on other Ewald spheres that are rotated about the origin with respect to the first one. When a sufficient number of spheres of data have been accumulated, one strategy is to interpolate the data to fill the frequency space on an appropriate grid of points. A three-dimensional Fourier transform then returns the desired values of the complex refractive index distribution of the sample.

This procedure has been applied using sound, radio waves, microwaves, and seismic waves, but not, apparently, using X-rays, presumably due to the difficulty in obtaining the phases. The computational methods that can be used are not limited to those based on interpolation and the generalized projection slice theorem as described above. A good deal of effort has been devoted to designing very general and powerful algorithms(22,23) which deal with limited amounts of data and noisy data and which allow prior knowledge to be used. The interest here is in understanding the conditions under which this kind of approach could work with X-rays. At first sight it seems that the whole procedure depends on the use of the Born approximation. However, cases where the Born approximation breaks down can be treated by a device called the Rytov approximation.(23,25,26) This method essentially provides an approximate way to calculate what the Born approximation to the scattered field *would* have been had a direct, forward calculation been done using it. From the point of view of implementation the Rytov and Born methods involve virtually the same amount of processing.

In Table 7.1 some optical data for some biological materials at an x-ray wavelength of 3 nm are given. These provide a feeling for whether the Born and Rytov validity criteria are met in soft x-ray imaging experiments. The criteria for both are based on the idea of a linear approximation to the wave equation for "weak" scattering of the incident wave. The exact calculation is very intractable and little progress has been made with it. Both approximations usually require that the scattering potential (i.e., the refractive index distribution) should be

Table 7.1. Optical Properties of Biological Materials at 3 nm

	Water	Protein	DNA	Lipid	Carbohydrate
Index real part (δ)[a]	0.0010	0.0015	0.0017	0.0012	0.0017
Index image part (β)[a]	0.000047	0.00057	0.00076	0.00038	0.00039
Phase change per wavelength ($2\pi\delta$) rad	0.0063	0.0094	0.011	0.0078	0.011
Attenuation per wavelength ($4\pi\beta$)	0.00059	0.0072	0.0095	0.0048	0.0049
Absorption length (μm)	5.1	0.42	0.32	0.36	0.61
Phase change per absorption length ($\delta/2\beta$) rad	10.7 (2.1 per μm)	1.3	1.1	1.62	0.2

[a]The complex refractive index is taken to be $1-\delta-i\beta$.

expressible as the sum of a background term that is real and a sample term that may be complex but is small compared to the background. This is well satisfied for X-rays because the refractive index is always equal to unity minus small correction terms. Both approximations also require that the scattered amplitude be small compared to the incident amplitude. This is also likely to be true for all soft x-ray diffraction experiments. The essence of the Born approximation is that each element of the sample is assumed to be illuminated with the *unmodified* incident wave. This requires that the total attenuation be small as well as the scattering, and this is certainly not satisfied by most soft x-ray imaging experiments. It is this difficulty that is resolved by the Rytov approximation because it requires only that the attenuation and phase change *per wavelength* should be small. This is a much easier requirement and, as shown in Table 7.1, is well satisfied for the kind of experiments envisaged.

The conclusion is that soft x-ray imaging experiments in or near the water-window spectral range (2.3–4.4 nm) do indeed satisfy the conditions for the use of the established procedures of diffraction tomography using the Rytov approximation. This opens possibilities for new types of experiments in the future.

7.6. Future Developments

The experiments reported above demonstrate resolutions of 40–100 nm with doses of 200 Mrad. However, they do not provide a very complete basis for projections about the future. First, the microdensitometer data were smoothed to diminish the size of the data set, and it is not presently known how much, if any, resolution was lost by this procedure. Second, no attempt was made to minimize the dose to the sample—the same resolution could perhaps have been obtained with less dose. It can be argued, based on spectral analysis of the unsmoothed microdensito-meter data, that information was recorded at about 20–30 nm, which is what should be possible with the dose used. It is also very difficult to estimate the ultimate dose that the sample can tolerate without loss of the interesting structures. The data(27,28) for the radiation tolerance of various organic materials in the electron microscope suggest a value around 1000 Mrad with about an order of magnitude variation in either direction for different materials. The endpoints used in these tabulations are not quite the same as in an x-ray imaging experiment, but they represent the closest available measurements. The types of holography experiment being considered are very much dependent on the properties of resist and the resolution limit set by this consideration is in the range 5–20 nm.

Remembering the sixth-power law referred to earlier, which gives the scaling of the needed dose with resolution for three-dimensional experiments, the above observations can be combined into a useful generalization about what might be achieved in the future. For a single hologram made with the currently used technologies, estimates of the ultimate dose-limited resolution are in rough agreement with those of the ultimate resist-limited resolution and both are in the region of 10–20 nm. Thus, it is reasonable to regard 10 nm resolution as a goal. If 10 nm resolution were achieved, then the depth resolution of single holograms would be useful and would be about 160 nm for 2.5-nm X-rays.

Some effort is presently being devoted to developing Fourier transform holography (Figure 7.1).[29,30] This approach does not depend on having a high-resolution detector and allows devices such as charge-coupled devices to be used instead of resist or film. The achievement of good resolution then depends on providing a coherent reference source of sufficiently small size. The responsibility for resolution is thus shifted from the detector to the condensing optics. Less progress has been made in this direction than in Gabor holography because it is considerably more difficult. There are some persuasive arguments, however, that it will be rewarding when it is implemented successfully, particularly with regard to dose reduction.

ACKNOWLEDGMENTS. The authors are grateful for much valuable help and advice from C. Dittmore, R. Feder, J. Grendell, D. Joel, R. Nawrocky, H. Rarback, and D. Sayre as well as the generous support of the staff of the National Synchrotron Light Source (NSLS) at Brookhaven National Laboratory. We also acknowledge the support of the National Science Foundation under grants BBS-8618066 and DMB-8410587 (CJ and JK) and the Department of Energy under contract DE-ACO3-76SF00098 (MRH and CJ). The work was carried out in part at the NSLS, which is supported by the Department of Energy under contract DE-ACO2-76CH00016.

References

1. D. Sayre, in: *X-Ray Microscopy: Instrumentation and Biological Applications* (P. C. Cheng and G. J. Jan, eds), pp. 13–31 and 213–223, Springer, Berlin (1987).
2. D. Gabor, "A new microscopic principle," *Nature* **161,** 777–778 (1948).
3. S. Aoki, Y. Ichihara, and S. Kikuta, "X-ray hologram obtained by using synchrotron radiation," *Jpn. J. Appl. Phys.* **11,** 1857 (1972).
4. S. Aoki and S. Kikuta, "X-ray holographic microscopy," *Jpn. J. Appl. Phys.* **13,** 1385–1392 (1974).
5. J. Kirz, "Specimen damage considerations in biological microprobe analysis," in: *Scanning Electron Microscopy II* (O. Johari and R. P. Becker, eds.), pp. 239–249, SEM Inc., AMF O'Hare (1979).
6. A. V. Baez, "A study in diffraction microscopy with special reference to X rays," *J. Opt. Soc. Am.* **42,** 756–762 (1952).
7. G. C. Bjorklund, S. E. Harris, and J. F. Young, "Vacuum ultraviolet holography," *Appl. Phys. Lett.* **25,** 451–452 (1974).
8. E. Spiller and R. Feder, "X-ray lithography," in: *X-ray Optics* (H.-J. Queisser, ed.), Topics in Applied Physics, Vol. 22, pp. 35–92, Springer, Berlin (1977).
9. D. M. Shinozaki, P. C. Cheng, and R. Feder, "Soft x-ray induced roughness in PMMA," in:

Proceedings of the XI International Congress on Electron Microscopy (S. Maruse, T. Imura, and T. Suzuki, eds), pp. 1763–1764, Japanese Society of Electron Microscopy, Kyoto (1986).
10. B. Niemann, "Detective quantum efficiency of some film materials in the soft x-ray region," *Ann. N.Y. Acad. Sci.* **342,** 230–234 (1980).
11. M. R. Howells, "Fundamental limits in x-ray holography," in: *X-ray Microscopy II* (D. Sayre, M. Howells, J. Kirz, and H. Rarback, eds.), Springer Series in Optical Sciences, Vol. 56, pp. 263–271, Springer, Berlin (1988).
12. C. Jacobsen, "X-ray Holographic Microscopy of Biological Specimens Using an Undulator," Ph.D. Thesis, State University of New York at Stony Brook (1988).
13. V. V. Aristov and G. A. Ivanova, "On the possibility of using holographic schemes in x-ray microscopy," *J. Appl. Cryst.* **12,** 19–24 (1979).
14. D. Attwood, K. Halbach, and K. J. Kim, "Tunable coherent X rays," *Science* **228,** 1265–1272 (1985).
15. D. Joyeux, S. Lowenthal, F. Polack, and A. Bernstein, "X-ray microscopy by holography at LURE," in: *X-Ray Microscopy II* (D. Sayre, M. Howells, J. Kirz, and H. Rarback, eds.), Springer Series in Optical Sciences, Vol. 56 pp. 246–252,Springer, Berlin (1988).
16. S. Aoki and S. Kikuta, "Soft x-ray interferometry and holography," in: *Short-Wavelength Coherent Radiation: Generation and Applications,* Am. Inst. Phys. Conf. Proc. **147,** 49–56 (1986).
17. M. R. Howells, C. Jacobsen, J. Kirz, R. Feder, K. McQuaid, and S. Rothman, "X-ray holograms at improved resolution: A study of zymogen granules," *Science* **238,** 514–519 (1987).
18. C. Jacobsen, J. Kirz, M. Howells, K. McQuaid, S. Rothman, R. Feder, and D. Sayre, "Progress in high-resolution x-ray holographic microscopy," in: *X-Ray Microscopy II* (D. Sayre, M. Howells, J. Kirz, and H. Rarback, eds.), Springer, Berlin (1988) in press.
19. L. Onural and D. Scott, "Digital decoding of in-line holograms," *Opt. Eng.* **26,** 1124–1132 (1987).
20. H. Rarback, S. Krinsky, P. Mortazavi, D. Shu, C. Jacobsen, and M. Howells, "An undulator source beamline for x-ray imaging," *Nucl. Inst. Meth. Phys. Res.* **A246,** 159–162 (1986).
21. R. J. Collier, C. B. Burckhardt, and L. H. Lin, *Optical Holography,* Academic, New York (1971).
22. A. C. Kak, "Tomographic imaging with diffracting and non diffracting sources," in: *Array Signal Processing* (S. Haykin, ed), pp. 351–423, Prentice-Hall, Englewood Cliffs, N.J. (1985).
23. A. J. Devaney, "Reconstructive tomography with diffracting wavefields," *Inverse Problems* **2,** 161–183 (1986).
24. E. Wolf, "Three-dimensional structure determination of semi-transparent objects from holographic data," *Opt. Comm.* **1,** 153–156 (1969).
25. V. I. Tatarski, *Wave Propagation in a Turbulent Medium,* McGraw-Hill, New York (1961).
26. A. J. Devaney, "Inverse scattering theory within the Rytov approximation," *Opt. Lett.* **6,** 374–375 (1981).
27. L. Reimer, *Transmission Electron Microscopy,* Springer, Berlin (1984).
28. R. Glaeser, "Radiation damage and biological electron microscopy," in: *Physical Aspects of Electron Microscopy and Microbeam Analysis* (R. Siegel and J. Beaman, eds.), pp. 205–227, Wiley, New York (1975).
29. W. S. Haddad, D. Cullen, K. Bowyer, C. K. Rhodes, and J. C. Solem, "Design of a Fourier transform holographic microscope," in: *X-Ray Microscopy II* (D. Sayre, M. Howells, J. Kirz, and H. Rarback, eds.), Springer Series in Optical Sciences, Vol. 56, pp. 284–287, Springer, Berlin (1988).
30. M. R. Howells and J. Kirz, "Coherent soft X rays in high-resolution imaging," in: *Free-Electron Generation of Extreme Ultraviolet Coherent Radiation,* Am. Inst. Phys. Conf. Proc. **118,** 85–95 (1983).
31. Using the latest soft x-ray undulator beamline (X1) at the NSLS, which began operations in late 1988, the hologram recording time has come down to about 3 minutes.

8

Prospects for NMR Microscopy

J. R. Mallard

8.1. The Development of Nuclear Magnetic Resonance Imaging

Early interest in the medical applications of magnetic resonance goes back to the late 1950s, when—it was then electron spin resonance (ESR)—differences were found in the signals obtained from tumors and normal tissue due to different concentrations of free radicals in them.[1] Preliminary forays were made into ESR imaging, but the frequencies of electromagnetic radiation required led to too much body absorption and scatter. The prospects for nuclear magnetic resonance (NMR) in medical research were first discussed in 1972.[2]

NMR imaging is the latest discovery in medical imaging, and is arguably as important as the discovery of X-rays. It forms images of the water in the body by virtue of the fact that the hydrogen nuclei of the water are spinning protons that behave as tiny bar magnets. When the body is placed in a magnetic field the protons try to line up to be parallel to the field direction. However, because their magnetism is so small and they are thermally agitated, they cannot all line up perfectly and about 1 in 10^6 of them (i.e. 10^{17} cm^{-3}) precess about the field direction (Figure 8.1). The vital factor that makes medical imaging possible is that this precession takes place at a frequency proportional to the applied magnetic field strength.

In proton NMR the protons are irradiated with electromagnetic radio waves at the same frequency as their precession. Energy is transferred to the protons and

J. R. Mallard • Department of Bio-Medical Physics and Bio-Engineering, University of Aberdeen, and Grampian Health Board, Aberdeen AB9 2ZD, United Kingdom.

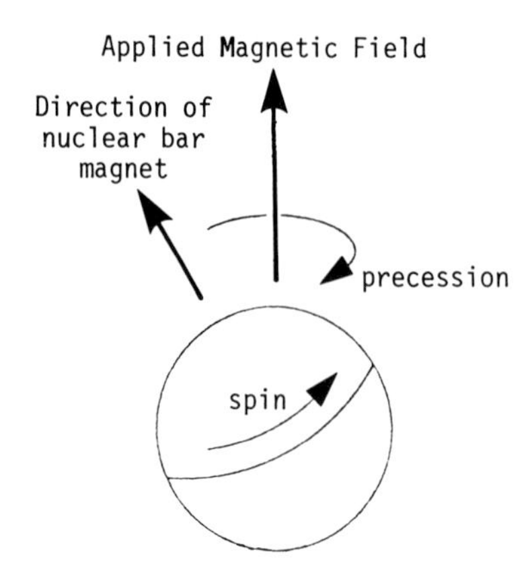

Figure 8.1. Model of spinning nucleus in static magnetic field. The spinning hydrogen nucleus is visualized with the direction of its magnetic moment indicated by an arrow. When placed within a standing applied magnetic field in the direction shown, the proton magnetic moment precesses around that direction. The important feature of the precession is that its frequency is linearly proportional to the applied magnetic field. In typical magnetic fields (e.g., 0.1 T) this frequency is in the radiofrequency band (4.24 MHz).

they are excited, so that they precess antiparallel to the magnetic field or at 90° to it. When the radiofrequency is switched off, they fall back from their excited states to their original states, radiating their surplus energy to the surroundings at the same frequency as their precession. This radiation, the NMR signal from the protons, can be detected by an aerial or coil of wire around the water sample. The water in the sample is thus detected and the intensity of the signal is related to the number of protons that radiate—that is, to the amount of water.

In NMR imaging the location of the water also has to be determined. This is done by applying a magnetic field gradient across the machine (Figure 8.2) so that on one side the standing magnetic field strength is less than on the other. The protons on the lesser side precess more slowly than those on the other side, and so their radiated signals are at a lower frequency. Thus, the frequency of the NMR signal determines the location of the water and the intensity tells how much water is in each position. This type of spectrum can be used with computer-assisted tomography (CAT) to form an image of a sample placed inside the machine.(3)

An early image of a mouse done by NMR(4) with CAT is shown in Figure 8.3. In addition to measuring the signal size (i.e., the proton density), which gives the outline of the animal, how rapidly the signal disappeared with time at each point was measured. This disappearance time from the antiparallel precession, the spin–lattice relaxation time T_1, is several seconds for pure water, while for soft tissues it is about ten times shorter. Each pixel of the image was color coded with the corresponding T_1 value. The region of the liver shows short T_1

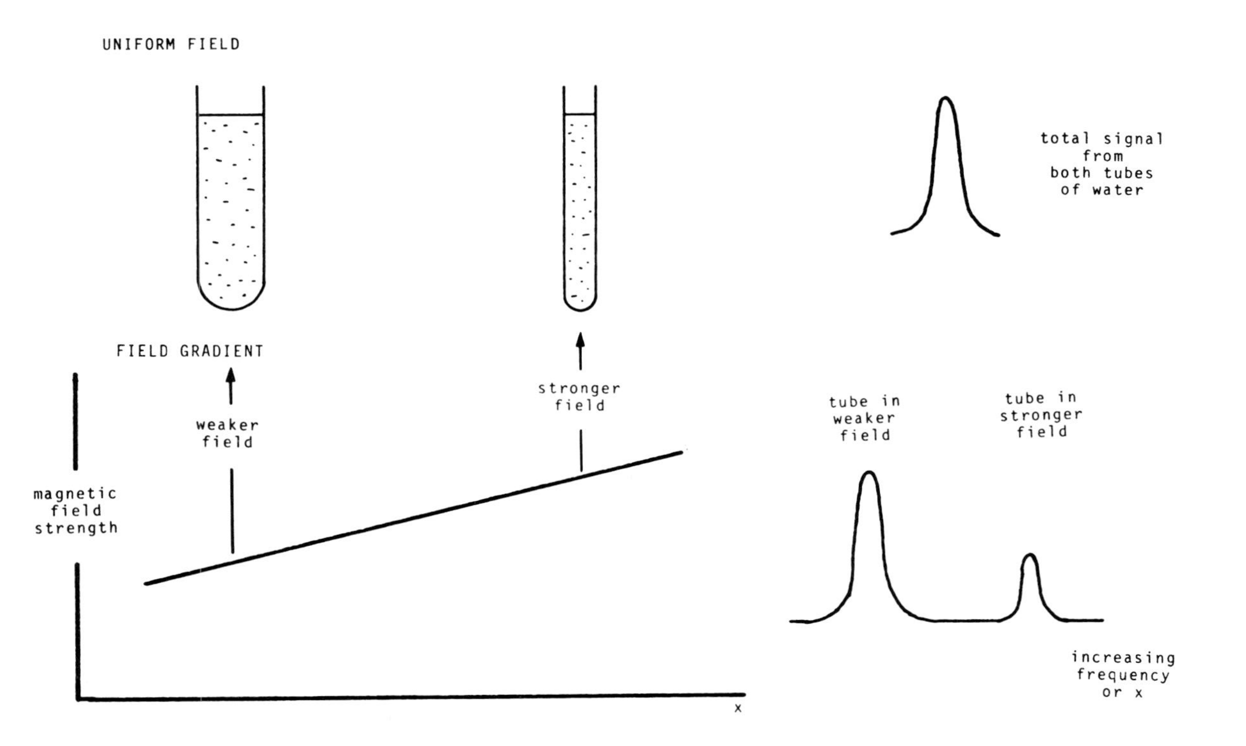

Figure 8.2. By applying a magnetic field gradient across the NMR machine the signals radiated from the hydrogen protons in the tube of water on the left, in the lower field, are at a lower frequency. The measured frequency of the signal locates the water, and the size of the signal is related to how many protons, that is, how much water, there is.

values, whereas in the head region the brain shows longer T_1. This was expected from earlier biological experiments on samples of animal tissues measured in vitro using an ordinary laboratory NMR spectrometer. The black region of maximum T_1 shows the edema that has formed at the neck fracture used to kill the animal to ensure that it was still during the exposure. This result came at about the same time as isotope tomography was being introduced.

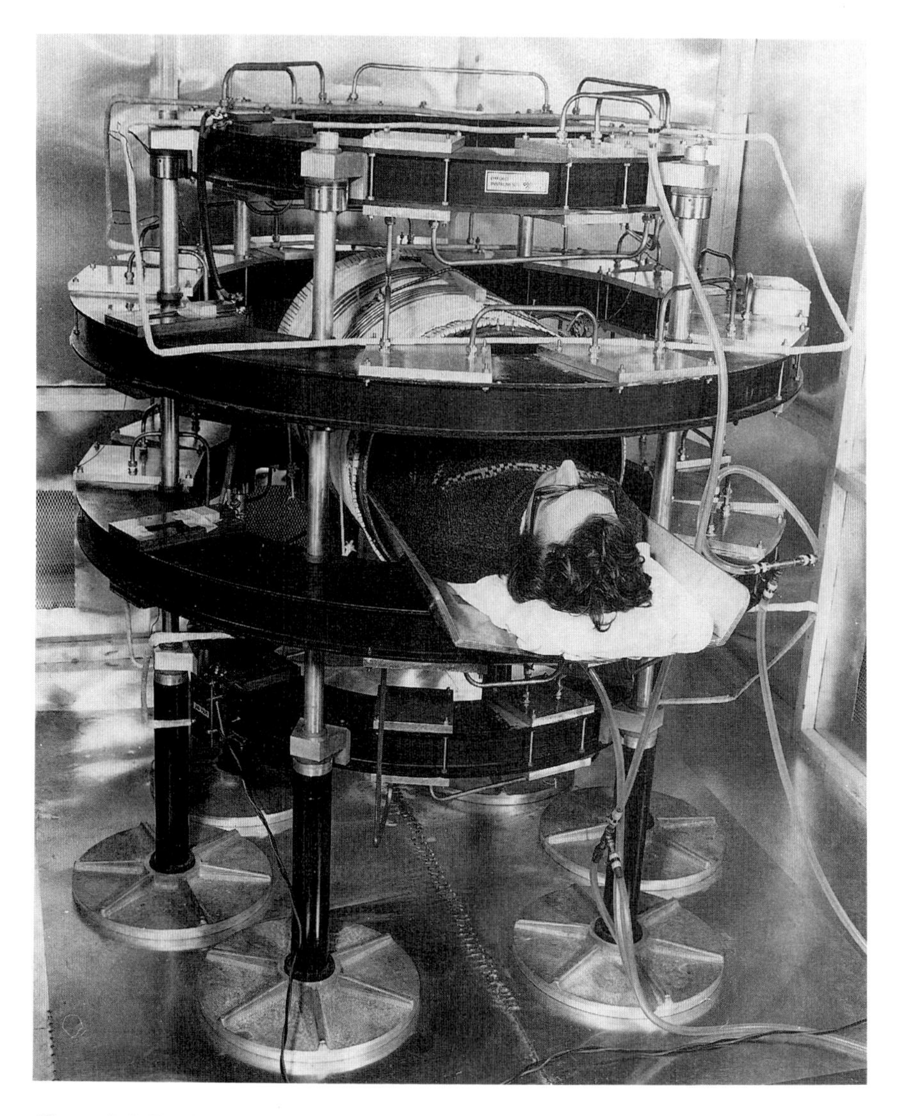

Figure 8.4. The first whole-body NMR imager used in Aberdeen diagnostically to study over 900 patients beginning August 1980. (From Ref. 5.)

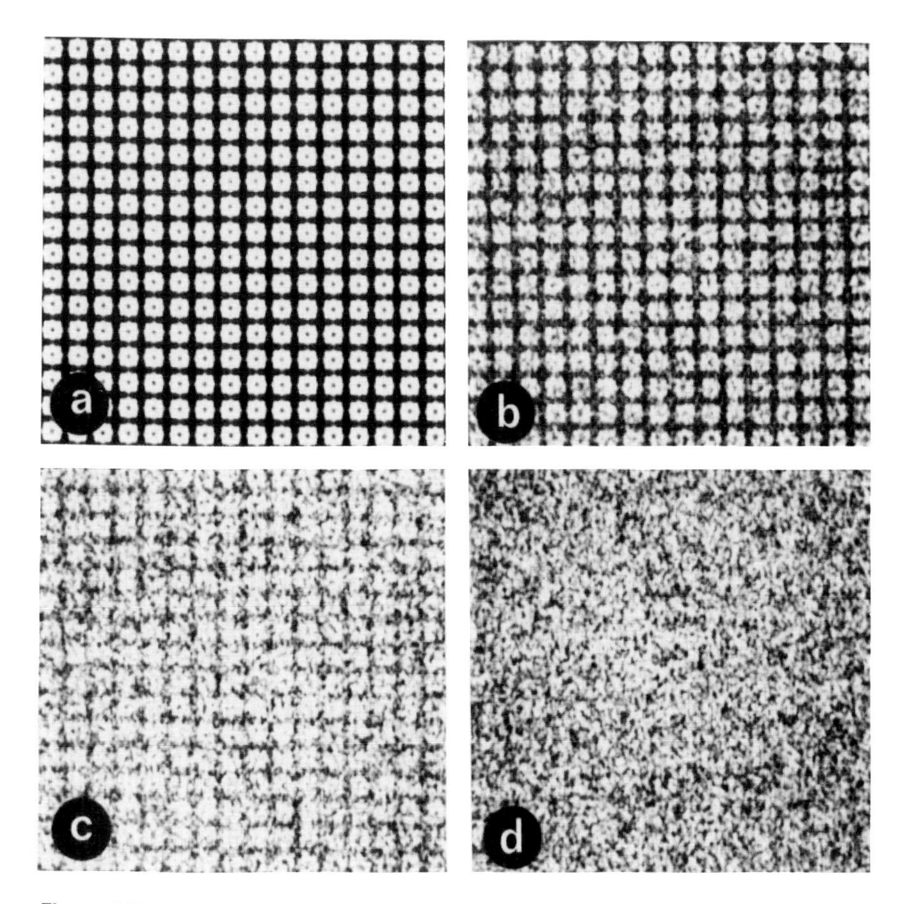

Figure 2.5. Effect of noise on fine structural detail: (a) a model two-dimensional crystal with an equal quantity of noise added (signal-to-noise ratio of 1). Fine details are easily made out. When the signal-to-noise ratio is decreased to 1/10 (b) most of the pattern is still visible, although fine details are harder to see. At a signal-to-noise ratio of 1/20 (c) only the position of the subunits is clear, and by a signal-to-noise ratio of 1/100 (d) even this feature cannot be made out. Electron micrographs usually resemble (b) and (c). Reproduced from Ref. 3.

alternate units is preserved only if the weak peaks that lie midway between the strong lattice peaks are included in the reconstructed image.[18]

2.3.3. Superposition Effects

The interference of patterns from different structural levels can sometimes make it difficult to interpret electron micrographs of regular biological arrays. Because the depth of focus in the electron microscope is large compared to

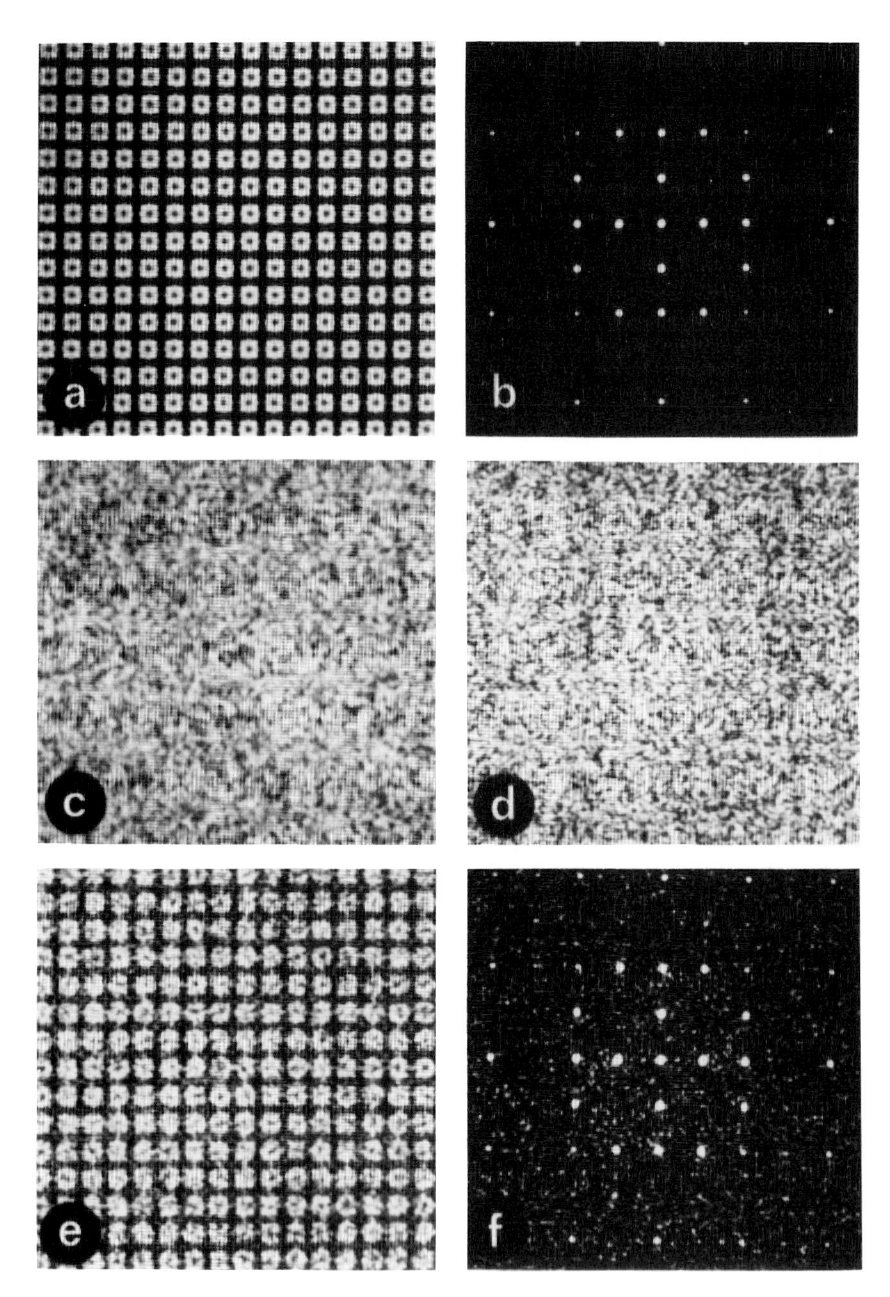

Figure 2.6. Fourier transforms from different types of object. A regular object (a) gives a transform (b) with sharp spots, whereas aperiodic noise gives (c) a pattern that covers the entire spectrum (d). Most objects are a mixture of signal and noise (e), and so give spots superimposed on a continuous background (f). Filtered images are produced by reconstructing the pattern using only the values of the Fourier transform at the spots corresponding to the regular transform (see Figure 2.2). Reproduced from Ref. 3.

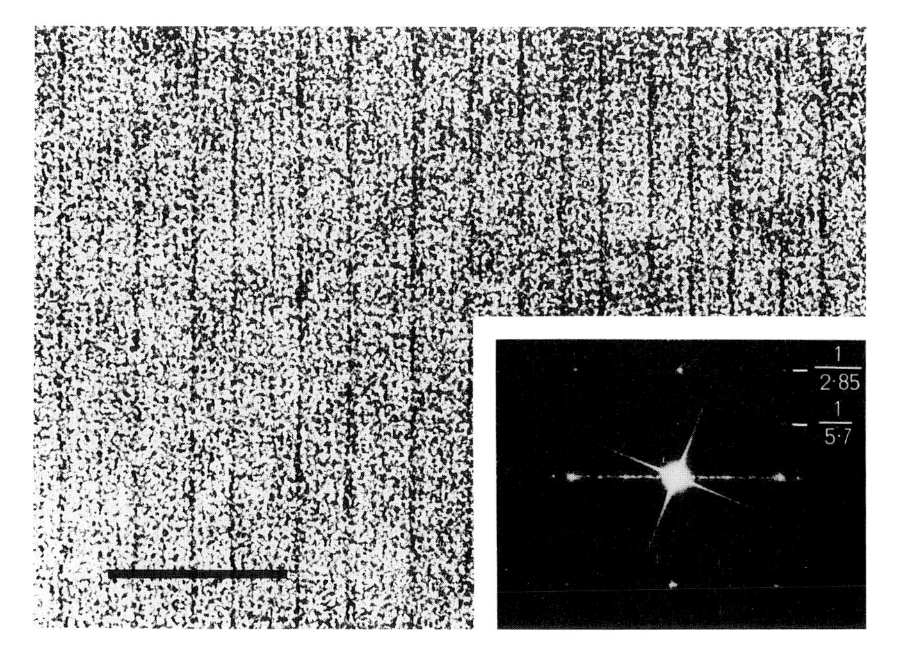

Figure 2.7. Electron micrograph of the sheath of *Methanospirillum hungatei* negatively stained with uranyl acetate.[18] The pattern has a fine repeat at about 2.8 nm vertically and horizontally, and optical diffraction (inset) shows corresponding strong spots at 1/(2.8 nm). However, there are also weak spots (arrows) at a vertical spacing of 1/(5.7 nm) indicating that successive subunits in this direction are different. Reproduced from Ref. 3.

specimen thickness, all levels of the object are in focus simultaneously. If two regular sheets overlap, their lattices will superimpose, producing a moire pattern as shown in Figure 2.9. If the sheets are rotated relative to each other, it is generally possible to separate the contributions of each in the Fourier transform, because most reciprocal lattice peaks from one pattern do not overlap with those from another. Thus, provided that the pattern can be indexed correctly, an image of a single sheet can be produced by Fourier synthesis. Note here that not only has

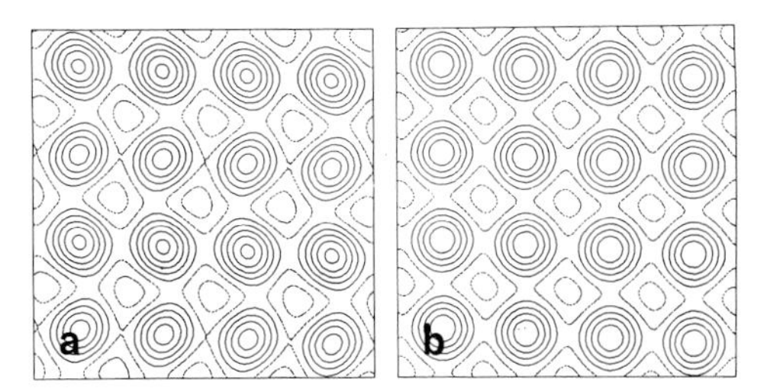

Figure 2.8. Reconstructed images of *Methanospirillum hungatei*. Images reconstructed using all spots in the transform (a) show a distinct difference between adjacent subunits in a vertical direction, but this feature is absent in images reconstructed omitting the 1/(5.7 nm) spots (b). Reproduced from Ref. 3.

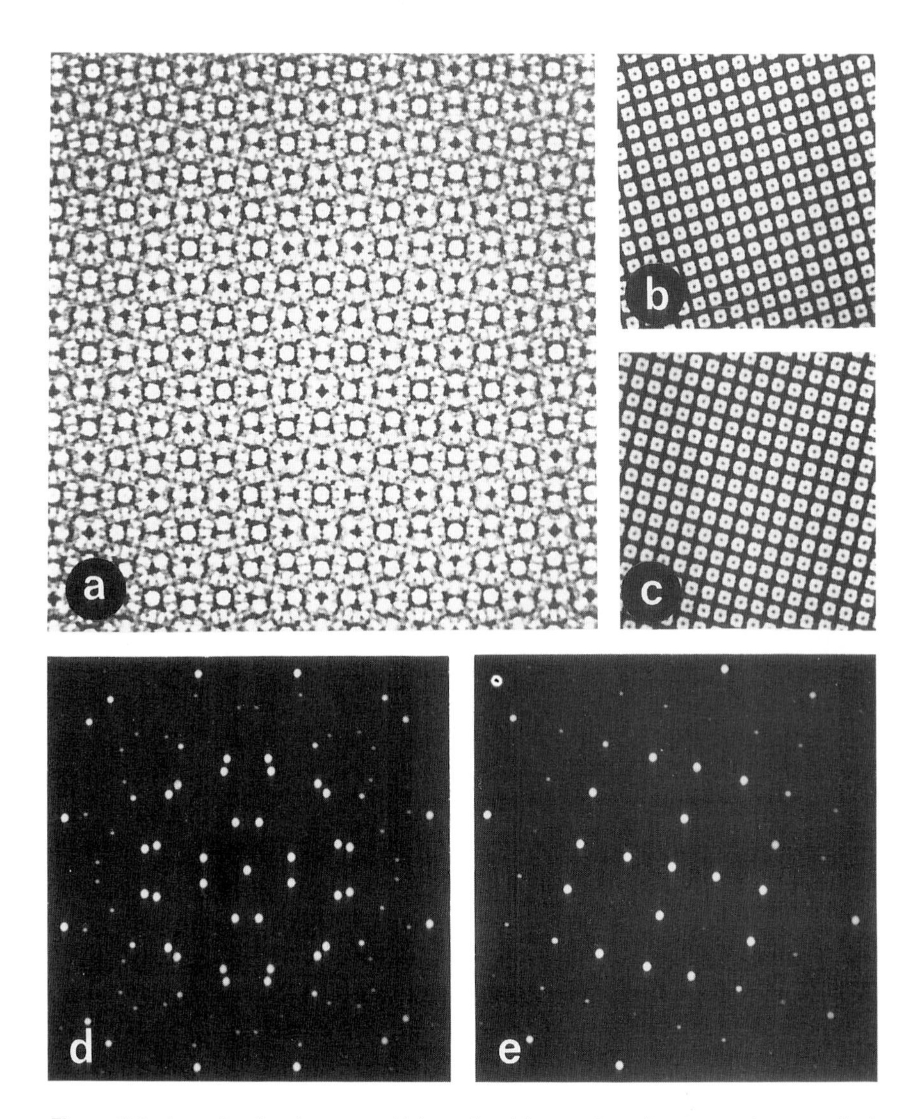

Figure 2.9. A rotational moire pattern (a) is produced by superimposing two regular arrays (b,c) that have been rotated relative to each other. The structure of the subunits can be easily made out in a single layer, but is confused in the moire pattern. The transform (d) of the superimposed layers contains spots from the transforms of each single layer, but these spots do not overlap. Thus, the transform of a single layer (e) can be identified and used to reconstruct a single layer [in this case, (c)]. Reproduced from Ref. 3.

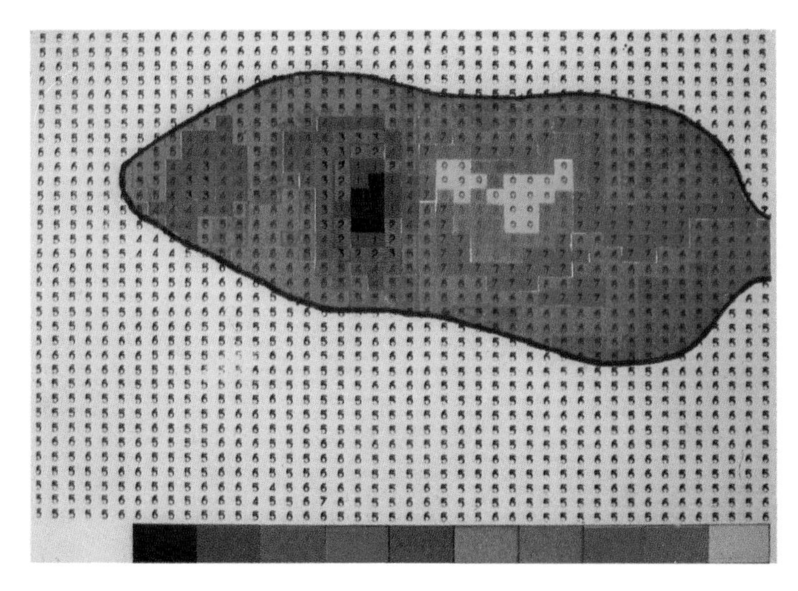

Figure 8.3. NMR image of a mouse (February 1974). The outline of the animal is seen from the proton density signal, while the coding of T_1 values shows the liver and brain with different values, and the black (longest T_1) region indicates edema at the neck fracture carried out immediately before the imaging was begun (1 hr exposure).

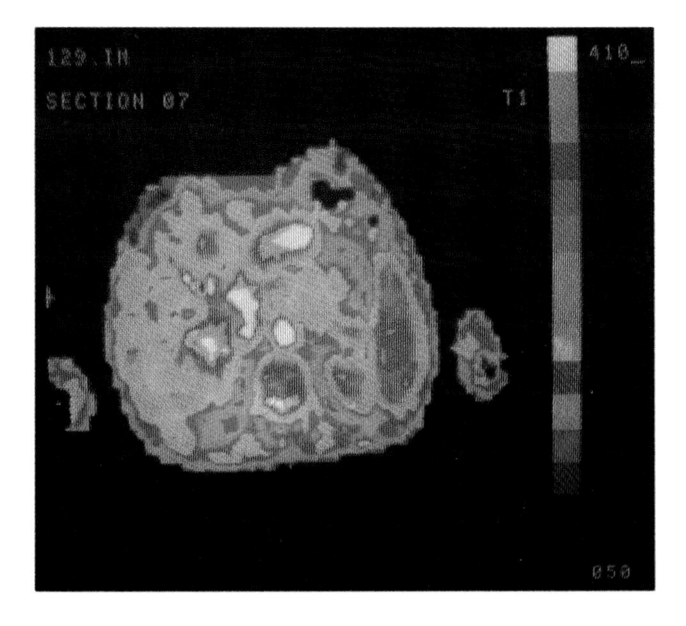

Figure 8.6. Transverse abdominal section by NMR of first patient investigated using the imager of Figures 8.4 and 8.5 in August 1980. The coding of T_1 values shows clearly the presence of multiple metastases in the liver (from carcinoma esophagus) with much longer T_1 values (seen as white for the large ones and red for the smaller ones) than the normal liver tissue seen as yellow ($T_1 \approx 140$ ms).

The next step was to build a human whole-body NMR imager (Figure 8.4) with measurement of T_1, which appeared to show pathology. This used a vertical resistive magnet of 0.04 T, made by Oxford Instruments Ltd., which was, at the time of conception, the highest field that could give the required uniformity in that configuration.[(5)] Figures 8.5a and 8.5b show diagrammatically how the image is formed from the data acquired from a transverse slice across the patient. The signals are coded in position across the patient by frequency, as explained above.

A major advance was the introduction of 2-D Fourier transform methods (spin warp).[(6)] In this, the signals are coded through the patient from back to front by phase, as a result of applying an additional magnetic field gradient through the patient from back to front. Figure 8.5a shows how a proton density image is obtained using a saturation recovery pulse sequence, and Figure 8.5b shows how a T_1 image is obtained by adding a radiofrequency inversion pulse prior to each alternate saturation recovery sequence (an inversion recovery sequence). From these two signals, the T_1 value in each pixel can be computed.

Figure 8.6 shows a transverse abdominal section of the first patient on whom this machine was used. It shows clearly multiple liver metastases from a carcinoma of the esophagus. It is a computed T_1 image in which the malignant tissue shows longer T_1 than the normal liver tissue. Another section (not shown here) highlighted a bone metastasis in the spine that had not been detected by other methods but was subsequently confirmed by a gamma camera ^{99m}Tc-diphosphonate bone image. Thus, the diagnostic use of NMR imaging was immediately demonstrated.

Figure 8.7 shows a liver section taken with an improved version of the NMR imager, with twice the field strength (0.08 T). It shows, on a computed T_1 image, very small multiple liver metastases from a colon carcinoma. Such small lesions are found because, at this low magnetic field strength, the contrast is maximized between the lengthened T_1 values of the metastasis and the value for normal liver tissue. This contrast is clear unequivocal biological contrast to which there is a biological meaning based upon the different utilization of the water in the tumor from that in normal liver. This has the potential to provide true identification of pathology, by different pathologies having different ranges of T_1 values.[(5)]

The diagnostic power of the NMR imaging method was very much enhanced when it became possible to form images in the coronal and sagittal planes as well as in the transverse plane. This cannot be done by x-ray CAT except by computer reconstruction from many transverse sections, which is very time consuming and entails loss of spatial resolution. The triaxial imaging of NMR imaging provides accurate three-dimensional localization of pathology.

The use of NMR imaging expanded rapidly in the early 1980s as commercial machines became available, and the whole field, together with its clinical utilization, has been reviewed by Mallard.[(7)]

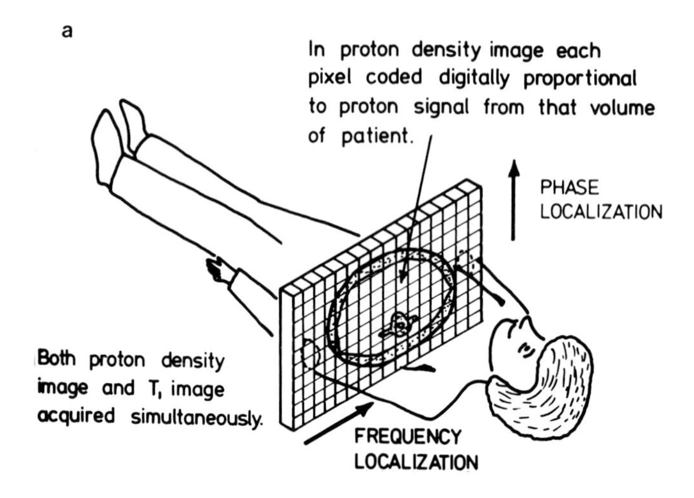

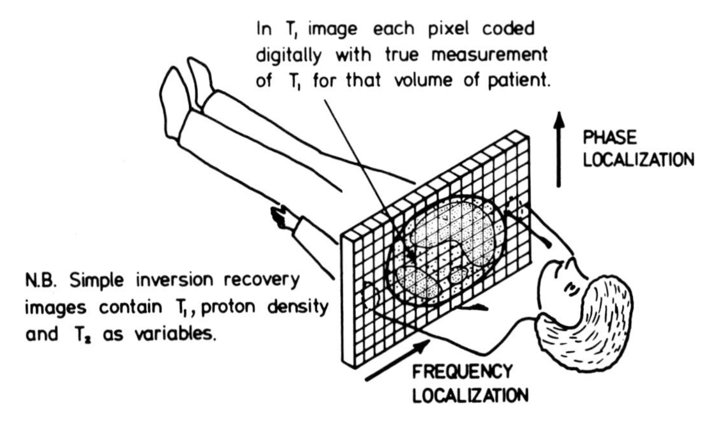

Figure 8.5. Diagrams showing how a transverse-section NMR image is obtained. Frequency coding positions proton signals across the patient, from left to right, while phase coding positions them through the patient from back to front. (a) Using a saturation recovery pulse sequence, the signals from each volume element (voxel) of tissue, which is displayed as a picture element (pixel), are related to proton density, number of proton spins, or water concentration. (b) Using an inversion recovery sequence interleaved with the proton density sequence, the T_1 value of the tissue in each voxel can be calculated and used to code each voxel on the T_1 image.

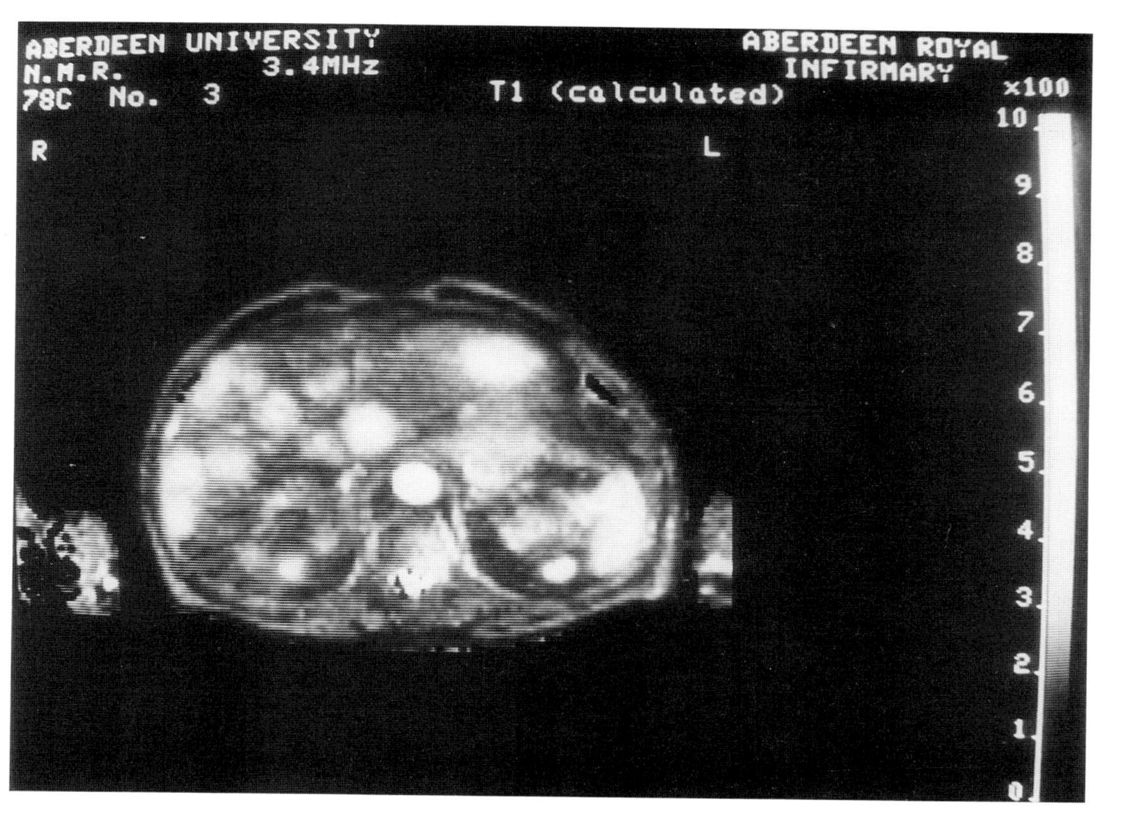

Figure 8.7. Transverse T_1 abdominal section of patient with carcinoma colon showing multiple large and very small metastases in the liver. The lumbar artery can also be seen clearly.

The Present Clinical Use of NMR Imaging (MRI)

It looks already as if NMR will prove to be the method of choice for detecting and studying demyelinating diseases. An example is seen in Figure 8.8, which is a transverse section of the head of a 61-year-old man with long-standing multiple sclerosis. This inversion recovery image shows extensive periventricular white matter demyelination. While the computed T_1 image shows these regions well, it also shows very clearly the normal white matter. The removal of the fatlike myelin sheath due to the disease lengthens the T_1 value of lesions, and provides an imaging contrast. Thus, inversion recovery images (which emphasize T_1) show small lesions from such diseases which are not detectable by other means. In addition, x-ray CAT usually needs contrast material and so it is invasive. It is interesting, however, that the lesions seen by x-ray CAT are sometimes in different positions from those seen by NMR imaging.[8] NMR imaging has a major potential in assessing the risk factors that influence a disease, in elucidating the time-course of the disease by sequential imaging, in improving our knowledge of the pathogenesis of the disease, and in determining the efficiency and mode of action of therapeutic agents.

Figure 8.9 shows a sagittal section of the head and neck of a 25-year-old man who had sustained a C7 vertebral fracture in 1979, and who has developed a syrinx in his cervical cord from C1/2 to C7. The vertebral fracture and the syrinx are clearly imaged. Figure 8.10 is a sagittal section of the spine of a 25-year-old man with severe back pain. The image shows an abnormal dehydrated disk at the L4/5 level; there is no obvious posterior disk prolapse and the other disks appear normal. Neither of these examples could have been detected by other methods.

NMR imaging can be useful in discriminating between different parts of tumors; for example, cysts can be discriminated from the malignant tissue.[7,9] Paramagnetic NMR contrast agents (Gadolinum–DTPA) have also been used to emphasize such tissue detail with success,[10] and much research is currently in progress to find the best substances and use them, particularly for the abdomen.[11–13]

The United Kingdom Medical Research Council has pursued a series of clinical evaluation projects at 9 centers and recently a review of progress to date was carried out.[14] The overall conclusions are summarized in Tables 8.1 and 8.2. The list of conditions and areas for which NMR imaging can now be accepted as the method of choice is impressive but, equally, the list where other imaging techniques are equivalent or better is disappointing. As the technique improves, becomes more refined and more commonly used, and extends in range, it is very likely that in a few years, because of its hazard-free nature, these lists will become more weighted in favor of NMR.

The value of NMR imaging is in its ability to provide better contrast resolution between different tissues, particularly soft tissues, than is possible with x-ray CAT. It offers the possibility to study tissue changes at the molecular level. Its

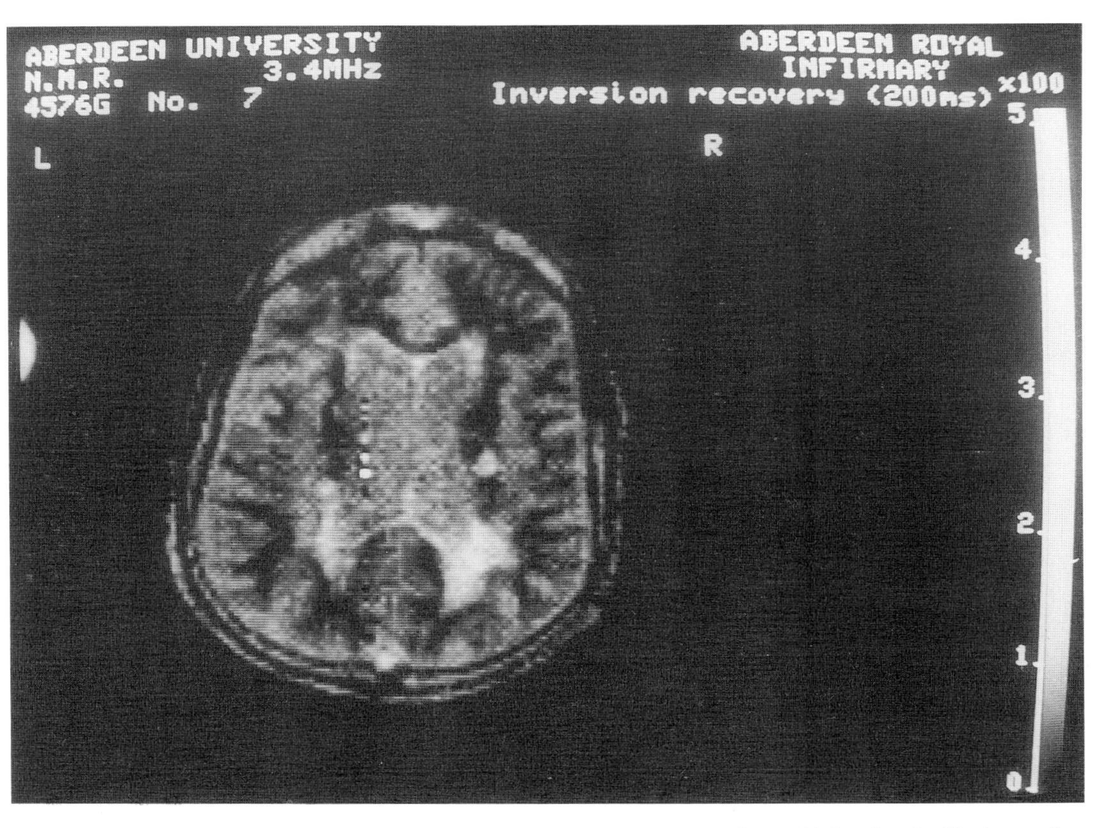

Figure 8.8. Transverse section of the head of a 61-year-old man with long-standing multiple sclerosis. The image shows extensive periventricular white matter demyelination.

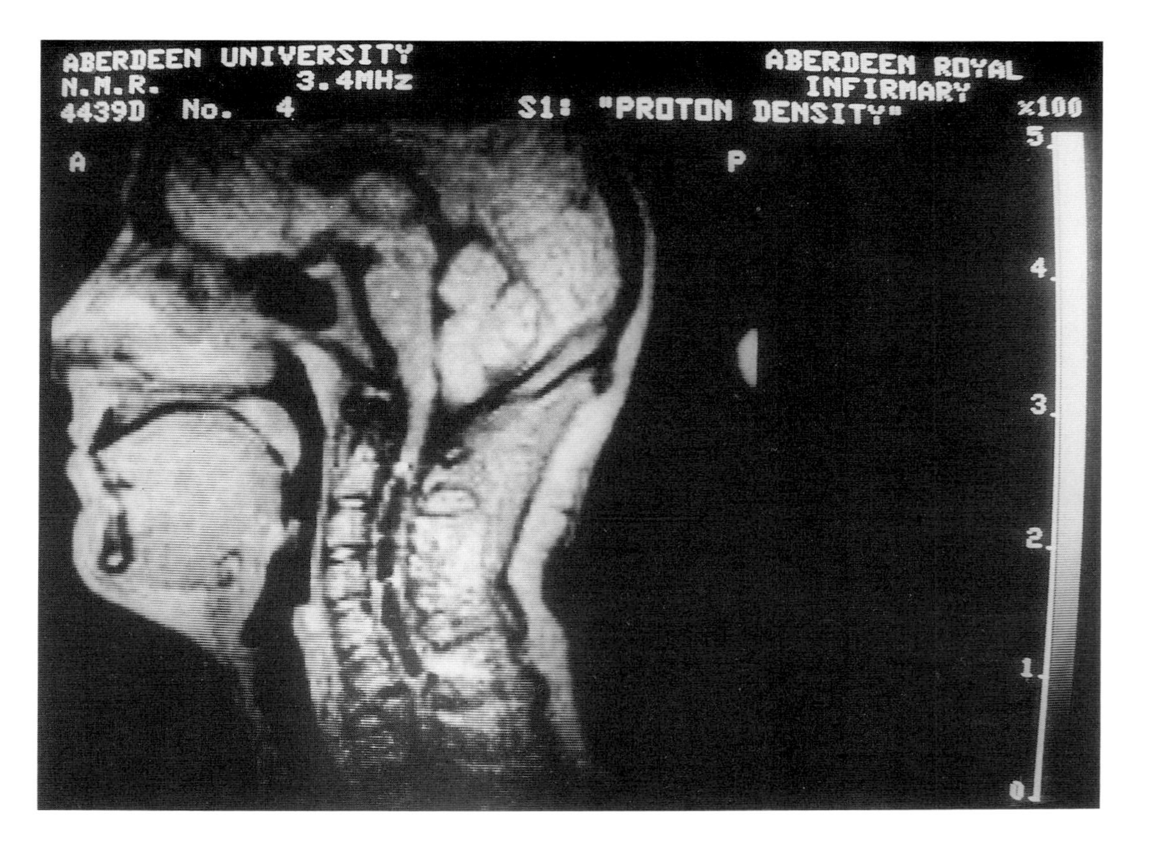

Figure 8.9. A sagittal section of the head and neck of a 25-year-old man who sustained a C7 vertebral fracture in 1979. A syrinx has now developed in the cervical cord from C1/2 to C7. Both the fracture and the syrinx are clearly imaged.

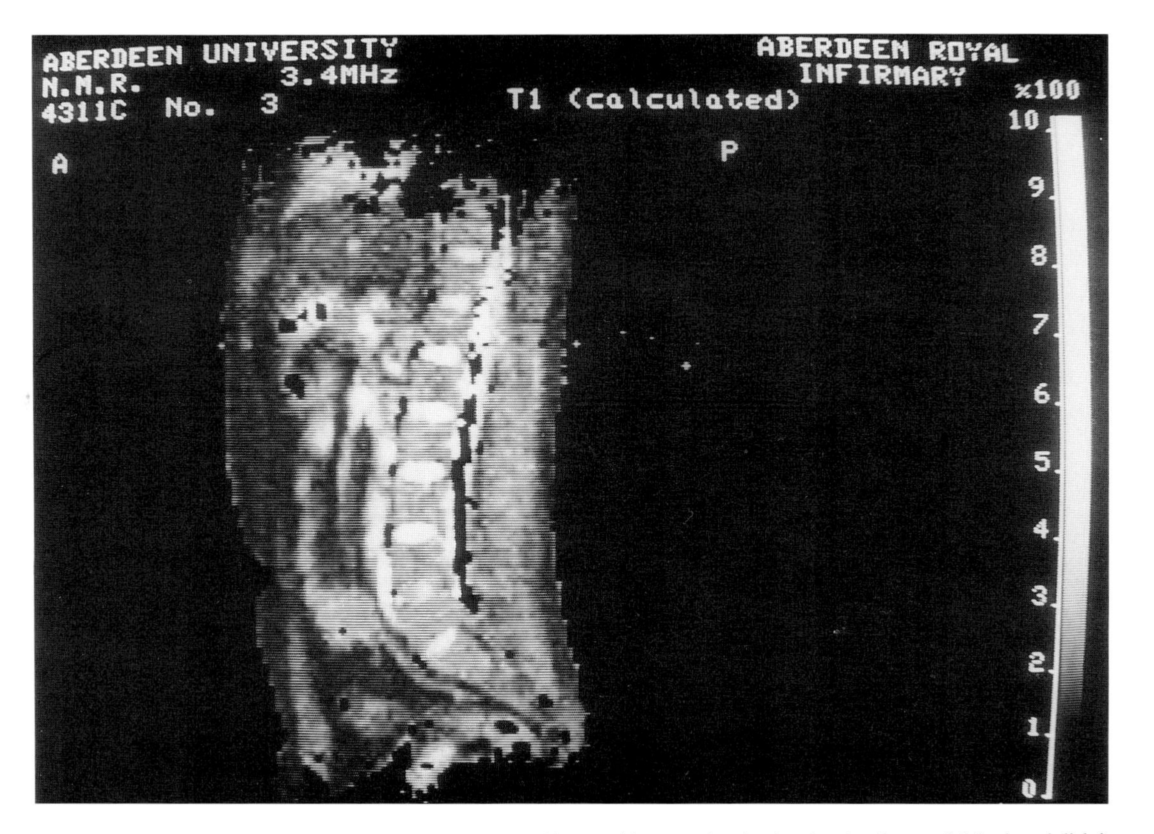

Figure 8.10. A sagittal section of the spine of a 25-year-old man with severe low back pain. An abnormal dehydrated disk is seen at the L4/5 level. There is no obvious posterior prolapse of the disk, and the other disks appear normal.

Table 8.1. Clinical Conditions for Which NMR Imaging is the Diagnostic Method-of-Choice[a]

Posterior fossa lesions
Foramen magnum lesions
Syringomyelia
Multiple sclerosis—multiple lesions
Acoustic neuromata
Lumbar spine and disk disease
Great vessels (cardiovascular disease)
Parotid gland tumors
Bone tumors
Pediatric imaging
Cerebral palsy

[a]United Kingdom Medical Research Council Review, 1986–1987.

Table 8.2. Clinical Conditions for Which Imaging Techniques Other Than NMR Are of Equivalent or Greater Diagnostic Usefulness[a]

NMR comparable or complementary to x-ray CAT or other diagnostic imaging	NMR not more useful than other imaging
Tumors	Lung
rectum	Spleen
prostate	Liver
cervix	Pancreas
uterus	Kidney
bladder	
breast	
cerebrum	
meninges	
pituitary gland	
Stroke	
Congenital heart disease	
Dementia	

[a]United Kingdom Medical Research Council Review, 1986–1987.

Table 8.3. Value of NMR as a Cross-Sectional Imaging Method

1. Good contrast between different tissues—particularly soft
2. No image degradation due to cortical bone
3. Images in any body plane—usually transverse, coronal, or sagittal
4. No ionizing radiation—can repeat test—safe for children
5. Both structural and functional information present in each image

value as a cross-sectional imaging method is summarized in Table 8.3. The real future of NMR imaging lies in developing much more the functional aspects, so that images emphasize abnormal changes of tissue function.

8.2. The Dependence of Image Quality on Spatial Resolution and Tissue Contrast: The Biological Basis of Tissue Characterization

As the static magnetic field is increased, the proton signal becomes stronger because more protons line up parallel to the field. The practical relationship between the signal size and the magnetic field strength is probably one of a square root or cube root, and thus a large increase in field strength is needed to provide a sizeable and useful gain in the signal-to-noise ratio.[7] Nevertheless, during the rapid development of NMR imaging by commercial manufacturers, the magnetic field has gradually increased from 0.1 T, through 0.5 T, and even up to 2 T using superconducting magnets. The images obtained at the higher field strengths can be acquired from much smaller volume elements (voxels) than at lower field (i.e., 512×512 or 256×256 instead of 128×128 images), and so they have finer spatial resolution and give more detailed anatomical images. They are more immediately striking and pleasing to a radiologist familiar with the fine detail of x-radiographs. An example, at 0.28 T, is seen in Figure 8.11 and one at 1.5 T in Figure 8.12. The claims of commercial manufacturers concerning the gains of using high fields has gradually become more muted, and recent literature from a major one includes Figure 8.13.

However, this is not the whole story. For an image to be of good quality it needs to have good contrast for the purpose of the image, that is, good contrast to show the tissue difference or abnormality of function that is sought. Finer spatial resolution alone will give only a more detailed portrayal of abnormal anatomy.

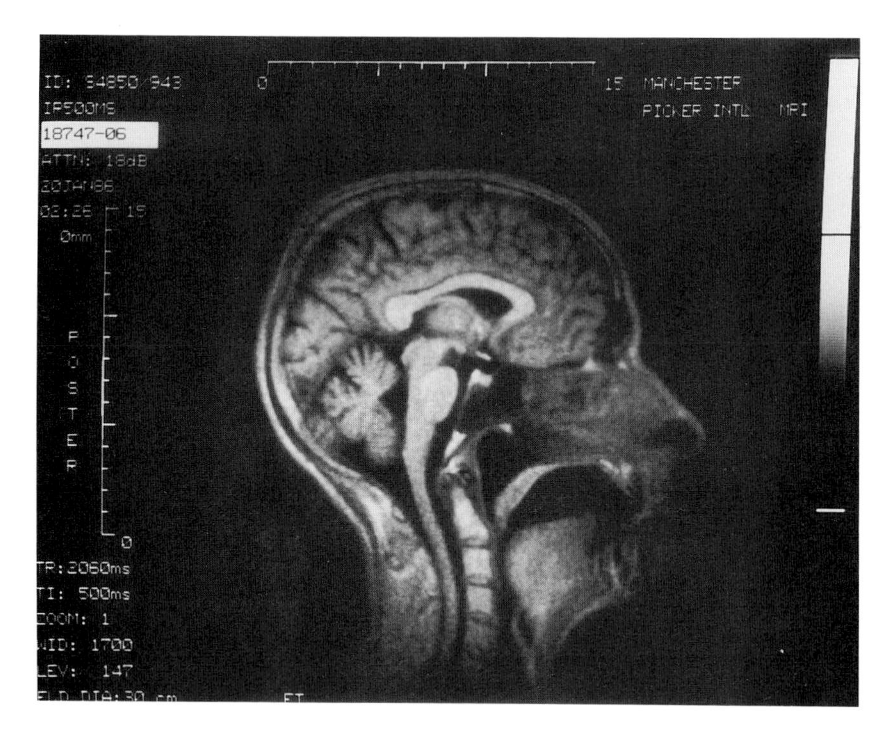

Figure 8.11. NMR image of normal head and neck taken at a magnetic field strength of 0.26 T. (Courtesy of I. Isherwood.)

If, when the field strength is increased to achieve this, the contrast is diminished, then the higher field image is not necessarily the better one.

The contrast in NMR images depends on four main factors:

1. Proton density (P) differences between tissues
2. Spin–lattice relaxation time (T_1) differences between tissues
3. Spin–spin relaxation time (T_2) differences between tissues
4. Flow velocity of fluid (e.g., blood, cerebrospinal fluid)

Proton density is difficult to measure consistently because it is an absolute quantity that has to be measured in very difficult circumstances. Proton density differences between tissue in the same slice can be measured if the signal homogeneity over the slice is good. This has to be measured with a test object. For grey brain, a contrast of 0.91 is obtained, and for white brain 0.75, compared with 1.00 for cerebrospinal fluid.[(15)] This 16% difference between the tissues corresponds well with the difference in water content of 13.5% as measured on goats.[(16)] The ratio of spin density between tissues and that of their water content normally show similarity except for tissues with a substantial lipid content.

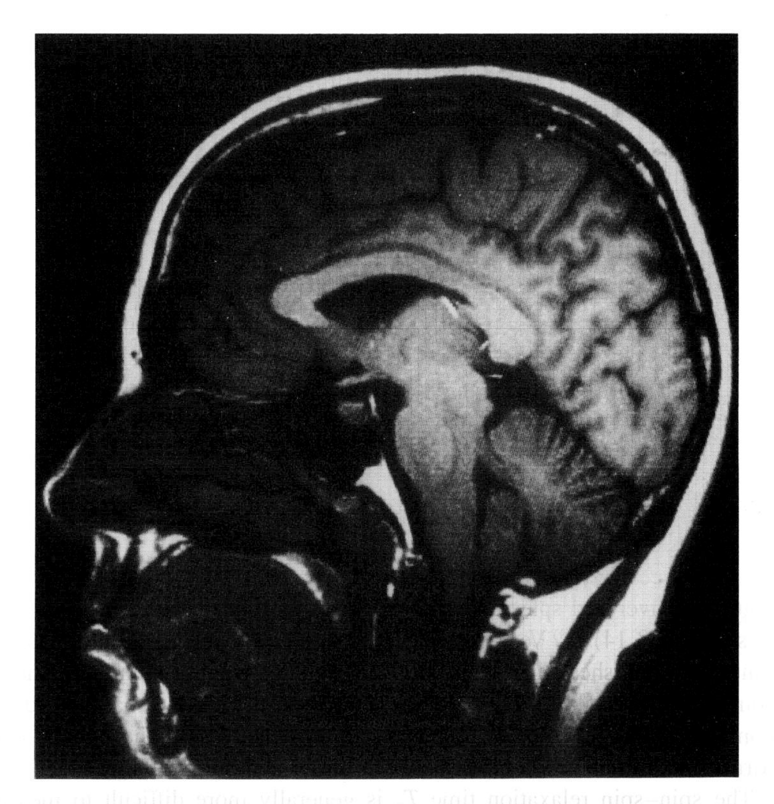

Figure 8.12. Sagittal image taken at 1.5 T of brain with enlargement of the pituitary gland (taken from manufacturer's brochure).

Free water has long T_1 and T_2 relaxation times, but, on becoming bound in hydration sheaths around macromolecules the relaxation becomes rapid. Such bound water is in fast exchange with free water; hence, when a measurement is made for each voxel, a weighted mean relaxation time is obtained. The variety of macromolecular composition in the cells causes differences in the relaxation times between different cell types.

In vitro studies of relaxation are valuable for a deeper understanding of disease processes, for helping interpretation of NMR images, and for assisting in the optimization of NMR imaging sequences. Relaxation times depend on a variety of biological and physical measurement conditions, and the considerable spread of values for a single tissue is largely due to variations of technique. Standardized methodology protocols and reporting procedures are absolutely vital for a more precise quantitization and understanding of these most important parameters.(17) As an example, tissue T_1 increases with increasing frequency used for the measurement (or increased static magnetic field strength), but each

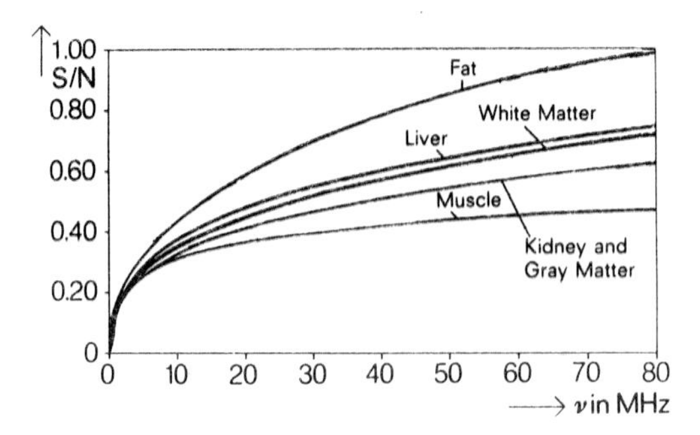

Figure 8.13. Graph taken from manufacturer's brochure of signal-to-noise versus frequency (or magnetic field strength) using a partial saturation sequence at 500 ms repetition frequency.

tissue type responds in a different manner: liver is much more frequency sensitive than spleen, whereas gray and white brain respond to a change in frequency in a similar manner. As a result of this important observation, image contrast based on T_1 differences between tissues is better at low frequencies (low magnetic field strength) for liver and spleen, but is not much affected by frequency for brain tissues (Figure 8.14).[18] Very importantly, the contrast between Yoshida sarcoma and muscle diminishes at a higher frequency;[18] this would lessen the chance of sarcoma detection at a higher field strength, even though the sharpness of the anatomical imaging, which is based on proton density, is higher. Other work confirms this diminished contrast between tumor and normal tissue.[7,19]

The spin–spin relaxation time T_2 is generally more difficult to measure

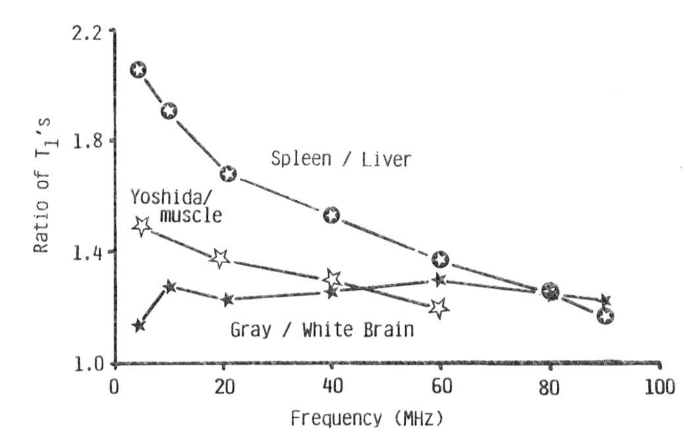

Figure 8.14. The ratio of T_1 values as a function of frequency. An image contrast that depends on T_1 differences is therefore greater at low frequencies and field, for spleen to liver, and for Yoshida sarcoma to muscle. Gray brain and white brain T_1 contrast, however, is virtually unchanged as the frequency (or magnetic field strength) is increased.

accurately, especially in imaging, but is much less frequency dependent. This is the underlying reason why very high field imagers use spin–echo sequences more frequently, because these are largely T_2 dependent. The contrast at the high field is therefore retained in spin–echo images, while that for T_1-dependent sequences is lower. It is still not proven that high-field imagers are superior to low- or medium-field ones for lesion detection. It will probably emerge that for some limited purposes the considerable extra cost is justified, but that magnetic fields of, say, 0.1–0.5 T are optimum for imaging. For imaging, two field systems would be useful at, say, 0.35–0.5 T for good spatial resolution, and 0.04–0.15 T for improved T_1 contrast. Such combinations would offer the best of both worlds for image quality.

Many types of biological factors affect relaxation time, including water content, age (especially in early life), tissue handling, and storage (for *in vitro* studies). The lipid content of the tissue is also an important factor; not only does it directly affect T_1 and T_2, but it also leads to nonexponential relaxation behavior that can distort the calculated values, if the calculation method assumes single exponential behavior.

High magnetic fields may be much more valuable for the *in vivo* NMR spectroscopy studies that cannot be carried out at low fields; an example is shown in Figure 8.15. The spectra result from the body phosphorus (^{31}P) resonating at slightly different frequencies when exposed to a magnetic field. This is because of the different magnetic environments of ^{31}P in inorganic phosphate, (P_i), in adenosine-triphosphate (ATP) and in phosphocreatine (PCr). Very high fields are also useful for imaging other nuclei such as ^{23}Na, which cannot be done at low fields; an example from a manufacturer's brochure is shown in Figure 8.16.

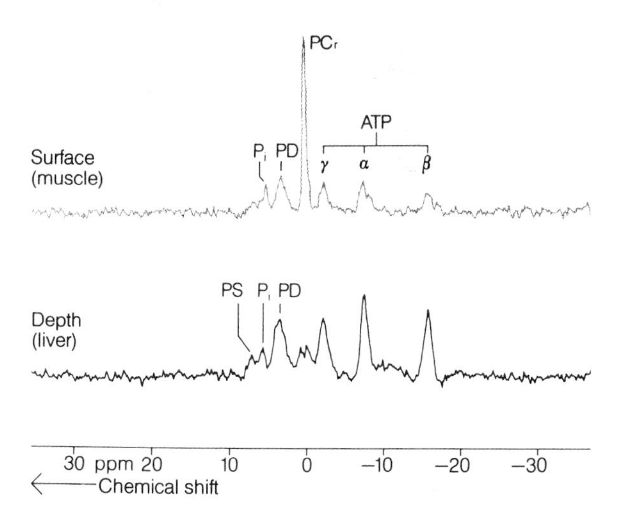

Figure 8.15. ^{31}P spectrum of superficial muscle, and of liver using depth selection techniques.

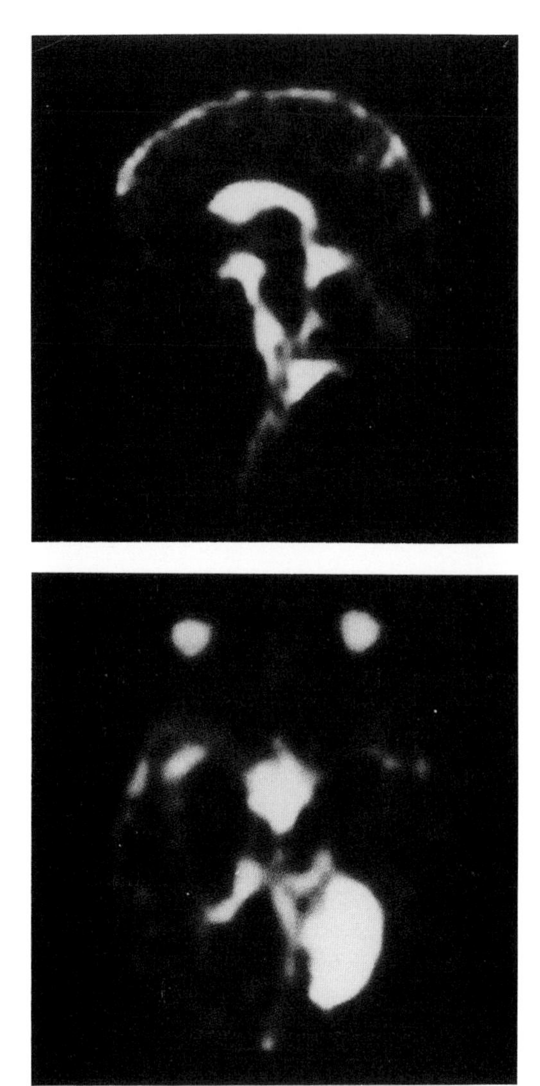

Figure 8.16. An 1.5 T image from manufacturer's brochure of ^{23}Na sagittal image of a normal brain (top) and of a cystic tumor in the brain (bottom).

8.3. Clinical Value of Tissue Characterization by NMR

Table 8.4 shows values of T_1 determined from some *in vivo* clinical images, both normal and pathological, at 0.08 T (3.4 MHz). Table 8.5 shows much more precise laboratory determinations of T_1 and T_2 for rat tissues, related to water

Table 8.4. Some Typical Spin–Lattice Relaxation Times of Human Tissues *in Vivo*[a]

Normal tissues	T_1 (ms)	Pathological tissues	T_1 (ms)
Lung	380–400	Pancreatitis	300–320
Medullary bone	230–250	Hepatitis	300–320
Cortical bone	0	Liver metastasis	400–500
Intervertebral disk	370–420	Hepatoma	400–500
Spinal cord	310–330	Simple serous cyst	800–1000+
Fat	130–160	Ascitic fluid	>1000
CSF	400–450	Renal carcinoma	360–380
White matter	260–270	Renal infarct	550–620
Gray matter	370–390	Lung carcinoma	320–360
Cerebellar cortex	340–370	Degenerate intervertebral disk	230–240
Eye		Anoxic edematous	
Lens	260–280	White matter	260–270
Globe	>1000	Gray matter	370–390
Sclera	420–460	Cerebral infarct	450–480
Blood	450–650	Cerebral glioma	520–570
		Active plaque multiple sclerosis	500–530

[a]Measured *in vivo* at 3.4 MHz (0.08 *T*).

content. Note the smaller values for T_2 and the smaller range of values compared with T_1; inherently T_1 should be more useful for clinical contrast as a result. In almost every case, the values for a single property vary sufficiently to cause overlap with other tissues; thus, tissue identification from a single NMR property is not possible. However, if more than one property is displayed (such as T_1 and water content), as in Figure 8.17[(16)] then better separation is achieved. If T_2 data is included as well, then even better separation results.

Unfortunately, this hopeful picture applies only to normal tissues. When tissues become abnormal they tend to have an even wider spread of NMR

Table 8.5. Tissue Identification by NMR *in Vitro*[a]

Tissue[b]	T_1 (ms)[c]	T_2 (ms)[c]	% H_2O[d]
White brain	326	85.2	78.4
Liver	146	42.2	71.8
Thigh muscle	236	50.2	77.4
Ventricle	260	59.0	78.2
Spleen	285	71.6	78.0

[a]Foster, 1986.
[b]Tissues taken from 32-day-old rats.
[c]At 2.5 MHz, 30°C.
[d]Drying-to-constant weight at 60°C.

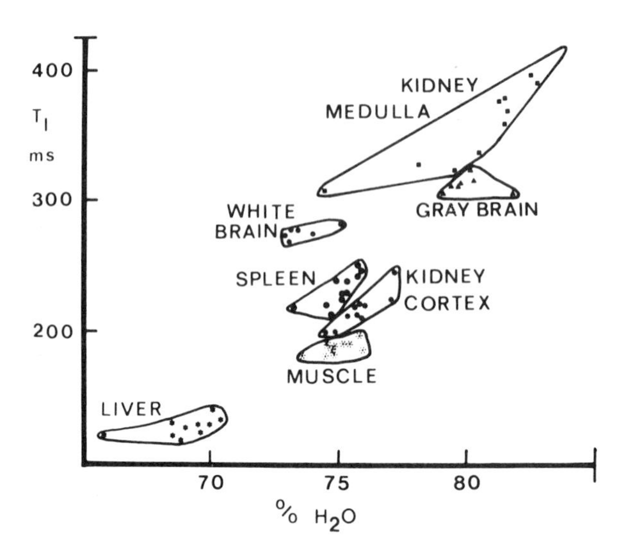

Figure 8.17. Separation of tissues by comparison of their spin–lattice relaxation times (T_1) with their water content. Only in a few cases is a complete separation and discrimination possible.

properties (see Table 8.4), presumably depending on the amount of abnormality in each local volume. The majority of abnormalities, however, are associated with an increase in relaxation times, and often with water content. It seems most unlikely that the majority of abnormalities can be identified or even distinguished from one another on the basis of these properties.

Knowledge of these properties, however, enables the imaging contrast to be enhanced for the clinical condition to be demonstrated on a particular patient. The type of pulse sequence, and its timing sequence, can be chosen to be related to the relaxation times of the tissues to be imaged. This can make all the difference between missing or detecting the abnormality. It also saves time by obviating the need to do a wide range of sequences on each patient section to be studied.

The problem of the partial volume effect, that is, the mix of tissues found within an imaging voxel, makes clinical abnormal tissue characterization even more difficult. This is perhaps particularly so for tumors. However, it may be that by measuring T_1 values for a big organ such as the liver, there would be a better chance of reducing partial volume problems, particularly if diffuse liver disease was studied. Consequently, the T_1 images of inflammatory disease of the liver, cirrhosis, fatty liver, and normal controls have been studied (Figure 8.18).[20] The controls show normal T_1 distributions while the diseases show an increasing spread of longer T_1 values outside the two-standard-deviation range from the normal average. Unfortunately, because each disease shows some patients with T_1 values within the normal range, it is not possible to definitively

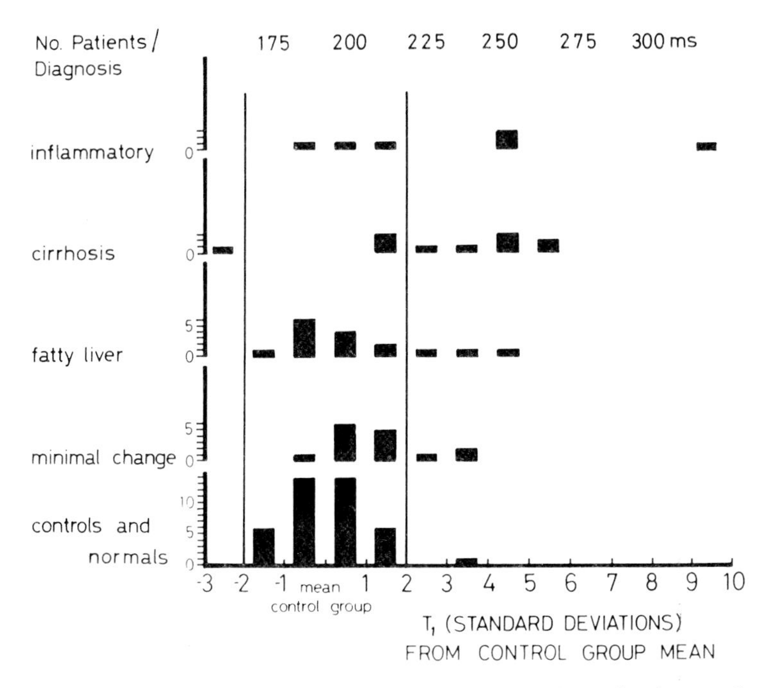

Figure 8.18. Histograms of hepatic T_1 values at 3.4 MHz averaged over liver images of normal patients and those with liver disease. The abscissa is in units of the standard deviation of the control group mean. (From Ref. 20.)

identify each disease in a given patient, but it is clear that T_1 tends to be elevated with disease. What can be said is that if the T_1 of a particular patient's liver mass is increased by two standard deviations above the normal range, then the liver is abnormal. There is then a high likelihood of cirrhosis or inflammatory disease. That, at least, is something that is useful in some circumstances. Also, if T_1 is below the normal range, it is a clear sign of the deposition of iron such as in hemochromatosis and cirrhosis.

A recent report suggests that the basic idea can have a real clinical value, even if in limited circumstances. It is very difficult to discriminate between postoperative fibrosis and early tumor recurrence using x-ray CAT. However, T_1 values seem to be successful for discriminating between carcinoma of the rectum and pelvic masses that are fibrous (Figure 8.19).[21] The tumor values of T_1 are higher than the fibrous ones probably because of greater total water content and because less of the water is bound to macromolecules such as proteins and nucleic acids. The T_2 values do not discriminate. Since tumor and fibrosis may be mixed in any pelvic mass, the pixel distribution of T_1 may give an even better discrimination and texture analysis of the images by computer needs to be explored.

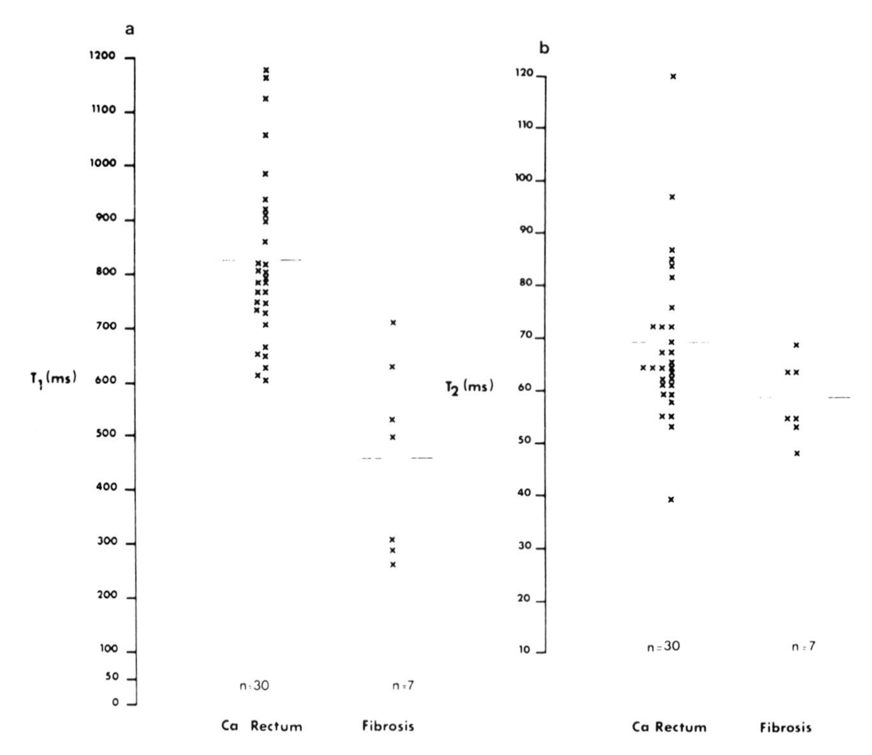

Figure 8.19. Quantitative MRI in rectal carcinoma. Histograms showing the distribution of T_1 (a) and T_2 (b) relaxation times for 30 patients with carcinoma of the rectum and 7 patients with pelvic fibrosis. Horizontal bars represent mean T_1 and T_2 values in each histogram. (From Ref. 21.)

T_1 measurement for tissue discrimination and measurement requires the imaging machine to be properly set up to give accurate measurement of T_1 or T_2. With care, errors in T_1 measurement can be low, and only rise to 5% when $T_1 >$ 500 ms. This gives an accurate measure of T_1 if the T_1 relaxation is a single exponential. Most tissues do have a monoexponential T_1 relaxation, but T_2 is usually multiexponential. As a result, T_2 measurements need multiple sampling of the relaxation in order to estimate the initial relaxation accurately. However, even though tissue T_1 is usually monoexponential, it must be borne in mind that each pixel measurement has come from a volume of tissue that is about one-tenth of a cubic centimeter. On the microscopic scale, that volume will not be uniformly filled with the same tissue, for example, tumor. It will have components of tumor and normal tissue, as well as fibrous, vascular, edematous, and necrotic tissues. This is known as the partial volume effect, and will tend to make the measured T_1 of that voxel a multiexponential one. A better understanding of this problem could lead to improved clinical tissue characterization.

8.4. Cine-NMR for Cardiac Imaging

Movie images of the beating heart can be made using ordinary NMR imagers. These are not real-time images, but the movement can be built up by a series of shots taken over many heartbeats using electrocardiogram gating to trigger the NMR pulse sequences. This allows movie recordings that show not only the movement of the heart muscles, but also visualization of the flow of the blood into and out of the chambers.(22,23) This method has now been improved using fast saturation recovery (spin–warp) sequences.(24,25) An R wave of the electrocardiogram initiates the data collection of the first line of each of 16 images. This is done by sequencing 16 complete pulse sequences each separated by 40–50 ms (Figure 8.20). The next R wave initiates 16 pulses which provide the data for the second line of each image, and so on. Sixteen complete images are acquired over 128 heartbeats, each image separated by 40–50 ms. Repetition of this builds up a complete movie of the heart action.

To acquire the data as rapidly as this, a fast saturation (spin–warp) recovery pulse sequence in conjunction with a field echo is used. Because the data collection occurs in bursts of 16 pulses, it is not continuous, and a 90° radiofrequency (rf) pulse is used to maximize the contrast between static tissue and flowing fluid. This is a fast-scan method, not dissimilar to the well-known FLASH method, but it does not use the reduced flip angle of that and other methods which only work for continuous-data collection.

The use of a field echo results in significant enhancement of the signal from protons flowing perpendicularly to the image slice. This is because of refreshment of in-plane spins by spins that have not experienced the previous burst of 16 rf pulses; this means that they give their maximum full recovery signal from the 90° pulse. If the flow is turbulent, however, then a loss of signal occurs instead of enhancement, because of the mixing of spins with widely different phase histories. These two effects provide the majority of the information in cardiac images because, for normal subjects, the flow provides enhancement that enables the chambers of the heart and associated vessels to be clearly defined. Also, the

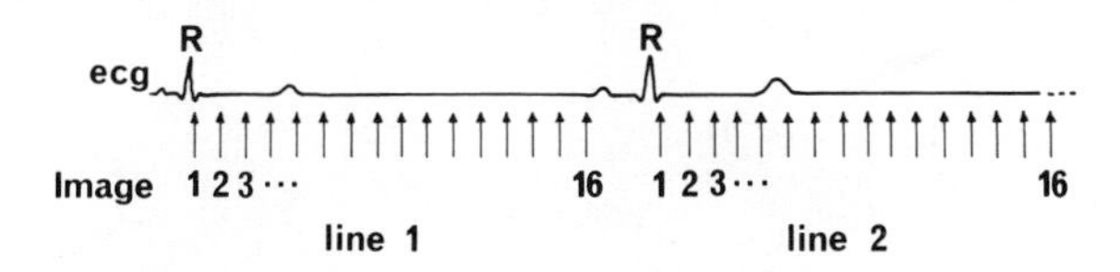

Figure 8.20. Cine-MRI for cardiac imaging. Each arrow represents the rf pulse starting a complete pulse sequence, 40 msec between each one. Each R wave triggers 16 pulses separated by 40–50 ms, each one collecting data to contribute a line to each of 16 images. The next R wave triggers the next line of the 16 images. After 128 heartbeats, 16 complete images have been acquired, each one separated by 40–50 ms.

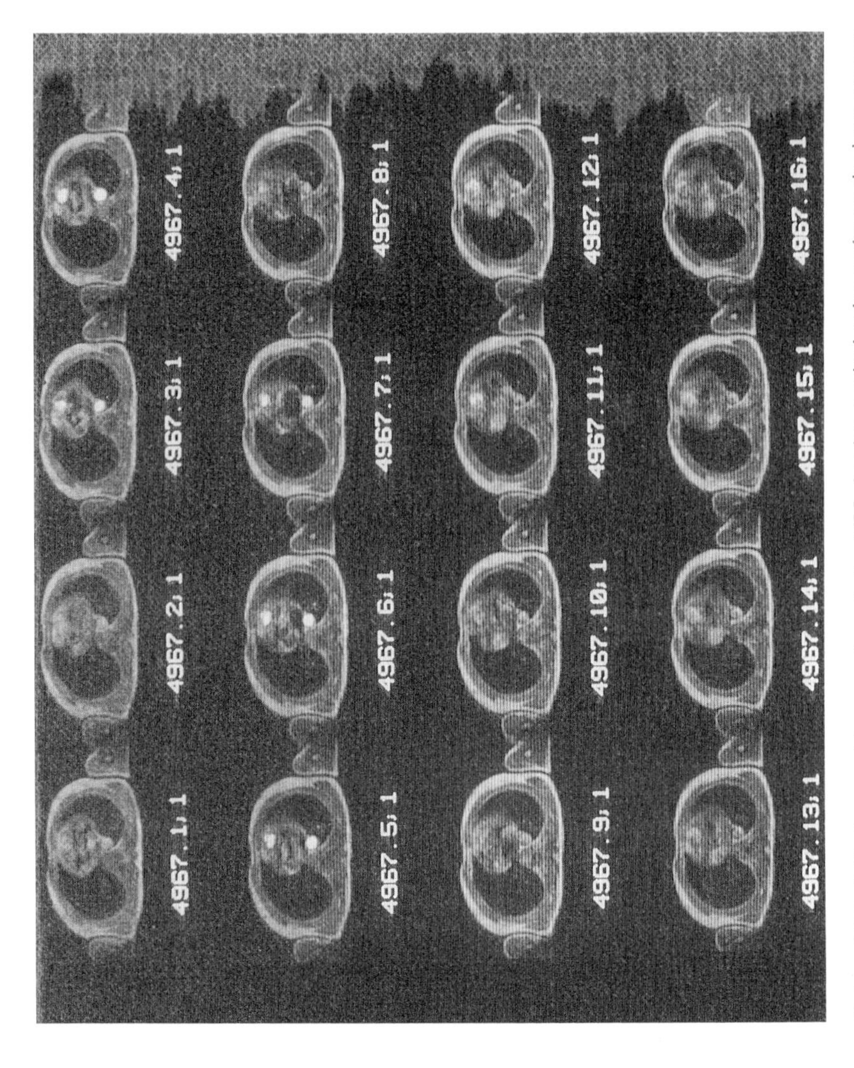

Figure 8.21. Stills from a video recording of cine-MRI showing a mitral value prolapse; the large amount of turbulence (black) is seen clearly.

degree of flow enhancement depends on the flow rate. Because the flow rates within the heart vary considerably during the cardiac cycle, the signal enhancement varies accordingly, which provides a visual movie display of cardiac function. This is now in regular use by cardiologists for conditions such as aortic stenosis, mitral valve prolapse, mitral regurgitation, hypertrophied arteries, and other heart conditions such as endocardial cushion defect (Gerbodie defect). Figure 8.21 shows stills from a video recording of mitral valve prolapse. It is hoped that NMR will compare favorably with x-ray ventriculography in the detection of abnormal intracardiac flow turbulence, cardiac wall motion, and calculation of ejection fraction. Because ventriculography requires general anaesthetic and injection of radio-opaque contrast, and as many of the patients are relatively poor anaesthetic risk, it would be a considerable advantage to be able to use a noninvasive technique such as NMR on this group. How best to add phase encoding is now being explored; this provides flow encoding, which will enable flow rates to be quantified.

8.5. NMR Angiography

Phase encoding has been used in NMR angiography. Flow influences the intensity of the detected signal in an NMR image by a paradoxical enhancement. By an appropriate pulse sequence it is possible to encode velocity information into the phase of the NMR signal. The resulting image displays information showing the direction and the measured velocity of the flow movement. Projective flow imaging produces images not unlike those from x-ray angiography.[26] The technique has been refined by using Fourier techniques;[27,28] an example is shown in Figure 8.22. This shows four images from an ECG-gated study to image the neck. The data was acquired 600 ms after detection of the R wave from a 200 mm square field of view. Flow velocity is imaged at 153 mm s^{-1}. The image at +153 mm s^{-1} shows flow in the carotids, and the carotid bifurcation is seen clearly. The image displaying −153 mm s^{-1}, flowing in the opposite direction, shows the jugular veins. These images are at low static magnetic field (0.08 T), but the power of the method is clearly demonstrated even at the diminished spatial resolution of this low field. Figure 8.23 shows an NMR angiogram at 1 T without flow encoding taken from a manufacturer's brochure. The improvement of resolution is clearly seen. In addition, pulsatile cerebrospinal fluid flow by NMR phase imaging has recently been demonstrated.[29]

8.6. Magnetic Resonance Imaging of Free Radicals and Oxygen Concentration

In the late 1950s the problems of ESR or electron paramagnetic resonance (EPR) to study surviving animal tissue samples were first addressed. The high

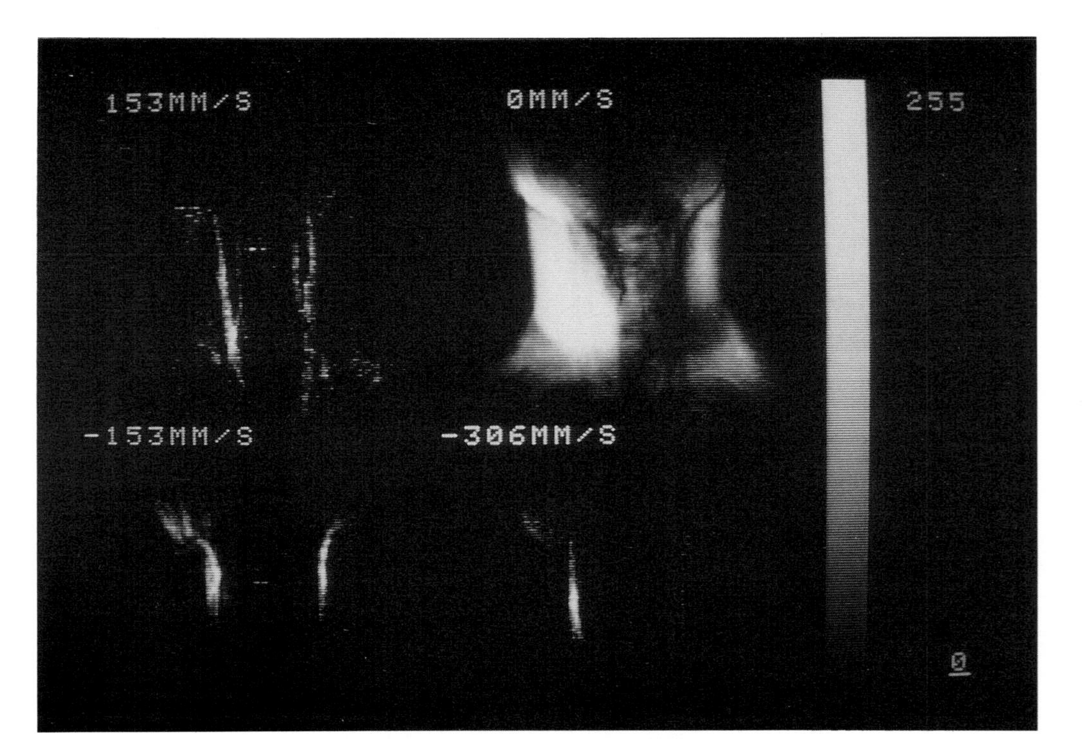

Figure 8.22. Projective angiography Fourier flow NMR image of neck (200 mm square field of view) at low field (0.08 T). The image at +153 mm s^{-1} shows the flow in the opposite direction in the jugular veins. (From Ref. 28.)

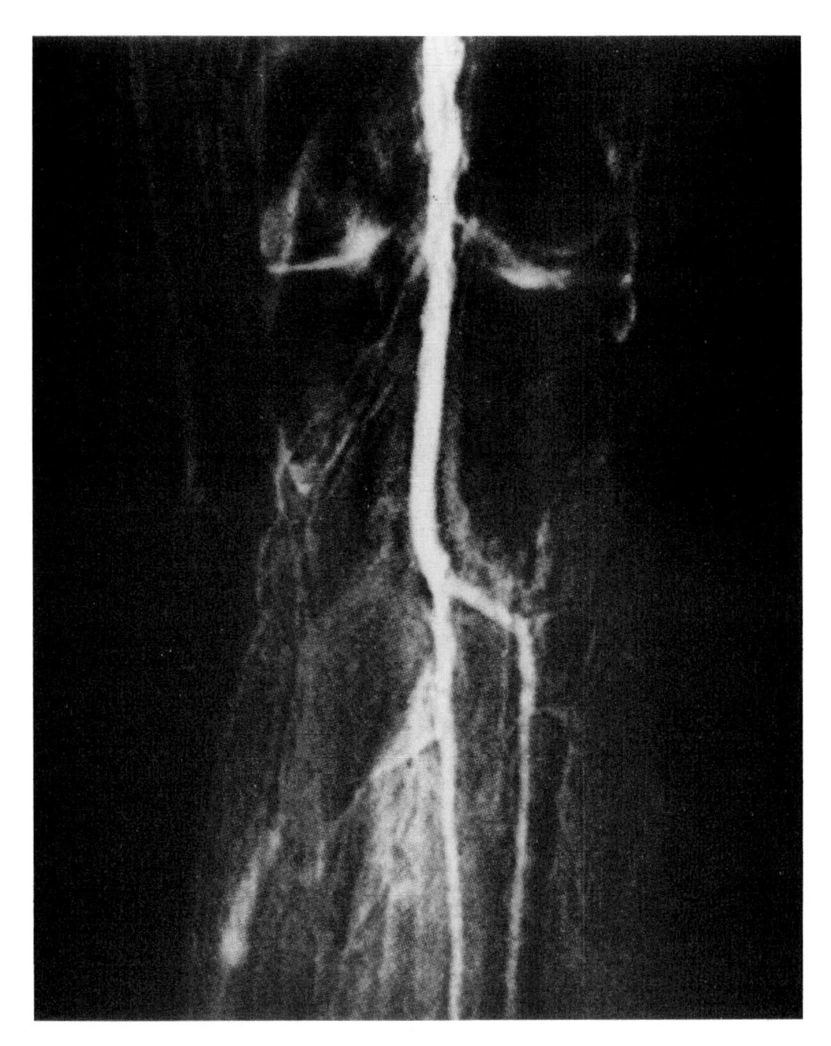

Figure 8.23. Angiograph of arteria poplitea at 1 T (from manufacturer's brochure).

water levels in the samples gave great problems, and special microwave cavities and sample holders had to be developed.[30] Using these cavities it was discovered that surviving tissues gave measurable signals at frequencies associated with the paramagnetism of the unpaired free electron. These signals could only come from free radicals, found later to be in the mitochondria. The sizes of the signals were different in normal tissue and tumors.[1,31,32] The free radical signal from normal rat kidney is compared with the much smaller one from spontaneous rat kidney carcinoma in Figure 8.24a and that from normal rat liver and rat liver carcinoma, where it appears to be greater, are compared in Figure 8.24b. These

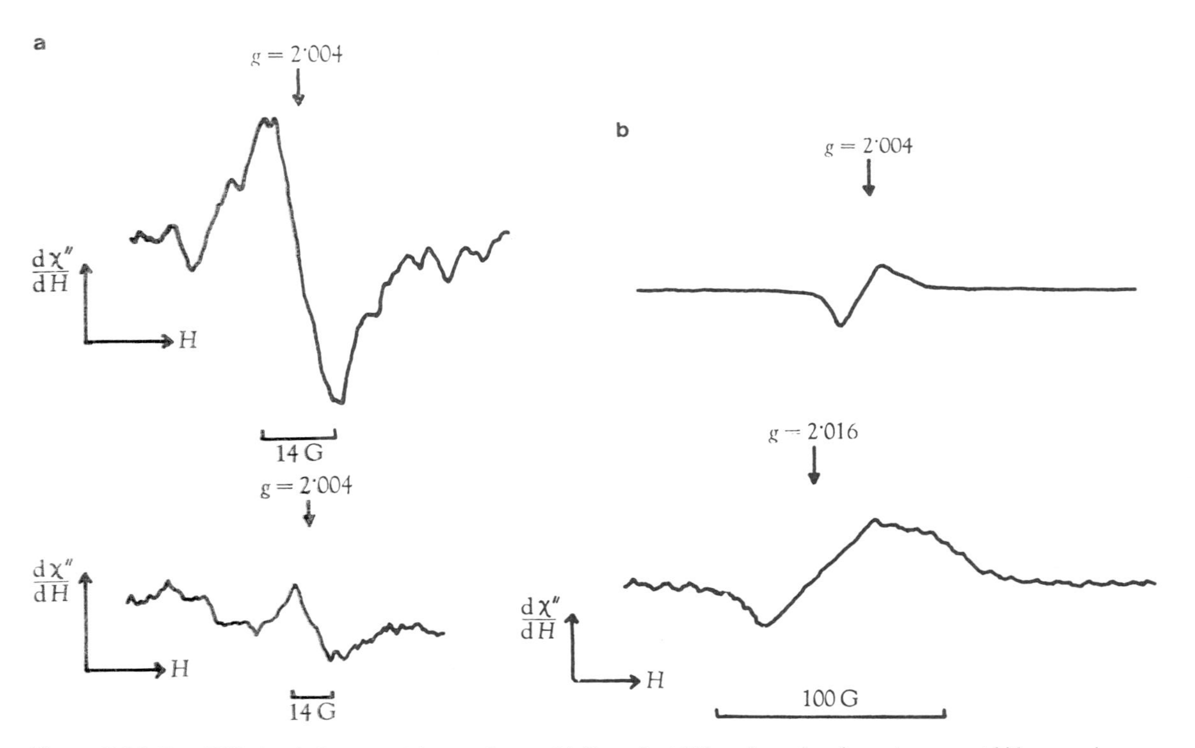

Figure 8.24. Free ESR signals from surviving rat tissues. (a) Normal rat kidney (upper) and spontaneous rat kidney carcinoma (lower). (b) Normal rat liver (upper) and rat liver carcinoma (lower) (same scale).

differences between the free radical concentration in the tumor to that in the surrounding normal tissue is thus natural contrast, which, if it could be detected in an imaging system, would lead to the display of tumor. Since, in addition, free radicals are known to play a part in carcinogenesis, imaging of free radicals could be very important.

This ESR work was done in the microwave region (950 MHz). In the late 1960s the possibility of *in vivo* ESR imaging at microwave frequencies was investigated,[33] but there is too much absorption and scattering to form a useful image.[34] The problem is now being reexamined, however, and EPR imaging may become possible at, say, 100 MHz[35,36] although it will be restricted to sample sizes of 5 mm or less.

It has been possible to form images of free radicals by a less direct method utilizing double magnetic resonance, which allows large aqueous samples to be imaged. When electron magnetic dipoles, such as free radicals, are close to the magnetic field of proton dipoles such as water (Figure 8.25a), then if the electron dipole is flipped through 180°, as it is during EPR, this will change the magnetic environment of the proton dipole and there will be an energy coupling between them. This coupling, which is a double resonance relaxation effect, increases the number of molecules in the upper molecular energy level. This greater population increases the number of protons that can take part in normal water proton NMR. The molecular energy distribution in normal water proton NMR is shown on the left in Figure 8.25b. The excitations between the populations of two upper bands are observed, giving a certain size of NMR signal. On the right, the unpaired electrons are irradiated at their EPR resonant frequency, the population in the upper band is increased, and the size of the water proton NMR signal is thereby increased. This enhancement of signal can theoretically be as high as 330 times at full saturation of the electron resonance.

This technique is called proton–electron double resonance imaging (PEDRI).[37] In PEDRI a conventional proton NMR image is collected while the EPR resonance of the paramagnetic solute (e.g., the free radicals) is irradiated. The NMR signal from those protons being relaxed by the paramagnetic solute is enhanced, and these regions exhibit a greater intensity of NMR image. Subtraction of images with and without the EPR irradiation provides an image that displays the distribution of the paramagnetic solute alone. The observed enhancement depends on the degree of saturation of the EPR resonance and therefore on the strength of the EPR irradiation; the concentration of the free radicals or paramagnetic compounds; the particular compound under study; the temperature of the sample; and the strength of the static magnetic field.

The first PEDRI image is shown in Figure 8.26. This was implemented at 0.04 T at an NMR frequency of 1.7 MHz and EPR frequency of 1.12 GHz. It shows a nitroxide free radical solution of 2.5 m*M* aqueous solution of TEMPOL in a 10 mm diameter glass tube. A signal enhancement of 6.9 times was observed, which gives a clearly increased image intensity of the tube within the

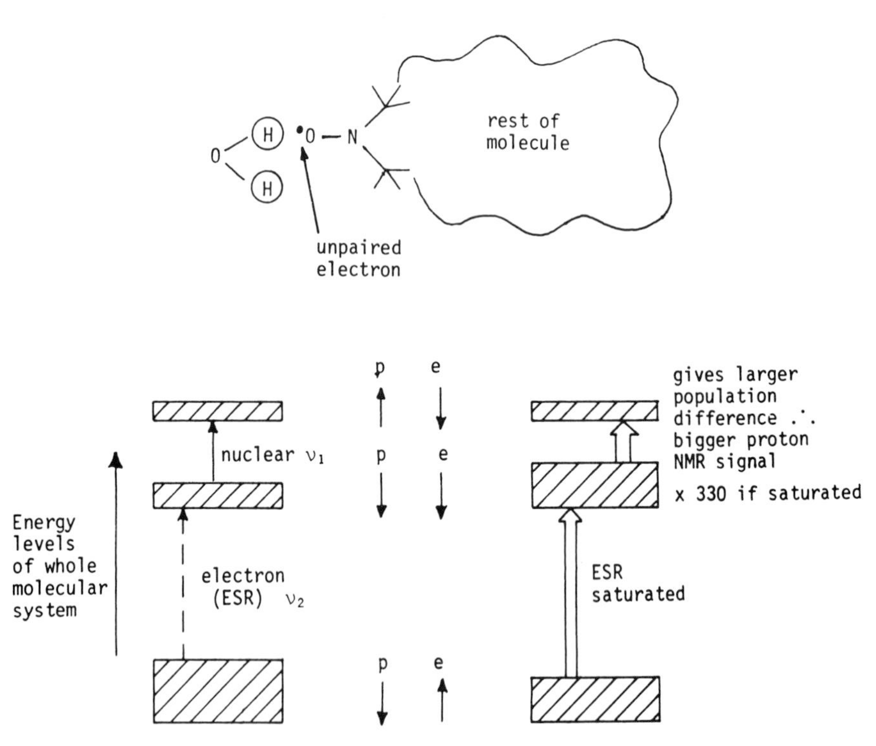

Figure 8.25. Molecular energy levels in PEDRI. (upper): There is an interaction between the proton of a water molecule in close proximity to the unpaired electron (nitroxide group) of a free radical which is caused by the magnetic field of one affecting that of the other. (lower left): Normal water proton NMR takes place by transitions between the two upper energy levels, unaffected by the electron energy levels below. (lower right): However, when the electrons are excited, as in EPR, the number of molecules in the middle level able to take part in the water proton NMR is increased. Thus, the NMR signal is increased when EPR excitation occurs simultaneously.

EPR resonator. Much larger enhancements have been observed using free radicals dissolved in organic solvents, but these would not be suitable for biological experiments. PEDRI images take 2 to 8 min, and images with aqueous TEMPOL concentrations as low as 1 m*M* have been obtained. This is about four orders of magnitude greater than the concentration of naturally occurring free radicals in tissues such as those shown in Figure 8.24. The goal of imaging these is not yet achieved, but significant progress has been made. Further exploration of the characteristics and optimization of PEDRI enhancement, and the targeting or encapsulation of free radicals, may lead to valuable new uses for this form of imaging contrast.

PEDRI could be the beginning of another new medical imaging tool. It is easily implemented using standard NMR hardware and software, requiring only the addition of a microwave source and antenna. Already, a use of great potential

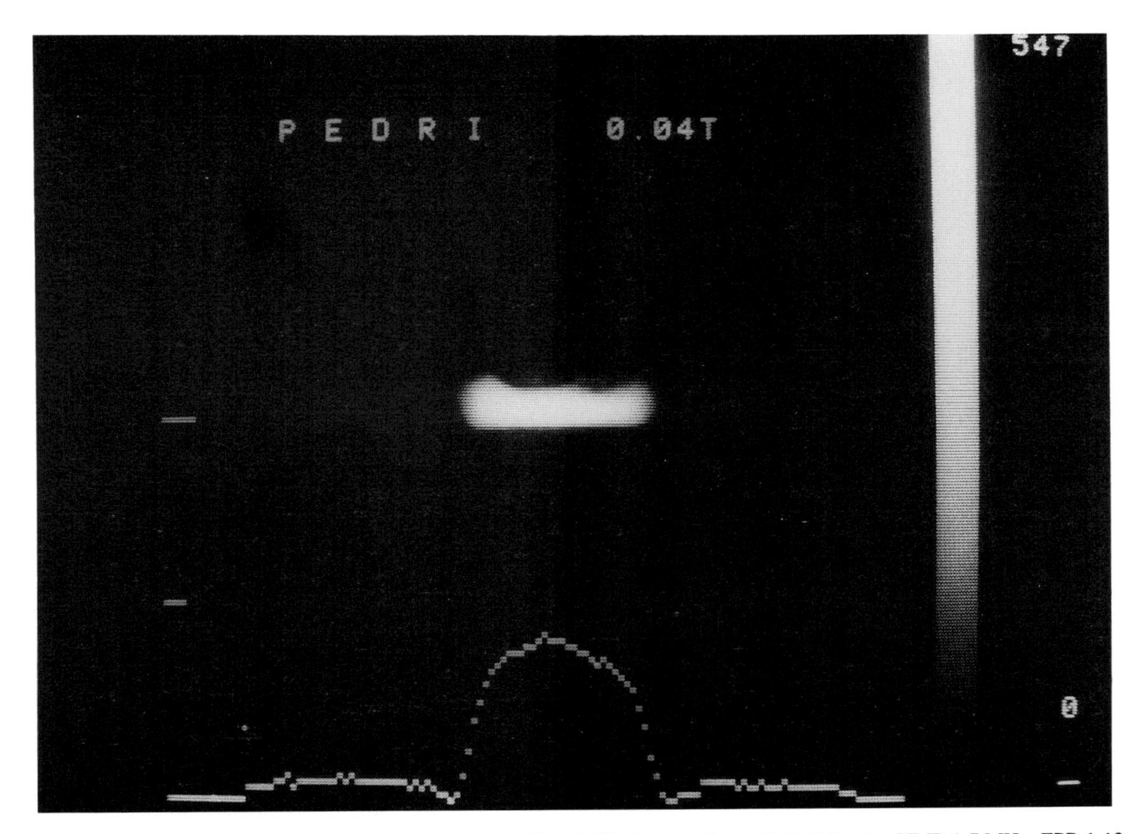

Figure 8.26. The first PEDRI free radical image (May 25th, 1987). Image taken at 0.04 T (proton NMR 1.7 MHz, EPR 1.12 GHz) of 10 mm diameter glass tube containing 2.5 m*M* aqueous solution of TEMPOL (4-hydroxy-2,2,6,6-tetramethyl-piperidine-1-oxyl) (Aldrich Chemical Co.) at room temperature.

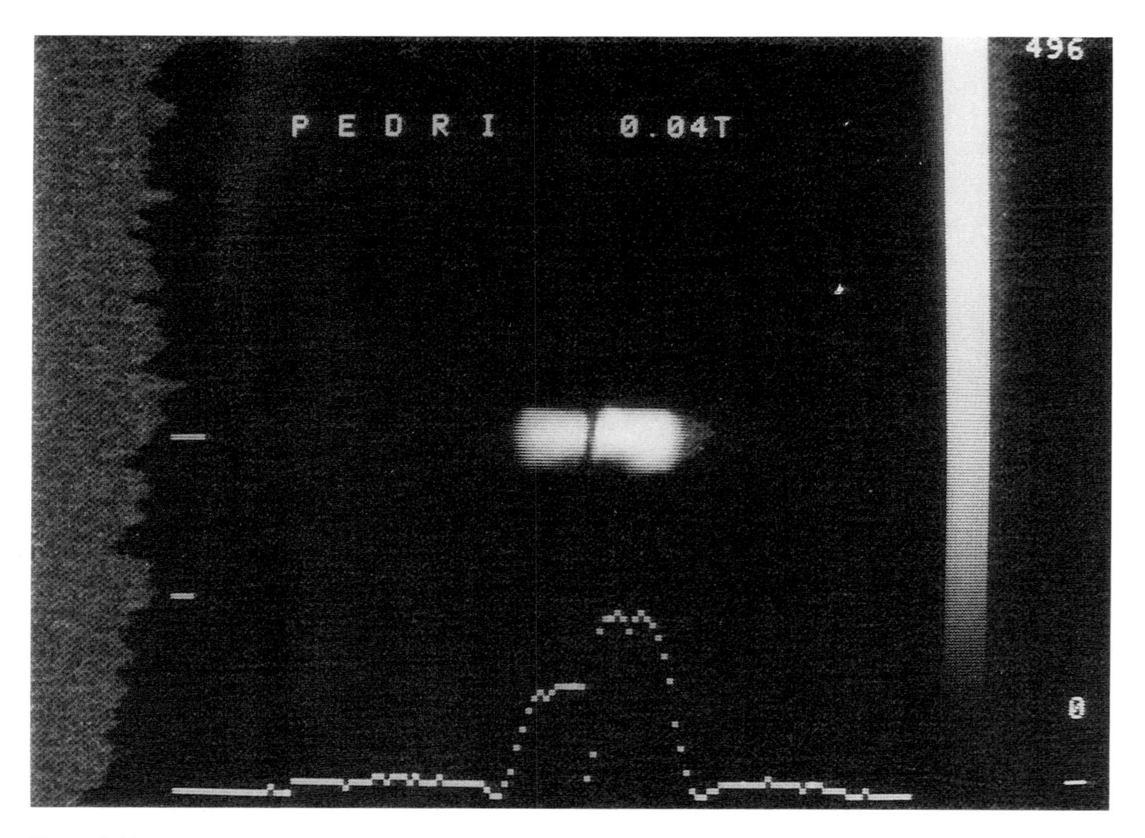

Figure 8.27. PEDRI image of two identical phantoms of 2.5 m*M* TEMPOL, the one on the left bubbled with air, and the one on the right with nitrogen. The right tube shows a 40% greater PEDRI enhancement due to a smaller oxygen content.

in measuring oxygen concentration has been demonstrated. Simultaneous images of two identical phantoms of 2.5 m*M* TEMPOL, one of which had been bubbled with air and the other with nitrogen are shown in Figure 8.27. The nitrogen-equilibrated sample exhibits a 40% greater enhancement than the air-equilibrated one. The EPR resonances of spin labels are broadened in the presence of dissolved oxygen, and the broad EPR line is more difficult to saturate which, in turn, results in a reduced enhancement factor for the air-equilibrated sample.

Using different irradiating frequencies the idea might be applied also to double nuclear resonances; for example, between ^{13}C and protons. Whether the double nuclear resonance effect will prove to be large enough to lead, for example, to the imaging of amino acids and sugars is currently being explored.

8.7. NMR Microscopy

A 1.5 T image of a plane across the breast, taken at maximum signal-to-noise using a surface coil, is shown in Figure 8.28. Clearly, the imaging of

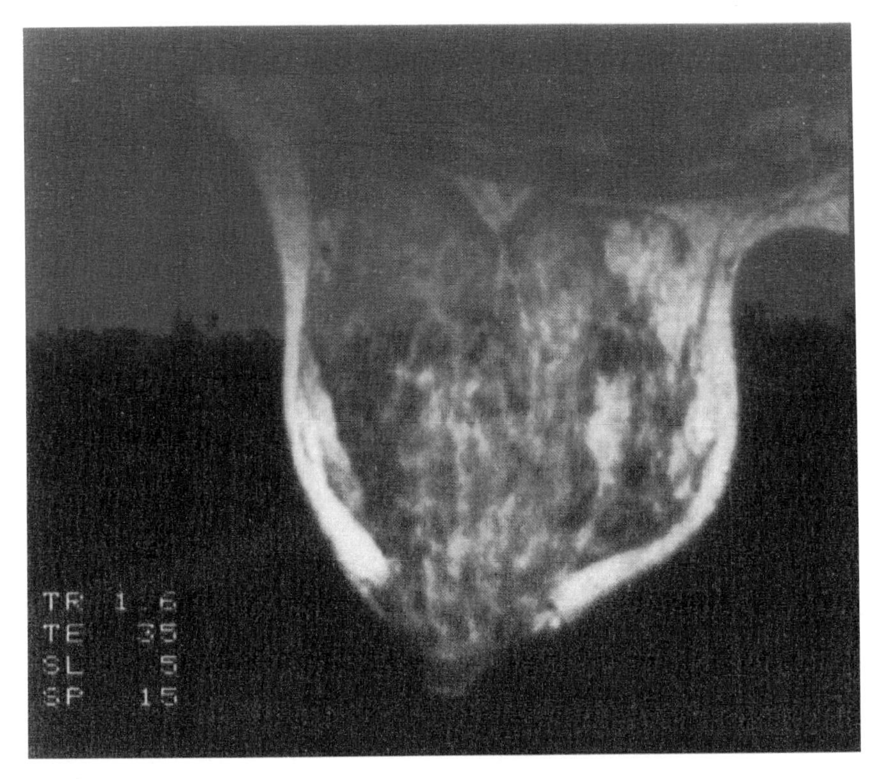

Figure 8.28. Breast tissue slice *in vivo* at 1.5 T with maximum signal-to-noise using a surface coil (from manufacturer's brochure). The anatomical detail is almost "microscopic."

microscopic structure such as blood vessels, fibrous tissue, and fat is being approached. The beginning of NMR microscopy is shown in Figure 8.29. This image is the first of a single cell by NMR imaging.[38] It is admittedly a large single cell, being the egg of the African clawed toad (*Xenopus*), which averages 1 mm in diameter. The nucleus can be seen separately from the cytoplasm showing that the nuclear water is different in some way from the cytoplasmic water. Also, the cytoplasmic water gives rather more signal in the lower part of the image than in the upper part. This image is 128 × 128 pixels, with a spatial resolution of 10 × 13 μm, but the slice thickness imaged is 250 μm, which is not truly a microscopic size. The image took 32 min 8 s to acquire at a magnetic field of 9.4 T (400 MHz).

An NMR microscope using 7.05 T (300 MHz) is planned in Aberdeen with the goal of achieving an imaging voxel of 10 × 10 × 10 μm. This will present enormous difficulties but it is a relevant goal biologically because the most commonly used human tumor cells (Hela) in culture are 13–30 μm in diameter with nuclei of 8–11 μm diameter, and human liver cells, for example, are 30–40 μm in diameter with nuclei of 12 μm.

Although such a microscope will have a very poor resolution compared with the optical microscope, which can resolve 0.1 μm, and the electron microscope (0.001 μm), NMR microscopy will have advantage since no special sample preparation will be needed. Any plane in the sample can be imaged; the technique is nondestructive and the sample can be alive. It may even be possible to develop an implantable probe such as the rf resonator, which would enable the observation of microscopic structures in a living animal rather than merely from tissue or cell preparations. If the resolution requirement could be relaxed to a voxel of, say, 30 μm, there would be a 27 times improvement in signal-to-noise, and this might lead to a smaller resonator to improve the sensitivity. A 10 μm voxel will only provide about 10^9 observable spins, and the statistical fluctuation is then high. The signal-to-noise ratio is only about 1.1 : 1, whereas about 20 : 1 is required. This can be achieved by repetitive acquisition of the signal, and signal averaging, but this will lead to data acquisition times of at least 10 min.

An NMR microscope will have a large number of uses: for example, living tissues could be studied at the cellular level. It will provide for the first time images and measurements of the intra- and extracellular distributions of water concentration, T_1 and T_2, respectively, and help in understanding the cellular role of water. Even more important it will provide cellular imaging and measurement of the diffusion of water, both intra- and extracellular. Regions of high and low diffusion can be related to cellular structures and membranes, and the effects of disease states on them studied. Perhaps, even more important again, microperfusion of water and NMR contrast-labelled materials (e.g., monoclonal antibodies) can be studied in functioning normal tissues, organs, lesions, and tumors. Changes in microperfusion during the breakdown of blood circulation in tumor development, the occlusion of blood vessels during atherosclerosis, changes

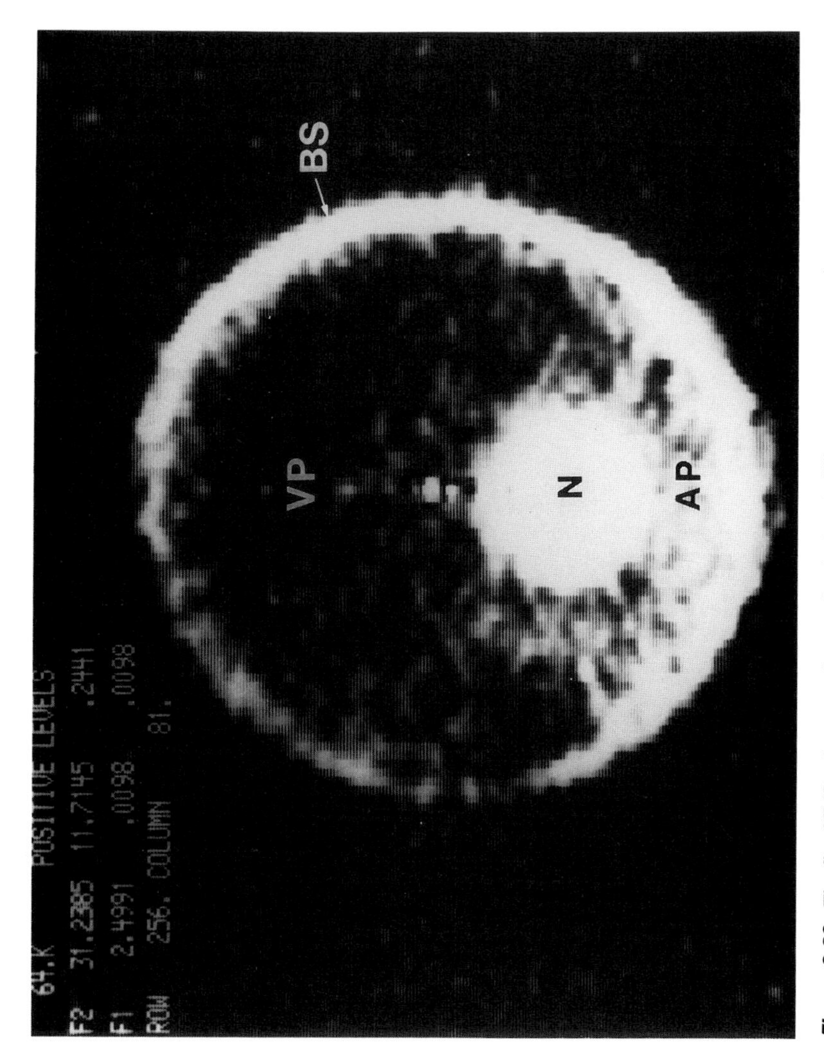

Figure 8.29. The first NMR microscope image of a single cell,[38] an ovum of *Xenopus* in which the nuclear water can be seen discriminated from cytoplasmic water. The imaging voxel is $10 \times 13 \times 250$ μm and the 128×128 image was acquired in 1928 s at 9.4 T.

during ischemia, infarction in brain and myocardium, and changes in the blood brain barrier—all can be studied.[39]

While studies can be carried out on living cellular preparations and tissues, and isolated organs, the development of an implantable probe, which is likely, may enable these cellular studies to be made in both the living plant and animal. This would enable continuous monitoring in an animal of the onset and development of diseases, the study of brain water in alcohol addiction, and other abnormal brain states. In a more general and fundamental way, it could lead to the validation or otherwise of small animal models in biomedical studies.

Some progress toward these goals has been made. An in-plane resolution of 4 μm at 2 T has been claimed,[40] but the slice thickness was not mentioned (although it is believed to be at least 200 μm), and the need for a compromise between minimizing the effect of diffusion (using a large bandwidth) and maximizing the signal-to-noise ratio (using a small bandwidth) has been stressed.[40] Microscopy at 7 T for rat brain studies has been described with a claimed resolution of 25 × 25 × 250 μm.[41] Implantable coils for microscopy at 2 T, surgically implanted around a rat kidney using inductive coupling to an external receiver coil have been reported, and a resolution of 117 × 117 × 1250 μm has been obtained.[42]

The difficulties of achieving a 10 μm voxel include the requirement for a readout magnetic field gradient over the sample volume of about 0.2 T m^{-1} and a selection gradient of 1 T m^{-1}, some 200 times greater than presently used for clinical imaging. The heat generated will be large (up to 50 W) so that cooling methods will be required. Also, the power requirement for the rf coil becomes very large, necessitating a very small, low-resistance coil. The practical difficulties of making these components are great and have not yet been solved.[43]

ACKNOWLEDGMENTS. This review could not have been written and collected together without the help and support of a large group of most generous people. I have tried hard to acknowledge this in the references. Especial recording of my gratitude to the following is given here: Dr. H. Deans, Dr. M. Foster, Mr. R. Hutcheon, Dr. J. M. S. Hutchison, Prof. I. Isherwood, Mr. R. Jones, Dr. D. Lurie, Miss E. McMaster, Mrs. M. Morrice, Dr. D. Norris, Mrs. H. Parry, Dr. T. Redpath, Mrs. D. Ross, Dr. F. W. Smith, Dr. P. Undrill, and my wife Fiona and daughter Katriona.

The help of the Medical Research Council, the University of Aberdeen, and the Grampian Health Board is gratefully acknowledged.

References

1. J. R. Mallard and M. Kent, "Differences observed between electron spin resonance signals from surviving tumor tissues and from their corresponding normal tissues," *Nature* **204**, 1192 (1964).

2. R. Damadian, K. Zaner, D. Hor, R. Dimaio, L. Minkoff, and M. Goldsmith, "Nuclear magnetic resonance as a new tool in cancer research: Tumors by NMR," *Ann. N.Y. Acad. Sci.* **222,** 1048–1076 (1973).
3. P. Lauterbur, "Image formation by induced local interactions: Examples employing nuclear magnetic resonance," *Nature* **242,** 190–191 (1973).
4. J. M. S. Hutchison, J. R. Mallard, and C. C. Goll, "*In vivo* imaging of body structures using proton resonance," in: *Proceedings of the 18th Ampere Congress* (P. S. Allen, E. R. Andrew and C. A. Bates, eds.), University of Nottingham, pp. 283–284 (1974).
5. J. Mallard, J. M. S. Hutchison, W. A. Edelstein, C. R. Ling, M. A. Foster, and G. Johnson, "*In vivo* NMR imaging in medicine: The Aberdeen approach, both physical and biological," *Phil. Trans. Roy. Soc. Lond* **B289,** 519–533 (1980).
6. W. A. Edelstein, J. M. S. Hutchison, G. Johnson, and T. W. Redpath, "Spin–warp NMR imaging and applications to human whole-body imaging," *Phys. Med. Biol.* **25,** 751–756 (1980).
7. J. R. Mallard, Wellcome Foundation Lecture, 1984. "Nuclear magnetic resonance imaging in medicine: Medical and biological applications and problems," *Proc. Roy. Soc. Lond.* **B226,** 391–419 (1984).
8. H. Deans, personal communication (1987).
9. F. W. Smith, "Magnetic resonance imaging of midline brain tumors using inversion–recovery sequences at 0.08 T (3.4 MHz)," *Magn. Reson. Med.* **5,** 118–128 (1987).
10. D. H. Carr, J. Brown, G. M. Bydder, R. E. Steiner, H. J. Weinmann, U. Speck, A. S. Hall, and I. R. Young, "Gadolinium–DTPA as a contrast agent in MRI," *Am. J. Roentg.* **143,** 215–224 (1984).
11. R. Felix, "The potential role of contrast media in magnetic resonance imaging," *3rd Congress. Eur. Soc. Magn. Res. in Med. Biol.*, University of Aberdeen, pp. 71–72 (1986). Published by European Society for Magnetic Resonance in Medicine and Biology, Geneva, Switzerland.
12. J. Sadler, personal communication (1987).
13. M. A. Foster, "Contrast manipulation," in: *NMR Proton Imaging Summer School Lecture Notes,* University of Aberdeen (1987).
14. Medical Research Council Review of Clinical Evaluation of Magnetic Resonance Imaging (1987). Available from Medical Research Council, London.
15. F. W. Wehrli, J. R. McFall, D. Shutts, R. Breger, and R. T. Herfken, "Mechanisms of contrast in NMR imaging," *J. Comp. Assist. Tomog.* **8,** 369–380 (1984).
16. M. A. Foster, "Tissue characterization by NMR," in: *Functional Studies Using NMR* (V. R. McCready, M. Leach, and P. J. Ell, eds.), pp. 147–166, Springer-Verlag, London (1986).
17. Concerted Research Project of the European Economic Community. Reported in *European Quarterly,* Nos. 1–15, published by the European Economic Community and available from F. Podo, Instituto Superiore di Sanita, 0016, Rome, Italy.
18. M. A. Foster and J. R. Mallard, "Biological basis and clinical value of tissue characterization by NMR," *Proc. 9th Ann. Conf. of IEEE Engineering in Medicine and Biology,* Boston, p. 1973 (1987).
19. R. M. Kroeker, E. R. McVeigh, P. Hardy, M. T. Bunskill, and R. M. Henkelman, "*In vivo* measurements of NMR relaxation times," *Mag. Res. Med.* **2,** 1–13 (1985).
20. G. R. Cherryman, A. P. Bayliss, P. W. Brunt, J. F. Calder, J. A. Harvey, J. K. Hussey, A. F. MacDonald, N. A. G. Mowat, E. M. Robertson, P. F. Sharp, J. J. Shrimankar, J. J. Simpson, T. S. Sinclair, F. W. Smith, J. Weir, and J. R. Mallard, "Magnetic resonance imaging of parenchymal liver disease: A comparison with ultrasound, radionuclide scintigraphy, and x-ray computed tomography," *Clin. Radiol.* **38,** 495–502 (1987).
21. R. J. Johnson, J. P. R. Jenkins, I. Isherwood, R. D. James, and P. F. Schofield, "Quantitative magnetic resonance imaging in rectal carcinoma," *Brit. J. Radiol.* **60,** 761–764 (1987).
22. D. G. Norris, "Phase-encoded NMR flow imaging," *Proc. 3rd Ann. Meet. Soc. Magn. Res. in Med.*, New York, pp. 559–560 (1984).

23. D. G. Norris and J. M. S. Hutchison, "Gated cardiac imaging using low-field NMR," *Phys. Med. Biol.* **31,** 779–787 (1986).
24. T. W. Redpath and R. A. Jones, "FADE—A new fast imaging sequence," *Magn. Res. Med.* **6,** 224–234 (1988).
25. M. J. Metcalfe, T. Redpath, S. Walton, and F. W. Smith, "Low-field cine-MRI in mixed aortic valve disease," *Magn. Res. Imaging* **7,** Suppl. No. 1, p. 92 (1989).
26. V. J. Weeden, R. A. Meuli, R. R. Edelman, S. C. Geller, L. R. Frank, T. J. Brady, and B. R. Rosen, "Projective imaging of pulsatile flow with magnetic resonance," *Science* **230,** 946–948 (1985).
27. D. G. Norris, R. A. Jones, and J. M. S. Hutchison, "Projective Fourier angiography," *Proc. 3rd Ann. Meet. Eur. Soc. Magn. Res. in Med. and Biol.*, University of Aberdeen, pp. 183–184 (1986). Published by European Society for Magnetic Resonance in Medicine and Biology, Geneva, Switzerland.
28. D. G. Norris, "NMR Flow Imaging," Ph.D. Thesis, University of Aberdeen (1987).
29. J. P. Ridgway, L. W. Turnbull, and M. A. Smith, "Demonstration of pulsatile cerebrospinal-fluid flow using magnetic resonance phase imaging," *Brit. J. Radiol.* **60,** 423–427 (1987).
30. P. Cook and J. R. Mallard, "An electron–spin resonance cavity for the detection of free radicals in the presence of water," *Nature* **198,** 145–147 (1963).
31. J. R. Mallard and M. Kent, "Electron spin resonance in surviving rat tissues," *Nature* **210,** 588–591 (1966).
32. J. R. Mallard and M. Kent, "Electron spin resonance in biological tissues," *Phys. Med. Biol.* **14,** 373–396 (1969).
33. J. M. S. Hutchison and J. R. Mallard, "Electron spin resonance spectrometry on the whole mouse *in vivo:* A 100 MHz spectrometer," *J. Phys. E: Sci. Instrum.* **4,** 237–239 (1971).
34. J. R. Mallard and T. A. Whittingham, "Dielectric absorption of microwaves in human tissues," *Nature* **218,** 366–367 (1968).
35. L. J. Berliner and H. Fujii, "EPR imaging of diffusional processes in biologically relevant polymers," *J. Magn. Reson.* **69,** 68–72 (1986).
36. M. M. Maltempo, S. S. Eaton, and G. R. Eaton, "Spectral–spatial two-dimensional EPR imaging," *J. Magn. Reson.* **72,** 449–455 (1987).
37. D. J. Lurie, D. M. Bussell, L. H. Bell, and J. R. Mallard, "Proton–electron double magnetic resonance imaging of free radical solutions," *J. Magn. Reson.* **76,** 366–370 (1988).
38. J. B. Aguayo, S. J. Blackband, J. Schoeniger, M. A. Mattingly, and M. Hintermann, "NMR imaging of a single cell: The NMR microscope," *Nature* **322,** 190–191 (1986).
39. M. A. Foster, personal communication (1987).
40. Z. H. Cho, C. B. Ahn, S. C. Juh, H. G. Lee, J. H. Yi, and J. M. Jo, "Some experiences on a 4 μm NMR microscopy," *Proc. Soc. of Mag. Reson. in Med. 6th Annual Meeting,* New York, p. 233 (1987). Published by Society of Magnetic Resonance in Medicine, Berkeley, California.
41. G. A. Johnson, S. A. Suddarth, P. B. Roemer, G. P. Cofer, W. A. Edelstein, and R. W. Redington, "MR microscopy at 7.0 T," *Proc. Soc. of Mag. Res. in Med., 6th Annual Meeting,* New York, p. 23 (1987). Published by Society of Magnetic Resonance in Medicine, Berkeley, California.
42. M. D. Hollet, G. P. Cofer, R. R. Maronpot, and G. A. Johnson, "Implanted RF coils for MR microscopy," *Proc. Soc. of Mag. Reson. in Med. 6th Annual Meeting,* New York, p. 467 (1987). Published by Society of Magnetic Resonance in Medicine, Berkeley, California.
43. J. M. S. Hutchison, personal communication (1987).

9

NMR Microscopy of Plants

P. G. Morris, A. Jasinski, and D. J. O. McIntyre

9.1. Introduction

The first papers on the subject of NMR imaging were published in 1973.[1,2] From the outset, the possibilities for the development of an NMR microscope were appreciated. Indeed, the seminal paper from the Mansfield[2] group was entitled "NMR 'diffraction' in solids?" Clearly, a method along the lines of x-ray crystal diffraction was envisaged. The subsequent development of the subject, however, was entirely devoted to clinical ends, the first prototype whole-body imaging systems generating results in 1977–1978.[3,4] Only within the last few years has attention returned to the development of a microscopic imaging system. Thus, in 1986 Aguayo *et al.* published NMR images of *Xenopus laevis* ova at various stages during oogenesis.[5] The resolution obtained was $10 \times 13 \times 250$ μm, and differences in the proton signal intensity from water were noted in the cell nucleus and the animal and vegetal poles of the cytoplasm. In the same year, Bone *et al.*[6] reported an NMR microscopic study of the development of a domestic chicken embryo. A more modest resolution of $200 \times 200 \times 1250$ μm was employed with their instrument, which was based on a 1.5 T whole-body imaging system. A three-dimensional imaging scheme was used. The same type of imaging system has been used to study changes in water distribution and binding in transpiring plants[7,8] and water transport in plants with light-stressed foliage.[9] The latter study involved observation of the root system *in situ* in the

P. G. Morris and D. J. O. McIntyre • Department of Biochemistry, University of Cambridge, Cambridge CB2 1QW, United Kingdom. ***A. Jasinski*** • Institute of Nuclear Physics, 31-342 Krakow, Poland.

surrounding soil. A more recent study using a higher field instrument (4.7 T) and a two-dimensional imaging scheme has achieved a resolution of $50 \times 50 \times 560$ μm in small excised sections of plant tissue.[10] Clear delineation of the xylem vessels was observed in 4-day-old maize seedlings. Mung beans were also followed during the germination process.

The physical basis of NMR imaging has been outlined in Chapter 8. Fuller treatments are also available.[11,12] The NMR microscope differs markedly from other imaging microscopes. Amongst the important distinguishing features are the following:

1. No specimen preparation (e.g., fixing, staining)
2. No restriction to surfaces (*in situ* observation at depth)
3. Control of slice orientation
4. Ability to perform 3-D (volume) imaging
5. Control of image contrast via NMR parameters
6. Control of resolution via strength of applied magnetic field gradients

Perhaps the most important of these points concerns the ability to work with biological samples in their natural state, that is, intact, hydrated, and at a physiological temperature. This is to be contrasted with electron microscopy where extensive sample preparation or the use of a cryostage is required. The technique is essentially noninvasive because the radiation employed is in the radiofrequency region of the electromagnetic spectrum and is therefore nonionizing. Biological time courses can therefore be followed; for example, differentiation in a developing plant, or the growth of a cultured embryo. Of great importance, too, is the ability to select a specific region within the object of study. Physical sectioning of samples is not therefore necessary. Against these many potential advantages must be weighed the one major drawback of the method, namely, an inherent lack of sensitivity. This arises because of the small magnitude relative to thermal quanta of the nuclear Zeeman energy splitting at practical magnetic field strengths. To take an example, at the highest commercially available field strength of 14.1 T, corresponding to a field of 600 MHz for protons, the fractional excess of protons in the low-energy state is only about 5×10^{-5}. This means that the temperature at which the thermal quanta have equivalent energy to the splitting is only ≈ 0.03 K. Thus, although in principle any desired resolution can be achieved by increasing the strength of the applied magnetic field gradients, there is a practical limit set by the sensitivity of the system. It is hard to predict absolute limits with any degree of confidence, but identification of the parameters of importance suggests ways in which matters might be improved. Mansfield and Morris[11] have given expressions [equations (6.39) and (6.42) in Ref. 11] for the signal-averaging time, t, required to achieve a given spatial resolution. These expressions can be inverted to give expressions for the in-plane resolution Δx. Thus, for cubic volume elements,

$$\Delta x \approx C \left[\left(\frac{S}{N} \right)^2 a^2 \frac{T_1}{tT_2} f^{-7/2} \right]^{1/6} \tag{9.1}$$

and for a slice-selected experiment,

$$\Delta x \approx C \left[\left(\frac{S}{N} \right)^2 a^2 \frac{T_1}{tT_2} f^{-7/2} \Delta z^{-2} \right]^{1/4} \tag{9.2}$$

where S/N is the desired signal-to-noise ratio in the final image, a is the radius of the receiver coil, T_1 and T_2 are the spin–lattice and spin–spin relaxation times, respectively, f is the NMR frequency, Δz is the slice thickness, and C is a constant. C can be predicted from theoretical considerations but values obtained in this way are only likely to give an order of magnitude impression. The real use of such expressions is that they enable extrapolation from a given situation to a new set of conditions. There exists considerable disagreement, however, concerning the frequency dependence of the NMR S/N. There is no problem with the signal; it increases as the square of frequency. The difficulty arises with the noise. Estimates of the strength of its frequency dependence vary from a fourth-root power to linear, depending on the nature of the sample and, to a lesser degree, the geometry of the receiver coil.(13) The most optimistic view has been taken in the derivation of equations (9.1) and (9.2), and these formulae apply for the case of an optimal imaging technique. For nonoptimal methods, not only will the value of C increase, but also the question of minimum performance time arises. Thus, only in the ideal situation will a single NMR record contain sufficient information to enable an image to be reconstructed. In the nonideal situation a number of such records ($n_z \times n_y$ in the case of three-dimensional Fourier imaging, for example) will be necessary before reconstruction is possible. This leads to the idea of a minimum performance time associated with a particular method. As will become evident, this can be quite substantial as a consequence of the generally long NMR relaxation times, particularly at the higher magnetic fields favored for microscopy. In the case of Fourier imaging, the method gives optimal sensitivity because signal from the entire sample is always observed, but many records are necessary for reconstruction, severely limiting the possibility of trading off image S/N against imaging time. The optimization of imaging parameters for maximum resolution has been considered in papers by Eccles and Callaghan.(14,15)

When designing an NMR microscope careful attention needs to be given to diffusion effects. These can certainly make substantial contributions to line broadening in the micrometer resolution range and dictate the minimum gradient strength that is required (see discussion in Ref. 11, pp. 174–175).

Unless there is some natural two-dimensional structure within the object of study (a series of parallel muscle fibers or a plant stem, for example), it makes little sense to have Δz larger than Δx—it is clearly absurd to speak of a resolution of 10 μm in a slice 1 mm thick. Thus, it is the isotropic case [equation (9.1)] that is generally the most appropriate. Putting in reasonable values for parameters leads to the conclusion that an isotropic resolution of 10 μm should be readily attained but that submicron resolution will be extremely difficult to achieve. It is likely that for the near future the resolution will be in the 1–10 μm range, that is, cellular resolution should be readily reached but subcellular resolution will be difficult.

NMR microscopy systems are slowly becoming available commercially as accessories to high-field NMR instruments. The system that has been constructed in Cambridge for use with a 400-MHz NMR spectrometer and preliminary experiences with it in the study of intact plant tissues are described below.

9.2. Experimental

The heart of the Cambridge NMR microscope is a Bruker AM400 NMR spectrometer fitted with a 9.4 T Oxford Instruments widebore (89 mm) superconducting magnet. To this standard NMR configuration, which is used for signal detection, processing, and for synchronization of imaging sequences, a purpose-built gradient controller, a gradient set, an image display system, and a probe-head for microscopy have been added.

The probe consists of a three-turn solenoid wound from high-purity copper wire on a 5 mm outer-diameter quartz tube. The length of the solenoid is also 5 mm, and it is mounted with its axis normal to the static field direction. Such transverse solenoids have long been known to confer a $\sqrt{3} \times$ S/N advantage over the saddle or "Dadok" arrangements usually employed with superconducting magnets. The coil is mounted parallel to the transverse magnetic field gradient used for slice selection in a two-dimensional Fourier imaging experiment (see below). The probe assembly is illustrated in Figure 9.1. Of the three tubular variable capacitors (0.6–6 pF), constructed from nonmagnetic material, one is used to parallel tune the coil to the resonance frequency of 400 MHz and the remaining two for a balanced network to match it to 50 Ω. The inherent Q of such circuits is high, typically ≈400. However, when loaded with a conducting sample such as physiological saline, it drops by typically an order of magnitude.

Whereas inspection of equation (9.1) would indicate that the smallest radius a should be used to achieve the highest filling factor for the coil, this becomes more problematic as the scale of the system is reduced to the point that the dimensions become comparable with the diameter of the wire from which the coil is constructed. The difficulty is the inhomogeneity in both B_0 and B_1 generated in the vicinity of the solenoid by the wires themselves. The effect on B_0 can be alleviated by adherence to cylindrical symmetry. It can also be completely eliminated by the use of zero-susceptibility material; this can often be achieved

Figure 9.1. The NMR probe used for microscopy experiments. The three-turn solenoid forms a 400 MHz parallel tuned circuit with the central variable capacitor.

as a composite by plating a weakly paramagnetic conductor onto a diamagnetic metal base. Such considerations are crucial in order to allow the maximum theoretical S/N to be reached. In this case the smallest overall bandwidth is required, and the applied field gradient is chosen to separate individual pixels by a frequency corresponding to the NMR linewidth. Ideally, this should be the natural linewidth, rather than one augmented by the static-field inhomogeneity caused by the coils. If these extreme limits are not required the effects can be overcome simply by increasing the gradient strengths.

The gradient controller has been described in detail elsewhere.[16] Essentially, it consists of three digital waveform generators corresponding to the G_x, G_y, and G_z gradient channels (see Figure 9.2). Each of these generators has its own memory which can be loaded with an appropriate gradient waveform from a small supervisory microcomputer (e.g., a BBC Model B). Waveforms are strobed out of the memory by clock pulses generated by the computer of the host spectrometer. They are routed though 8 $\times$ 8 bit fast (85 ns) digital multipliers (e.g., ADSP 1080). This is a particularly convenient way in which to implement amplitude and phase encoding. It also enables the sequence of image acquisition to be readily altered to allow rapid coarse imaging to be conducted (with subsequent high-resolution infill, if indicated) or acquisition in a bit-reversed sequence to enable preprocessing in a fast Fourier transform reconstruction procedure.

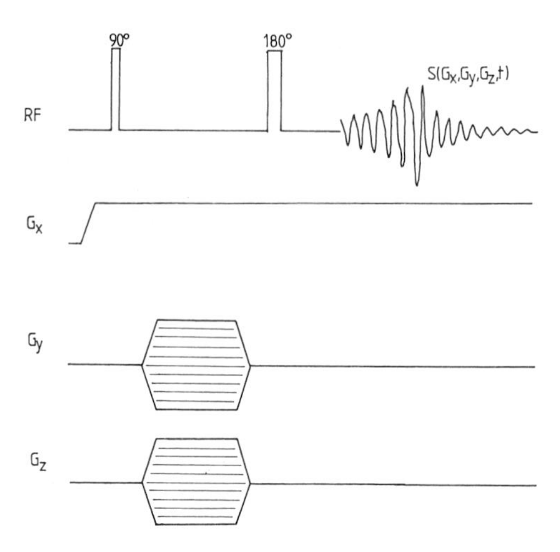

Figure 9.2. A three-dimensional Fourier imaging (3DFI) scheme with amplitude phase encoding.

The imaging sequences that have been used are essentially those based on Fourier imaging methods,[17,18] using gradient amplitude encoding. This method was originally described under the name spin–warp imaging.[19] In the present case a full three-dimensional form requiring phase encoding in G_y and G_z has been used, as shown in Figure 9.2. The sequence repetition time T_R is largely determined by the spin–lattice relaxation times of the nuclear spins. In the case of tissue water these may typically be in the range 0.2–1 s at clinical imaging frequencies. Spin–lattice relaxation times do increase with frequency[12]; in contrast, the spin–spin relaxation time is relatively independent of frequency since low-frequency components are also strongly influential in determining this relaxation time. Thus, for this type of experiment, the minimum performance time referred to above is $n_y \times n_z \times T_R$. Typically, n_y might be 64 and n_z 32. Taking $T_R = 1$ s gives a minimum performance time of 34 min. This can only be reduced if decreased resolution in the phase-encoding dimensions (smaller n_y, n_z) is acceptable. Herein lies the temptation to depart from isotropic resolution.

As the number of phase-encoding steps becomes smaller the image becomes relatively less well-represented by the Fourier analytical procedure, and there is greater leakage from pixel to pixel. As an alternative to three-dimensional Fourier imaging (3DFI) a two-dimensional experiment can be carried out. Unless a projection is acceptable (rarely), this requires the selection of a slice using a slice-selective pulse.[12] Imaging times are thereby reduced by a factor n_z. If a small number of slices is required such methods can be faster than a three-dimensional experiment with comparable z-resolution (slice thickness). This is possible because a true selective pulse will interact only with those spins lying in the selected slice, leaving those in adjacent slices unperturbed and hence capable of excitation by other selective pulses applied during the T_R relaxation recovery interval. In this way several such slices may be imaged in the time it takes to image a single slice. This is known as multislice imaging. For this scheme to work properly any 180° refocusing pulses (see Figure 9.2) must also be selective. This can be problematic since selective 180° refocusing pulses are difficult to design due to the nonlinearity of the NMR spin system.[20] For rapid time courses two-dimensional Fourier imaging experiments will be the order of the day, whereas when three-dimensional structure is important or longer time courses are involved the full three-dimensional experiment is appropriate.

In the present case the two- or three-dimensional data set is held to 24-bit accuracy. This information is normally truncated to 8 bits before conversion to a gray scale for image display. Real-time windowing facilities are available to set upper and lower intensity masks and to multiply the data in the selected range. Commercial packages with much more sophisticated display options are also available.

9.3. Results and Discussion

To assess the resolution achievable by the system described, a simple capillary tube phantom has been examined. The tube had an internal diameter of 1.6 mm and was filled with tap water. It was imaged using a solenoidal transceiver coil and a 3DFI sequence with $n_x \times n_y \times n_z = 128 \times 64 \times 32$. Gradient strengths and phase-encoding periods were selected such that the resolution per image point was 40 μm in each of the x-, y-, and z-dimensions. A two-dimensional plane normal to the capillary axis has been selected for display and is shown in Figure 9.3a. This image can perhaps be better appreciated in the stacked plot of Figure 9.3b, which shows the actual Fourier amplitudes prior to conversion to a gray scale. Vertical deflection rather than intensity corresponds to signal strength. The observed S/N is 150:1. Extrapolation of these results via equation (9.1) leads to a lower resolution limit for this system of about 10 × 10 × 10 μm. At present it is not possible to operate in this limit because of the rather modest power available from the gradient amplifiers. This situation will be remedied shortly through the provision of a smaller gradient set requiring less drive current to achieve the same gradient strength. Of course, subject again to gradient limitations, resolution in any dimension can be traded off against another provided the overall voxel volume is maintained.

Plant tissue studies have been performed using the basic system described. Four consecutive 1.7 mm NMR sections through an agricultural lupin seedling near the root/tip junction are shown in Figure 9.4. These four slices are the central ones from a full 3DFI set of 32. The resolution in the plane of the slice is approximately 70 μm. During development discrete vascular bundles eventually fuse to give an annular transport system. This is the bright ring visible at about

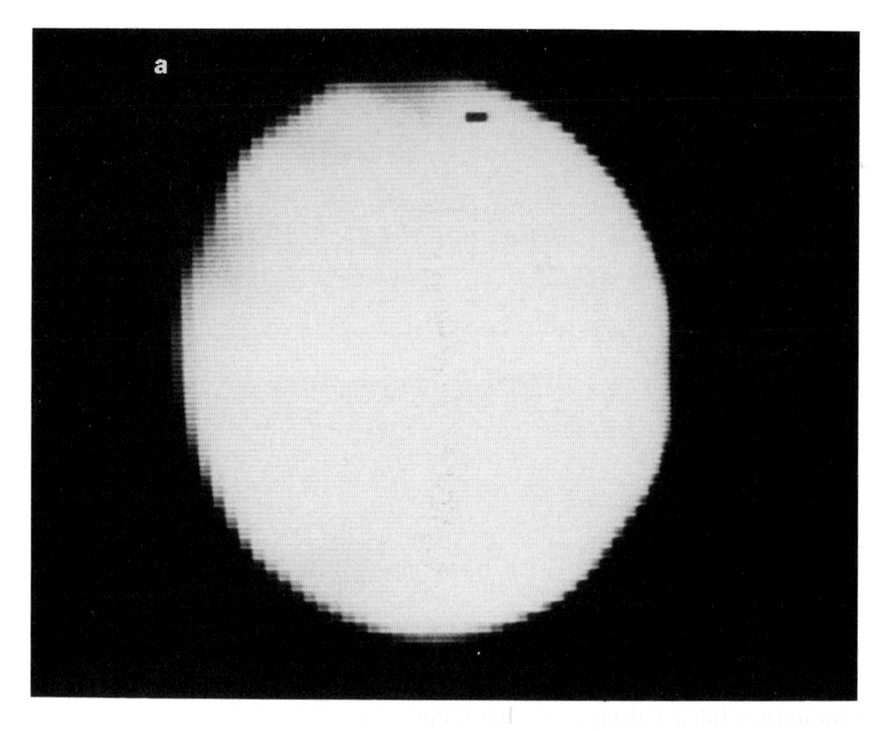

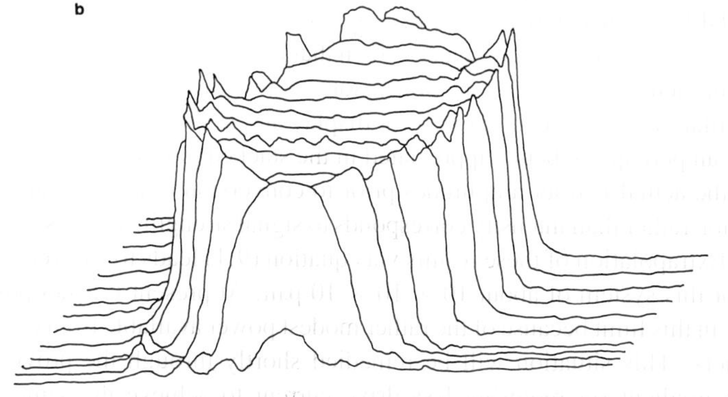

Figure 9.3. (a) Cross-sectional image through a 1.6 mm inner-diameter capillary phantom. Selected from a three-dimensional isotropic image data set. Resolution: 40 μm. (b) Stacked plot showing line intensities through a central cross section.

one-third the stem radius in each cross-section. Evidence of the discrete structures can also be seen (top and bottom, left and right of each annulus).

Similar image arrays of somewhat larger plant structures are shown in Figures 9.5 and 9.6. All of these were obtained with a more conventional Dadok coil mounted on an 8 mm inner-diameter cylindrical former. Figure 9.5 shows

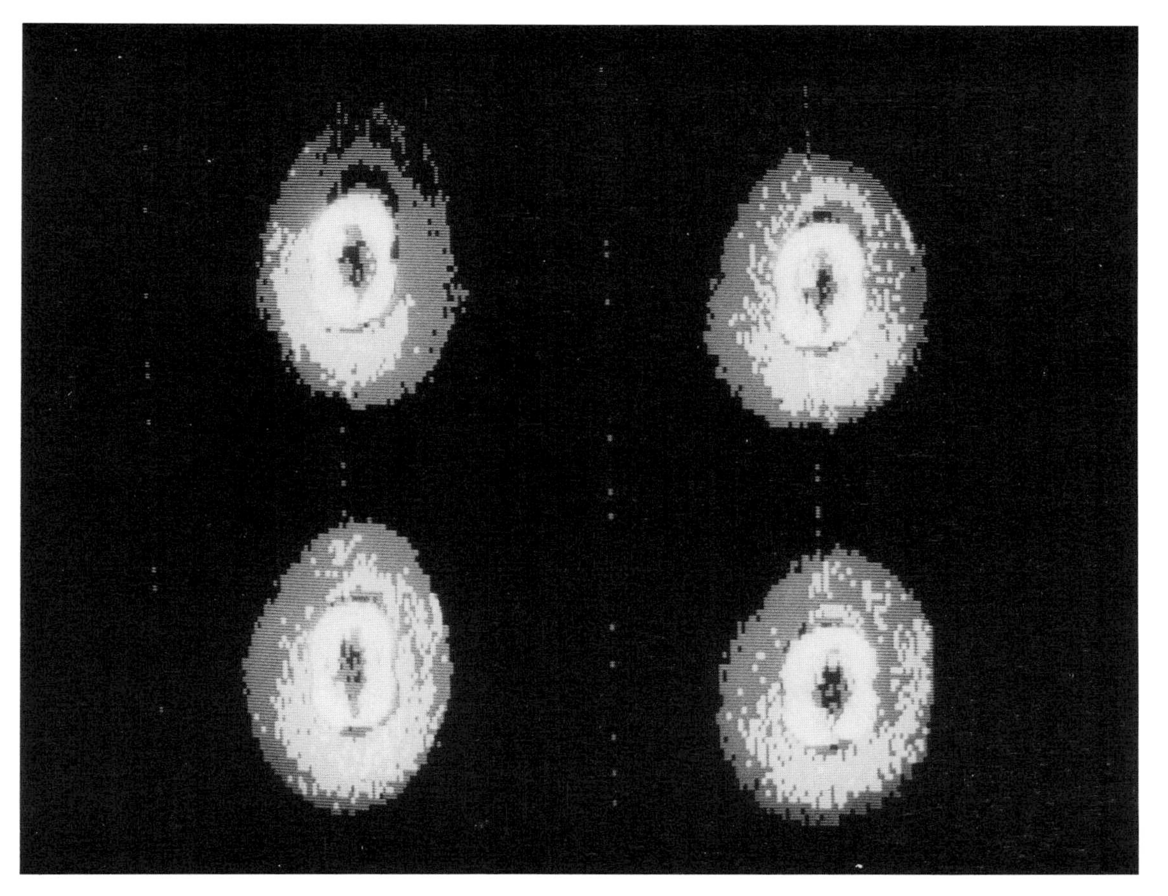

Figure 9.4. Four adjacent cross sections through an agricultural lupin seedling near the root/tip junction.

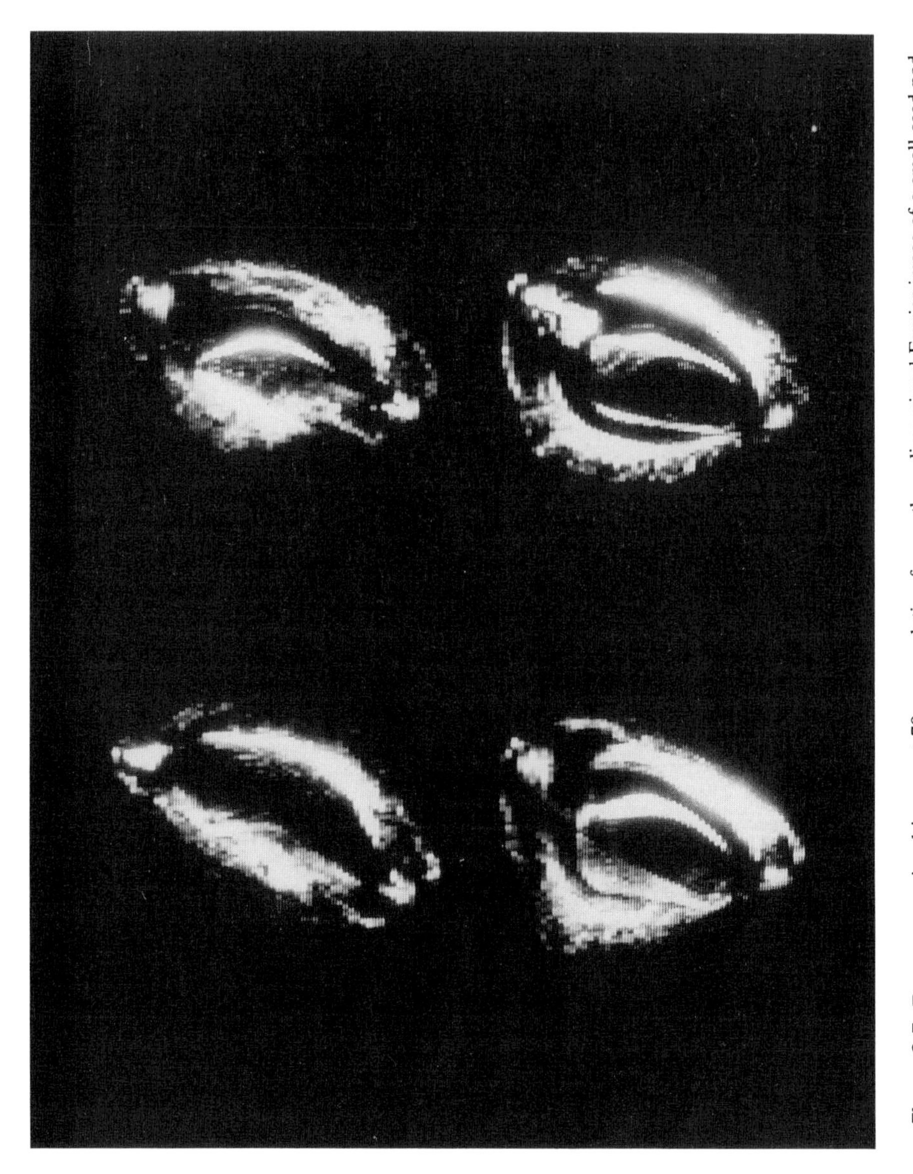

Figure 9.5. Four cross-sectional images at 70 μm resolution from a three-dimensional Fourier image of a small seed pod.

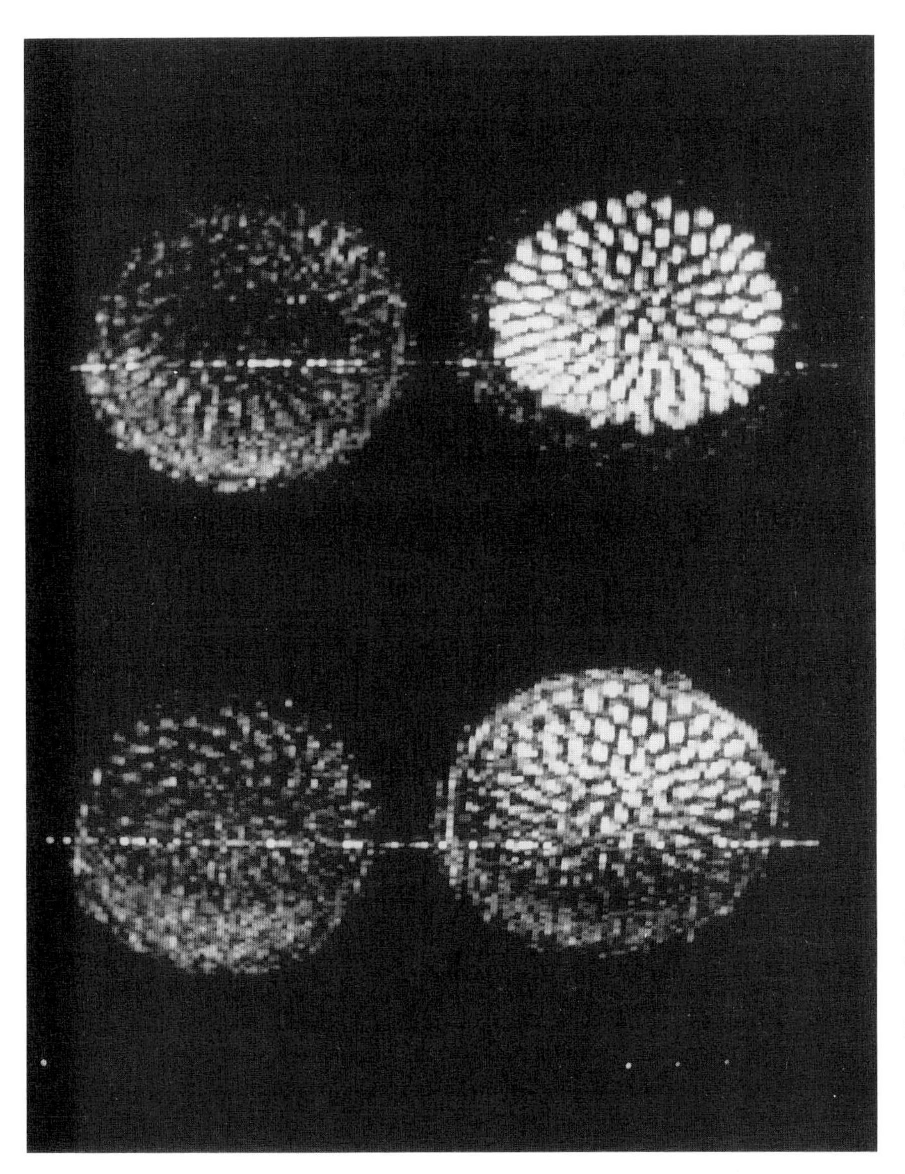

Figure 9.6. Four cross-sectional images at 70 μm resolution near the base of a thistle flower head.

consecutive sections through a small seed pod. Top left is above the level of the seed, which is traversed sequentially from upper right through lower left to lower right. Figure 9.6 shows cross-sectional images through the head of a small thistle. The fine hairlike structures of the flower contain little water at their extremities, but as they enter the base they coalesce into discrete fleshy bundles forming a characteristic pattern.

9.4. Conclusions

The development of a prototype NMR microscope which, in its present form, is capable of an isotropic linear resolution of 40 μm has been described. With the provision of higher-strength gradients this should improve to some 10 μm in each dimension, giving a voxel size of 10^3 μm^3. Resolution in one dimension can be traded against resolution in any other dimension in any convenient way provided that this voxel volume is not diminished.

The application of this system to plant tissue in a noninvasive manner has been discussed. The availability of higher-field gradients should enable cellular resolution to be reached with the system described, permitting plant differentiation processes to be studied. Another possibility concerns the observation of chemical species other than water. One candidate would be sugar transport—plant sugars are present at concentrations approaching *M* and are therefore within the scope of such experiments. It will not then be possible, however, to maintain 10 μm isotropic resolution. With 1% of the signal intensity, the resolution will be reduced by $100^{1/3}$, that is, by ≈4–5 times, but this still represents an attractive proposition. Chemical-shift imaging methods will be required to retain the spectroscopic information[12] and solvent suppression will be necessary to reduce the level of the water signal to manageable proportions. Germination studies are another possibility. Here the entry of water into the seed triggers a cascade of early biochemical processes leading to the development of the seedling. Provided some rudimentary form of chemical shift resolution is maintained simultaneous images can be obtained of the distribution of mobile lipids via the strong —CH_2— resonance some 3 ppm upfield of the water signal. This enables the time course of lipid mobilization to be determined.

At the present stage of development there is clearly a wide range of interesting studies possible in developmental biology and embryology. As the resolution approaches the 1 μm level an accelerated growth in this form of microscopy can be anticipated. The inorganic potential of this method should also not be ignored.[21,22]

ACKNOWLEDGEMENTS. P. G. M. and A. J. are grateful to the British Council for a link award enabling exchange visits to be made between their respective institutions. We are also grateful to Richard Hodgson for construction of the solenoidal coil as part of an undergraduate research project.

References

1. P. C. Lauterbur, "Image formation by induced local interactions: Examples employing nuclear magnetic resonance," *Nature* **242,** 190–191 (1973).
2. P. Mansfield and P. K. Grannell, "NMR 'diffraction' in solids?," *J. Phys.* **C6,** L422–L426 (1973).
3. R. Damadian, M. Goldsmith, and L. Minkoff, "NMR in cancer: XVI. FONAR image of the live human body," *Physiol Chem. Phys.* **9,** 97–100 (1977).
4. P. Mansfield, I. L. Pykett, P. G. Morris, and R. E. Coupland, "Human whole-body line-scan imaging by NMR," *Br. J. Radiol.* **51,** 921–922 (1978).
5. J. B. Aguayo, S. J. Blackband, J. Schoeniger, M. A. Mattingly, and M. Hintermann, "Nuclear magnetic resonance imaging of a single cell," *Nature* **322,** 190–191 (1986).
6. S. N. Bone, G. A. Johnson, and M. B. Thompson, "Three-dimensional magnetic resonance microscopy of the developing chick embryo," *Invest. Radiol.* **21,** 782–787 (1986).
7. J. M. Brown, G. A. Johnson, and P. J. Kramer, "*In vivo* magnetic resonance microscopy of changing water content in *Pelargonium hortorum* roots," *Plant. Physiol.* **82,** 1158–1160 (1986).
8. G. A. Johnson, J. Brown, and P. J. Kramer, "Magnetic resonance microscopy of changes in water content in stems of transpiring plants," *Proc. Natl. Acad. Sci. USA* **84,** 2752–2755 (1987).
9. P. A. Bottomley, H. H. Rogers, and T. H. Foster, "NMR imaging shows water distribution and transport in plant root systems *in situ,*" *Proc. Natl. Acad. Sci. USA* **83,** 87–89 (1986).
10. A. Connelly, J. A. B. Lohman, B. C. Loughman, H. Quiquampoix, and R. G. Ratcliffe, "High-resolution imaging of plant tissues by NMR," *J. Expt. Bot.* **38,** 1713–1723 (1987).
11. P. Mansfield and P. G. Morris, *NMR Imaging in Biomedicine,* Academic Press, New York (1982).
12. P. G. Morris, *NMR Imaging in Medicine and Biology,* Oxford University Press, (1986).
13. D. I. Hoult and R. E. Richards, "The signal-to-noise ratio of the nuclear magnetic resonance experiment," *J. Magn. Reson.* **24,** 71–85 (1976).
14. C. D. Eccles and P. T. Callaghan, "High-resolution imaging: The NMR microscope," *J. Magn. Reson.* **68,** 393–398 (1986).
15. P. T. Callaghan and C. D. Eccles, "Sensitivity and resolution in NMR imaging," *J. Magn. Reson.* **71,** 426–445 (1987).
16. T. A. Frenkiel, P. G. Morris, and A. Jasinski, "Apparatus for generation of magnetic field gradient waveforms for NMR imaging," *J. Phys. E* **21,** 374–377 (1988).
17. A. Kumar, D. Welti, and R. R. Ernst, "NMR Fourier zeugmatography," *J. Magn. Reson.* **18,** 69–83 (1975).
18. A. Kumar, D. Welti, and R. R. Ernst, "Imaging of macroscopic objects by NMR Fourier zeugmatography," *Naturwissenschaften* **62,** 34 (1975).
19. W. A. Edelstein, J. M. S. Hutchison, G. Johnson, and T. W. Redpath, "Spin–warp NMR imaging and applications to whole-body imaging," *Phys. Med. Biol.* **25,** 751–756 (1980).
20. J. T. Ngo and P. G. Morris, "General solution to the NMR excitation problem for noninteracting spins," *Magn. Reson. Med.* **5,** 217–237 (1987).
21. L. D. Hall, S. Luck, and V. Rajanayagam, "Construction of a high-resolution NMR probe for imaging with submillimeter resolution," *J. Magn. Reson.* **66,** 349–351 (1986).
22. L. D. Hall, V. Rajanayagam, and C. Hall, "Chemical-shift imaging of water and *n*-dodecane in sedimentary rocks," *J. Magn. Reson.* **68,** 185–188 (1986).

10

Confocal Optical Microscopy

A. Boyde

10.1. Introduction

Although confocal microscopes were invented a long time ago, they have taken a long time to get off the ground. Until recently, this family of methods of microscopy was in the hands of the developers. However, many different manufacturers have now appeared on the scene with instruments ready for purchase.

The first recorded invention of a confocal scanning optical microscope appears to be that of Minsky,[1] who built the instrument but published no results obtained with it. The next microscope to be described and which was demonstrated to work from its inception was the Tandem Scanning Reflected Light Microscope (TSRLM) of Petráň and Hadravský.[2] This microscope appears to have been difficult to construct and align in the first instance, but, nevertheless, appealed to one of its American allies, M. D. Egger, who was associated with the development of two further variants. A scanning device was constructed on essentially similar principles to that in the TSRLM, but slits were used as apertures instead of holes, thus making the instrument much simpler to align.[3] In a second device, a laser was used as a source of illumination, the specimen being irradiated in only one point at a time, and the image being built up by classical scanning techniques. In this case, scanning was achieved by moving the objective lens.[4] Peculiarly, the early success of all these microscopes was ignored

A. Boyde • Department of Anatomy and Developmental Biology, University College London, London WC1E 6BT, United Kingdom.

even by the biological microscopy community, yet biologists were closely associated with their invention and construction.

The further history of the development of the instrument was that of a continued development of the TSRLM in Pilsen, Czechoslovakia, and of improvements to the confocal scanning laser microscopes (CSLMs), with particularly notable achievements in Amsterdam in the group led by Brakenhoff (with some of his collaborators later moving to Heidelberg),[5,6] and by Sheppard and Wilson in Oxford.[7–9] The work of these groups has led directly to at least two of the currently available commercial instruments. Later starters in the field[10–13] have introduced the scanning beam from a "black box" separate from an already existing light microscope, putting the scanning beam in, and taking the necessary reflection or fluorescence signal back out, through the vertical phototube of a conventional (trinocular head) light microscope.

10.2. Basic Principles

The basic physical principle of any scanning microscope is that the sample is scanned with a radiation that illuminates one point in, or on, the specimen at one time. A resultant signal from the interaction of the radiation probe with the sample is collected, processed, and used to reconstruct an image when the illuminating probe is scanned point-by-point over the sample (for example, in a pattern something like a television raster).

Scanning optical microscopy was invented by Young and Roberts in the late 1940s.[14] They acquired a scanning spot of light by taking the image of a cathode ray tube spot as a source of illumination for a light microscope. They were thus able to illuminate one point in the field of view at a time. This procedure proves to be physically equivalent to imaging only one point at a time, which can be done much more simply by putting the image into a TV camera. Nevertheless, there are many reasons why point-by-point irradiation of the sample proves to be a better way of proceeding. The further development of unitary-beam scanning optical microscopes has strongly depended on the advantages of the laser, giving a powerful source of monochromatic radiation that can be concentrated into one spot.[4–13]

Most of the real specimens looked at with light microscopes are either translucent (as, for example, most geological and biological samples are) or, if they are surface reflective, do not have flat surfaces. In both cases, light interacts with the sample over a considerable vertical range and is reflected (or fluorescent light emanates) from a large vertical "slice." It does not help too much if the specimen is illuminated very brightly at one focus level, that is, the focal plane of the objective lens, because light still interacts with the sample in depth and any of this reflected or fluorescent light can enter a normal image forming system.

10.3. The Principle of the Confocal Microscope

The principle of the confocal scanning microscope is to eliminate the scattered, reflected, or fluorescent light from out-of-focus planes. This can be done by combining the principle of illuminating only one spot in the focal plane at a time with that of imaging only one spot in the focal plane a time (Figure 10.1). If an aperture is placed in the illuminating beam so that it is imaged (in the common case of reflection and epifluorescence modes by the objective lens) in the focused-on plane, and if, in the imaging system, a conjugate aperture is placed such that only the light from the focused-on plane in the specimen will pass this second aperture, then the scattered, reflected, or fluorescent light from out-of-focus planes will be strongly discriminated against. In effect, this gives a microscope that will image only the very thin layer on which the beam is focused.

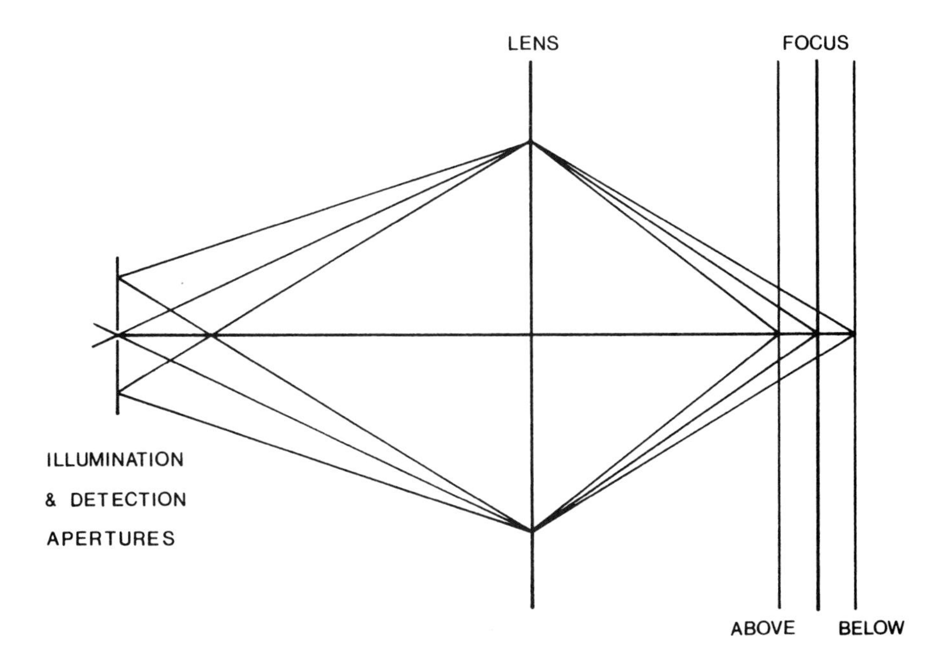

Figure 10.1. Diagram illustrating the confocal principle; an aperture (or apertures in a TSM) is located at the center of the line labelled "illumination and detection apertures." The lines emanating from that point represent the ideal focused confocal rays which are reflected from the desired plane of focus and pass back through the confocal pinhole. Rays reflected in front of (from the plane labelled "above") or behind (from the plane labelled "below") the focused-on plane are intercepted by the solid material surrounding the aperture.

The tubelength of standard RMS objectives is 160 mm. This means that an image plane lies 160 mm behind the objective lens. For a reflection microscope configuration with illumination through the objective, a small aperture in this intermediate image plane will be imaged in the focused-on plane of the specimen. If another identical aperture is placed 160 mm behind the objective, the returning imaging light from the focused-on plane will be brought to a focus at the level of the aperture. It would normally be necessary to use a beamsplitter to separate the illuminating and imaging apertures and rays. The description thus far explains how it is possible only to illuminate a discrete spatial spot and to collect the rays from that spot separately from other scattered rays into a detector aperture. How to make this into a confocal scanning microscope is described below.

10.4. Multiple-Aperture Array (Tandem) Scanning Microscopes

In the real-time, direct-view confocal microscopes, the apertures are moved physically in the plane of the intermediate image plane (Figures 10.2 and 10.3).[2,15–18] A large number of apertures are used, and the specimen is illuminated with a large number (e.g., 1000 or more) of scanning beams simultaneously (tens of thousands during one frame interval). Thus, these microscopes have an inherent and major advantage in the speed of reconstruction of the image. The illuminating and detecting apertures have to be scanned in tandem, hence the name "tandem scanning microscope" (TSM, Figure 10.2).

The apertures in most existing TSMs are holes near the edge of a spinning disk. (Other methods of scanning aperture arrays are possible and are currently being introduced into the field.) In the original and usual configuration,[15–18] one pair of holes on opposite sides of one diameter, and at the identical radial distance from the rotation center of the disk, functions to produce one scanning line (in the brightly illuminated focused-on plane) in the image as the disk rotates. In models built so far, it has been usual to allow roughly 1% of the aperture field to be open; the solid parts block off the remainder of the illuminating light. For a perfect mirror reflector lying at the focal plane of the objective, all of the 1% illuminating light would pass back through the conjugate detector-side apertures.

It is also possible to use only one side of a rotating-aperture disk for both illumination and observation. The advantage is that such a system requires no alignment, but the problem is that 99% of the illuminating light stands a chance of being reflected back from the top of the disk and into the viewing channel. This reflected light can be suppressed by using crossed polars, with one plane of polarization in illumination and one plane in observation, and a quarter-wave

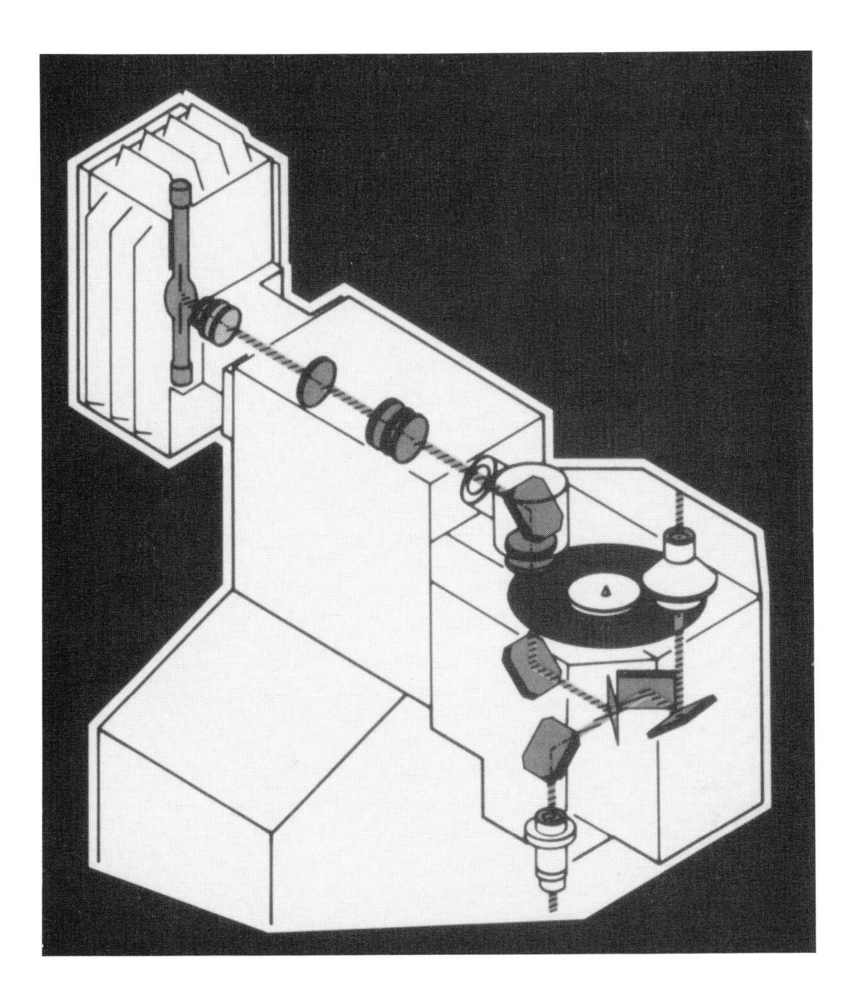

Figure 10.2. Layout of a recent tandem scanning microscope (TSM), the "Confocal 2002" instrument manufactured by JZD Komorno, Pilsen, Czechoslovakia (figure taken from the manufacturer's literature, with permission). Light from a mercury arc lamp is reflected to pass a field lens placed close to the 10 cm diameter, 1% transmissive aperture disk, with tens of thousands of holes, typically $\approx$30 μm in diameter. Light passing through the disk is reflected from a beamsplitter and is then reflected downward to enter the 160 mm tubelength RMS objective. Light reflected in the specimen passes back through the same lens, off the same final mirror, passes the beamsplitter and suffers two more reflections before reaching the observation side of the disk. The last optical component is a Ramsden-type eyepiece used to observe the image in the scanning disk.

Figure 10.3. Photograph of the Tracor Northern TSM being used to examine the internal structure of the bone of a human skull. The standard specimen stage has been removed and a precision jack is used for coarse focusing. Fine focus in this instrument is achieved with a piezoelectric device moving the objective lens.

plate somewhere in the optical train above or below the objective lens. These means of suppressing the reflected light from the disk are not good enough to allow a one-sided configuration of the scanning-disk microscope to be used for looking at specimens, such as living biological tissues, which have an inherently low level of reflection. Kino and Xiao have put together a one-sided tandem scanning microscope and demonstrated that it is very effective in examining good reflectors.(19)

10.5. Single-Beam (Laser) Confocal Scanning Microscopes

Several different methods of scanning the beam have been adopted in the confocal single-beam (nowadays of laser origin) scanning microscopes (CSLMs). In the original microscope of Minsky (who used a conventional source), the optical beam was stationary and the specimen was scanned past it by vibration and screw-driven translation mechanisms.[1] Similar mechanisms included vibrating piano wires driven by loudspeakers in the earlier versions of the microscopes of Sheppard and Wilson.[7–9] Some very fine engineering is necessary to develop a high-resolution mechanical scanning stage, but this approach has been pioneered with excellent results by Brakenhoff and co-workers.[5,6] Needless to say, if the specimen is scanned past a stationary optical beam, the frame speed must be limited to the rate at which the specimen can be moved, which will depend on its mass. Whether or not a specimen remains stable when scanning at speed is another problem. The method would not be suitable for live biological cells in a fluid medium, for example. Thus, although the unitary-beam, specimen-scanning approach has the advantage of being on-axis and avoids all off-axis optical deformations, other methods of scanning had to be adopted to overcome the problem of either massive or floppy specimens and to increase the frame speed to make it possible to examine moving specimens. The lens-scanning approach of Egger *et al.*[4] was no faster than the specimen-scanning approaches adopted later.

Several different microscopes have been constructed in which the beam is scanned by mirrors (commonly galvanometer mirrors), perhaps combined with high-speed rotating polygon mirrors for scanning the line axis. In practice, the potential advantages of using the higher scanning speed of the polygon mirrors have been outweighed by other engineering problems, and these have been abandoned in later production models of microscopes stemming from laboratories that first adopted this approach.[11,12]

The scanning mirrors can be positioned inside the microscope, or can be in a package physically separate from the real microscope (Figure 10.4).[10–13] The external scanning system utilizes the fact that there are many conjugate scanning (image) planes in the microscope, one of which is located above the eye-piece at the so-called "eye-point." The appeal in such a system is that one of the normal operational modes of the conventional microscope can be used to locate an interesting field of view in a specimen, before switching to confocal reflection or fluorescence mode. The disadvantage would be that the specimen would have to be configured to fit a conventional microscope.

TSMs can also be used simultaneously with conventional illumination or observation modes. For example, the specimen could be placed on an inverted optical microscope and viewed in transmitted light using the TSM to illuminate the inverted microscope. Translucent specimens can be placed on a conventional microscope illuminator base, and viewed via the TSM using conventional (non-confocal) transmitted bright-field or dark-field illumination.

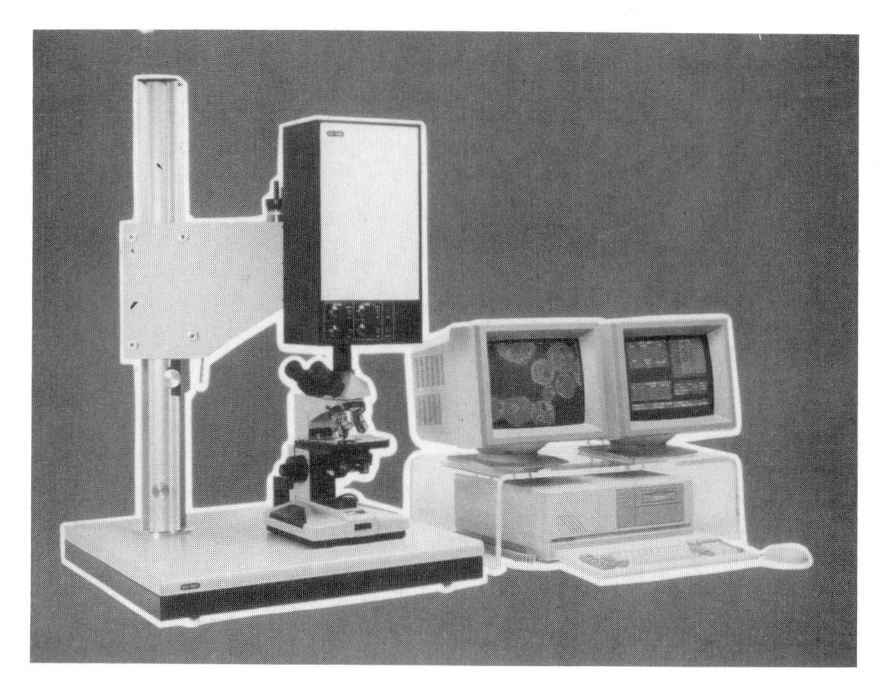

Figure 10.4. Photograph of the Bio-Rad Lasersharp confocal imaging system being used in conjunction with a Nikon Optiphot light microscope (photograph from manufacturer's literature).

10.6. Results and Applications

The operating principle of the tandem scanning reflected light microscope (TSM, or TSRLM) has just been described. It was invented to look inside live brain cells in their natural location.[15] To look at cells in live tissue it is necessary to have bulk samples, which cannot easily be examined by transmitted light microscopy, so the configuration is that of a reflecting light microscope. The difficulties of using reflected light microscopy to look at live biological tissue are that normally absolutely nothing but a pink fog will be seen. For dry bone (Figure 10.5) and teeth (Figure 10.6) the situation is the same, except that the fog is white, even if the sample is stable with time.[16–18]

The confocal or tandem scanning principle (the terms are synonymous) eliminates the fog, which is due to the halo of scattered or fluorescent light from all elements in the sample in front of and behind the plane of focus; an image of a very thin focal plane is seen which changes as the focus is changed. The effect is dramatic, and all the confocal microscopes show the same thing.

TSMs presently have much higher frame speeds that CSLMs, far higher than the eye and brain can appreciate, so that the image formation is, to all

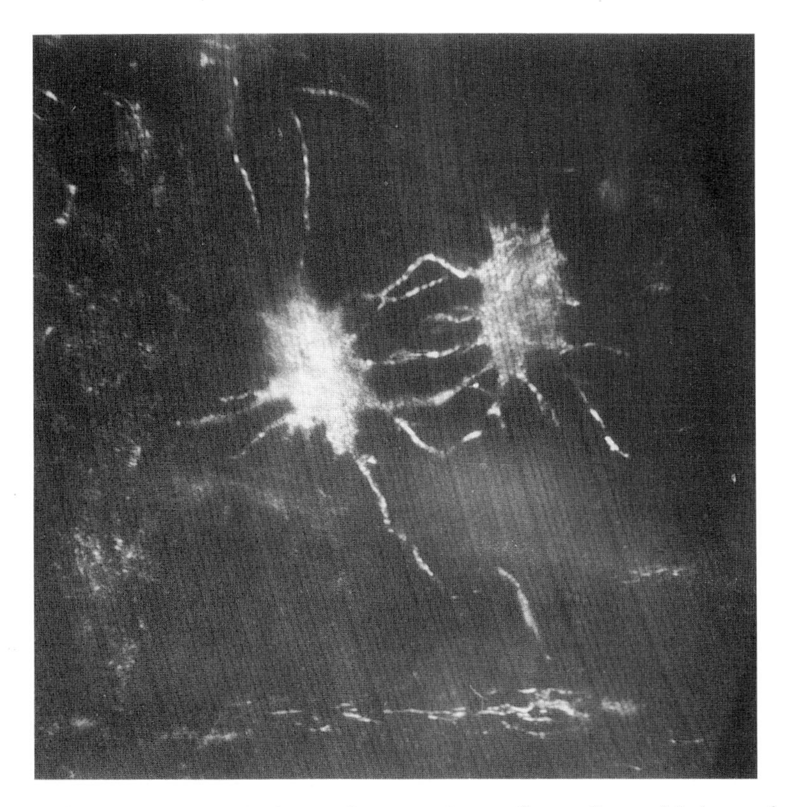

Figure 10.5. Confocal reflection image of *osteocyte lacunae* (bone cell spaces) in human skull. (Pilsen TSM. 100/1.3 oil. 10 × eyepiece. Mercury arc lamp.)

intents and purposes, as instantaneous as in a conventional microscope. When using the TSM, conventional controls are handled and the image looks like that given by a conventional microscope. The problem is that biologists are not accustomed to working with reflected light images. With CSLMs there is a delay while a frame builds up on a conventional TV display. The fastest scanning rates for full resolution are about one frame per second, but at reduced pixel resolution it is possible to increase this frame speed to, say, four frames per second for rapid scanning to locate a plane of interest in a sample. Many of these instruments take much longer than 1 s to make a full-resolution single frame.

Scanning can have different meanings. So far it has been taken to mean the point-by-point reconstruction of an image; not all points, even in the TSM, are scanned and imaged at exactly the same moment. Scanning also has the meaning, however, of looking over a lot of material to find what is wanted, or to obtain a general overview of what a particular kind of sample shows. For both reasons, real-life situations dictate that this mode of scanning should be as rapid

Figure 10.6. Confocal reflection image of enamel prisms in fossil human ancestor tooth imaged through tooth surface. (Pilsen TSM. 100/1.3 oil. Mercury arc lamp.)

as possible. Finding a good field to record should be done rapidly, and it is next to impossible to have a good impression of a range of appearances in a range of samples if each frame takes several seconds. That delay is convoluted with the need to look at many focus levels per field of view (remember that all confocal scanning microscopes give only a shallow depth of field). High temporal resolution is, therefore, a factor to be looked for when looking at real microscopical problems. The TSMs, which are the only confocal microscopes that work without image reconstruction on a display monitor, are presently the winners in temporal resolution, and far larger sample volumes and numbers of samples can be handled in such an instrument than in any of the CSLMs. It is therefore necessary to consider what advantages the CSLMs offer.

The main reason why the unitary-beam CSLMs are slow is that the most rapid mechanical scanning systems using mirrors are not fast enough to match even existing TV frame speeds. This problem can be overcome by using acousto-optical deflection devices to move the scanning beam and so avoid mechanical moving parts altogether. This approach has been adopted by Draaijer and Houpt[20] who have built a system that scans a 512 × 512 pixel matrix at 50 frames per second.

10.6.1. Surface Imaging

The need for high-speed scanning will never be as great if the purpose of a microscopy is to characterize the morphology of a surface. Then there is no need to scan in depth (in z) beyond the limited range of heights at which the surface can appear. When examining a specimen with a reflective surface in any of the confocal instruments, including the TSMs, focus is easily found because only when the microscope is in focus is there any significant returning signal, and both signal and contrast reach sharp maxima. This demonstration will easily convince anyone who has not seen such a microscope in operation that it works very differently. In examining a reflecting surface with a normal reflection microscope, there is more or less the same intensity of light for a wide range of distances above and below focus. In the confocal microscopes the returning light from out-of-focus planes is largely intercepted by the solid parts of the detector aperture (or aperture array in the case of the TSM), and the surface only comes into view when it is in focus. Contrast is dramatically higher than in the case of the conventional microscope.

For the most detailed inspection of surface morphology, the highest resolution and the highest contrast will be the main demands. It will not matter if it takes several seconds to examine one particular focal plane. To satisfy the demands of highest resolution, the ideal kind of confocal microscope would have one beam, one wavelength, and on-axis optics. Utilizing the fact that focus is determined by finding the highest brightness or contrast, the specimen can be scanned mechanically in the z direction as well as in the x and y directions. Since the signal level can change significantly for a displacement of a fraction of a micrometer (less than the wavelength of light) these microscopes are eminently suitable for surface mapping at very high resolution.[8]

Special ultraclean, highly precise confocal scanning laser microscopes have been built for the inspection of semiconductor devices, and are able to examine the largest wafers presently used in that industry. Every kind of confocal microscope can be used to advantage in examining the surfaces of semiconductors. However, the TSM, which has to be illuminated by a conventional light source, presents its own interesting characteristic advantage. When the TSM is used to look at a reflecting surface, the problem of chromatic aberration of the objective lens is encountered (Figure 10.7). Even the best lenses have an element of longitudinal chromatic aberration, which means that the microscope will not be equally confocal for all wavelengths simultaneously. With a uniform white spectrum and a tilted reflective surface all the colors of the rainbow would be seen across the surface as each color satisfied the best compromise in being confocal.

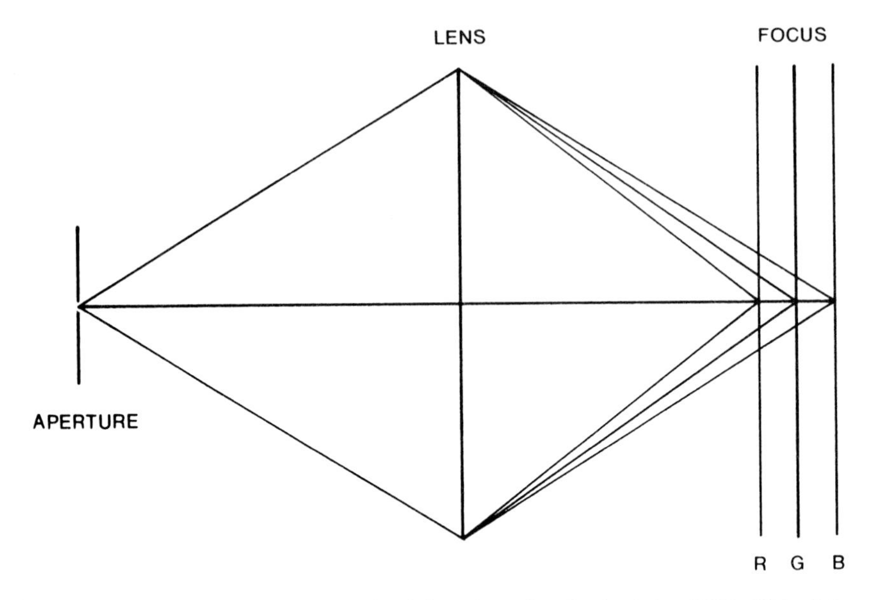

Figure 10.7. Diagram showing the origin of the color-coding for depth in a TSM. White light passing an illuminating aperture is focused at different distances (e.g., at the planes labelled "R," "G," and "B" for red, green, and blue, respectively) in front of the objective lens because of the longitudinal chromatic aberration of that lens. Returning light through the pinhole is maximal in blue, green, and red for reflectors lying at distances indicated by the blue, green, and red planes.

Figure 10.7 shows a grossly exaggerated view of what happens to the image of a plane reflector in the TSM. A single image of a nonflat surface recorded in color at one focus level in the TSM gives rise to a colored map of the specimen surface, features in the same color having the same height, that is, distance from the objective.[17] Pseudo-color-coded contour maps can, of course, be reconstructed from the information obtained from x, y, z scanning with the other kinds of confocal scanning optical microscopes.

10.6.2. Extended-Focus or Range Images

The shallow depth of field of all the confocal scanning microscopes can sometimes be represented to be a disadvantage. It may be required to combine the improvement in resolution in this kind of microscope with an increase in the depth of field. This can be simply achieved in any of these instruments by making use of the fact that the signal level is very low for out-of-focus planes. Thus, if images representing different focal planes in the specimen are acquired, these simply have to be added together to make an in-focus image for all of these focus planes.[9,11] By through-focusing when photorecording a single frame in the TSM it is therefore possible to acquire an extended-focus or range image (Figure 10.8). The same maneuver can be done by summing or processing images in a video framestore or CCD camera with the TSM, or with the laser instruments.[5,10,11,21] If it seems peculiar to take an instrument designed to produce a shallow depth of field, and then to operate it to increase the depth of field, it should be remembered that the confocal instrument will only image the depth slice that has been chosen. The conventional microscope image will have confusing contributions from layers that are out of focus above and below the depth slice chosen for the range image.

10.6.3. Stereo Imaging

Reconstructing two extended-focus views through the same depth slice produces a stereoscopic view of that slice. One of the most exciting vistas for the application of all the confocal microscopes is that they are able to extend three-dimensional imaging to the limits of resolution in optical microscopy.[5,10,11,21–24] In the case of the TSM, it is only necessary to scan through focus mechanically while recording two separate photographic exposures with z axes inclined at a suitable angle. Alternatively, the TSM or the CSLM image can be recorded via a TV camera and processed in a framestore. One view through the stereo slice can be reconstructed simply by adding together all of the individual focal plane images, and the other is reconstructed by shifting each incoming image sideways by a small increment to simulate an oblique view through that depth volume of the sample. It is also possible to shift each plane of the summed image laterally but in opposite senses to give rise to symmetrical oblique viewing (Figures 10.9 and 10.10). Operating the TSM or the CSLM in this way it is possible to produce stereo images after only a single pass through the specimen; this allows the inspection of the stereoscopic view soon after focusing is completed.

The confocal laser scan microscopes will doubtless all be equipped with framestores and, depending on the size of memory available, with a range of image-processing routines for stereo reconstruction or for any other manipulations. Brakenhoff's group, in particular, have shown some splendid demonstrations of high-resolution stereo images in reflection and fluorescence. They have shown the advantages of selecting the most significant information in a particular line of sight through a three-dimensional volume and only displaying that piece of information, ignoring any other features that would overlap in a particular line of sight. Two or more discrete sets of information, such as reflective features or fluorescence from one or more fluorophores, can be combined in one microscopic image by pseudocolor processing.

The TSM can also be used to produce direct, photographically recorded color-coded stereo pairs.[24] These are achieved by changing the color of the illuminating light at the same time as changing the focus level of the instrument,

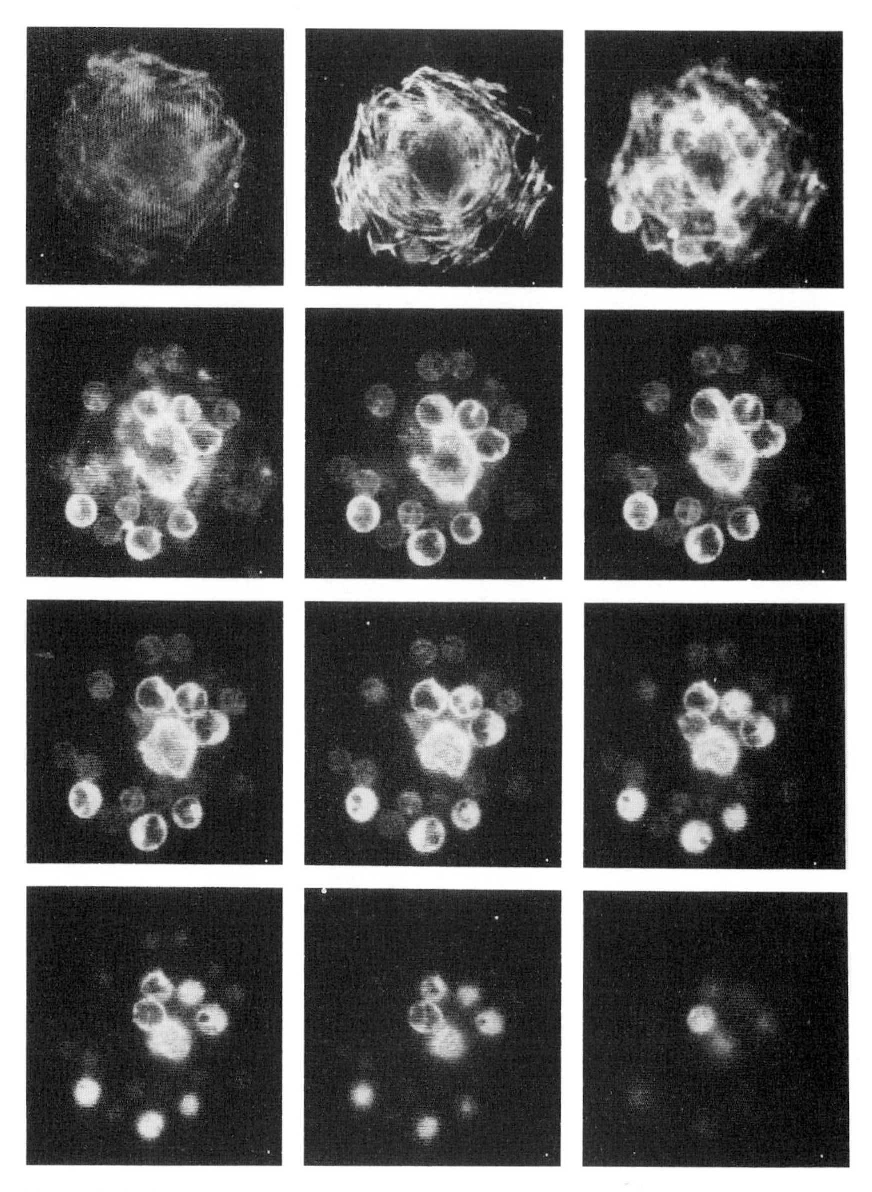

Figure 10.8. Bone cell immuno-labelled with FITC (fluorescin)-tagged antitubulin antibody (preparations by Louise Taylor and Sheila Jones). Prominent blobs were probably caused by formaldehyde fixation and do not represent living condition. This set of confocal fluorescence images was recorded at 1 μm focus intervals in the Tracor Northern TSM using a Wright Instruments Ltd. cooled CCD camera.

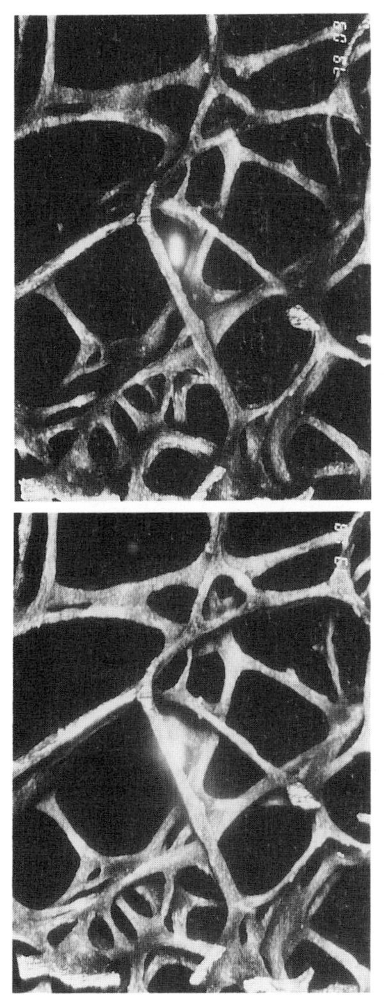

Figure 10.9. Stereo pair synthesized from one through-focal set of reflection images of human osteoporotic trabecular bone using the Bio-Rad Lasersharp CSLM with a Nikon 2/0.018 planapochromatic objective. Confocal microscopes present advantages in stereo imaging even at such a very low magnification. (Preparation by J. A. P. Jayasinghe.)

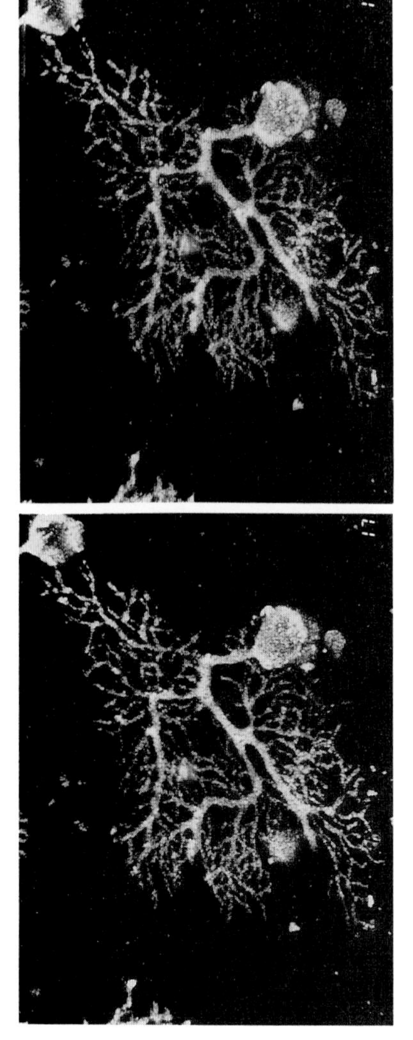

Figure 10.10. Stereo pair synthesized from one through-focal set of reflection images of a Golgi-stained preparation of mouse cerebellum showing the dendritic tree of one Purkinje cell. (Bio-Rad Lasersharp. 60/1.4 oil. Preparation by Martin Berry.)

thus color coding features that lie at the same depth in the same color. When this method of coding for depth is combined with the effects of stereoscopic parallax, the two are additive and offer a considerable advantage in 3-D interpretation.

Three-dimensional images reconstructed from any confocal microscope have a different geometry to the three-dimensional images normally acquired from light optical systems. They have the geometry of parallel projection, distinct from perspective or central projection geometry.[22] The geometry for reconstruction from *x*, *y*, and *z* parallax coordinate measurements in such image pairs is simple.[25] This has been demonstrated for many years in scanning electron microscopy and special measuring, and analytical equipment has been evolved to deal with this case.[25,26] This is an important advantage of confocal microscopy, which extends even into the range where no advantage in resolution in *x*, *y*, and *z* is necessary over what could already be achieved with, for example, a low-powered stereo binocular light microscope. Systems for recording stereo pairs in the TSM or CSLM work even at magnifications so low that the sample could as well be looked at with two different camera orientations using a macro lens on a 35-mm camera (Figure 10.9). However, the three-dimensional photogrammetric measurements in the latter case are much more difficult to calibrate and calculate.

10.6.4. Stereology

Stereology is the science and art of reconstructing three-dimensional information from two-dimensional sections. Most of stereology is based on examining single flat, polished surfaces through bulk samples, or one of a pair of sections.[27] Stereologists are enthusiastic about confocal scanning optical microscopy because they can now do their job really well. Not only can they look into a translucent three-dimensional sample at a depth, but the set of sections they obtain by through focusing are perfectly in register.[28] Anyone who has attempted a three-dimensional reconstruction from a set of histological sections will know the difficulties in achieving exact register, particularly if the sections are distorted from one to the next in the series.

10.6.5. Image Processing

Any kind of image processing and image analysis possible with any other form of image can, of course, be applied to confocal light microscope images. Measurements of area and intercept methods of edge enhancement, and so on, are all applicable. Fast Fourier transforms can be applied to determine periodicities. A mask can be applied to a FFT and, through a reverse FFT process, used to remove certain spatial frequencies and enhance others. A fuller description of image processing techniques has been given in Chapter 2, and in the references quoted here.

10.6.6. Fluorescence

All the confocal microscopes work in fluorescence as well as in reflection. Biologists tend to think that they have a monopoly in the application of fluorescence microscopy, but this is not so; there are many ways in which it can be used in materials science. The applications in cell biology, however, are the most critical because the smallest structures have to be characterized by the use of immunocytochemical fluorescence probes (Figure 10.8).

The disadvantage of conventional epifluorescence is that detail is obscured by the halo of fluorescent light coming from features above and below the desired plane of focus. A dramatic improvement is secured by eliminating this out-of-focus information. Three-dimensional information can be obtained by through focusing and mental processing, but better results are obtained by computer image processing. The laser-based confocal scanning microscopes would certainly have an advantage in applications where only fluorescence imaging of stable, prepared (dead) biological samples was concerned—the high intensity of illumination available from the laser would give rise to a much higher intensity of fluorescent light. Furthermore, since the returning signal can be handled point-by-point, even very low levels of fluorescence can be successfully imaged using a sensitive photomultiplier. In the case of the TSM, low-fluorescence signals can be handled with a low light level video camera or a cooled CCD camera. Splendid examples of applications of fluorescent tagging techniques in cell biology using CLSMs have been recorded by Brakenhoff and his group, by Carlsson and co-workers in Sweden, and by White and co-workers in Cambridge.

10.7. Comparisons

The TSMs are configured to look at even very large samples (see Figure 10.3), and they form an image in real color as well as real time. The on-axis laser-based instruments in which the specimen is scanned mechanically past a fixed beam would be the most "confocal," and show the highest resolution in x, y, and z. Resolution will be limited by the choice of wavelength of the illuminating laser. The price of this marginal improvement in resolution is the cost of increased time in scanning each frame. As noted above, such instruments have been conspicuously successful in examining surfaces of solid specimens and in fluorescence applications in biological work.

The deflected optical beam laser-based instruments potentially have the advantage that they can be used to look at bulk samples, although none have been configured for this purpose to date. They can have much higher frame speeds than the mechanical-specimen scanning variety of microscope, though still much slower than the TSMs.

At present, a TSM would be chosen for work with large samples, with live samples to study fast interactions (e.g., cutting and fracturing phenomena or rapid dye-penetration experiments) or, if it is necessary to search a lot of areas, or a lot of specimens, and at a lot of depths, or if real color (in natural reflection or fluorescence) is a requirement. The advantage of the TSM in fluorescence is that any fluorophore in the visible wavelength range can be chosen. The microscope will also work in the infrared or ultraviolet if the image is detected via a suitable TV camera and filters.

A mechanically scanned sample, laser-based instrument would be chosen if ultimate resolution in either reflection or fluorescence was the sole requirement. They would never be chosen for bulk samples, living samples, or samples that change with time.

A scanned-beam type of microscope, ideally the kind in which the beam is injected into an existing microscope from a free-standing black box (Figure 10.4), would be satisfactory if it is needed to combine standard methods of optical microscopy in transmission and reflection with the advantages of confocal imaging. This kind of microscope could be used for slowly moving live biological objects. The high intensity of illumination of a sensitive biological sample may be a problem with any of the laser-based instruments.

ACKNOWLEDGEMENTS. The greater part of the text of this chapter was first published in *Microscopy and Analysis* and is reproduced with the kind permission of the publisher, Mrs. Jeanne Gordon.

References

1. M. Minsky, Microscopy Apparatus, United States Patent Office. Filed Nov. 7, 1957, granted Dec. 19, 1961. Patent No. 3,013,467 (1961).
2. M. Petráň, M. Hadravský, M. D. Egger, and R. Galambos, "Tandem-scanning reflected light microscope," *J. Opt. Soc. Am.* **58,** 660–664 (1968).
3. P. Davidovits and M. D. Egger, "Scanning laser microscope for biological investigations," *Appl. Opt.* **10,** 1615–1619 (1971).
4. M. D. Egger, W. Gezari, P. Davidovits, M. Hadravský, and M. Petráň, "Observation of nerve fibers in incident light," *Experientia* **25,** 1225–1226 (1969).
5. G. J. Brakenhoff, P. Blom, and P. Barends, "Confocal scanning light microscopy with high-aperture immersion lenses," *J. Microsc.* **117,** 219–232 (1979).
6. H. T. M. Van der Voort, G. J. Brakenhoff, J. A. C. Valkenburg, and N. Nanninga, "Design and use of a computer-controlled confocal microscope for biological applications," *Scanning* **7,** 66–78 (1985).
7. T. Wilson, "Imaging properties and applications of scanning optical microscopes," *Appl. Phys.* **22,** 119–128 (1980).
8. D. K. Hamilton and T. Wilson, "Three-dimensional surface measurement using the confocal scanning microscope," *Appl. Phys* **27,** 211–213 (1982).
9. T. Wilson and C. Sheppard, *Theory and Practice of Scanning Optical Microscopy,* Academic, London (1984).
10. K. Carlsson, P. E. Danielson, R. Lenz, A. Liljeborg, L. Majlof, and N. Aslund, "Three-

dimensional microscopy using a confocal laser scanning microscope," *Opt. Lett.* **10,** 53–55 (1985).
11. K. Carlsson and N. Aslund, "Confocal imaging for 3-D digital microscopy," *Appl. Opt.* **26,** 3232–3238 (1987).
12. J. G. White, W. B. Amos, and M. Fordham, "An evaluation of confocal versus conventional imaging of biological structures by fluorescence light microscopy," *J. Cell Biol.* **105,** 41–48 (1987).
13. W. B. Amos, J. G. White, and M. Fordham, "Use of confocal imaging in the study of biological structure," *Appl. Opt.* **26,** 3239–3243 (1987).
14. J. Z. Young and G. Roberts, "A flying-spot microscope," *Nature* **167,** 231 (1951).
15. M. D. Egger and M. Petráň, "New reflected-light microscope for viewing unstained brain and ganglion cells, *Science* **157,** 305–307 (1967).
16. M. Petráň, M. Hadravský, J. Benes, R. Kucera, and A. Boyde, "The tandem scanning reflected light microscope. Part I: the principle, and its design, *Proc. Roy. Microsc. Soc.* **20,** 125–129 (1985).
17. A. Boyde, "The tandem scanning reflected light microscope. Part II: pre-Micro '84 applications at UCL," *Proc. Roy. Microsc. Soc.* **20,** 131–139 (1985).
18. M. Petráň, M. Hadravský, and A. Boyde, "The tandem scanning reflected light microscope," *Scanning* **7,** 97–108 (1985).
19. G. Q. Xiao and G. S. Kino, "A real-time confocal scanning optical microscope," *Proc. 4th Int. Symp. on Optical and Optoelectronic App. Sci and Eng.,* The Hague (1987).
20. A. Draaijer and P. M. Houpt, "A real-time confocal laser scanning microscope (CSLM)," *Scanning* **10:** 139–146 (1988).
21. I. J. Cox and C. J. R. Sheppard, "Digital image processing of confocal images," *Image Vision Comput.* **1,** 52–56 (1983).
22. A. Boyde, "Stereoscopic images in confocal (tandem scanning) microscopy," *Science* **230,** 1270–1271 (1985).
23. R. W. Wijnaendts van Resandt, H. J. B. Marsman, R. Kaplan, J. Davoust, E. H. K. Stelzer, and R. Stricker, "Optical fluorescence microscopy in three dimensions: microtomoscopy," *J. Microsc* **138,** 29–34 (1984).
24. A. Boyde, "Color-coded stereo images from the tandem scanning reflected light microscope (TSRLM)," *J. Microsc.* **146,** 137–142 (1987).
25. A. Boyde, P. G. T Howell, and F. Franc, "Simple SEM stereophotogrammetric method for three-dimensional evaluation of features on flat substrates," *J. Microsc.* **143,** 257–264 (1986).
26. H. F. Ross, "A new comparator for SEM stereophotogrammetry," *Scanning* **8,** 216–220 (1986).
27. C. V. Howard, "Real 3-D measurements in microscopy using geometrical probes," *Microsc. Anal.* **2,** 15–17 (1987).
28. A. J. Baddeley, C. V. Howard, A. Boyde, and S. A. Reid, "Three-dimensional analysis of the spatial distribution of particles using the tandem-scanning reflected light microscope," *Acta Stereol. Suppl. II* **6,** 87–100 (1987).

11

Acoustic Microscopy in Biology

An Engineer's Viewpoint

M. G. Somekh

11.1. Scope

This chapter describes the instrumental principles behind the acoustic microscope and its actual and potential applications to biological science. We will show that the motivation for the development of the acoustic microscope and the basic physics of ultrasound are closely related. We will describe the historical developments in acoustic imaging leading to the present configuration. The mechanism and peculiarities of image formation in the microscope will be described because they are an important reminder that information from the microscope must be interpreted with some circumspection. We will also describe the properties of tissue that may be measured in the microscope, giving appropriate examples. The effects of nonlinear wave propagation are discussed briefly both as an under-used form of contrast and as a possible source of error in measurements where it is not taken into consideration.

It should be made clear that the author is an engineer and not a biologist, and is therefore concerned with instrumentation rather than applications. The approach will be to describe what properties and features can be imaged and measured in the microscope. Interpretation of results, from a biological point of view, will not be attempted. Instead, we will try to provide the necessary back-

M. G. Somekh • Department of Electrical and Electronic Engineering, University of Nottingham, Nottingham NG7 2RD, United Kingdom.

ground so that biologists can appreciate whether the method is applicable to their problems.

11.2. Basic Physics of Ultrasound

To understand the motivation leading to the development of the acoustic microscope it is necessary to appreciate the basic physics of ultrasound. The reasons for using ultrasound to form images depend on the manner in which sound waves interact with the matter under investigation.

When creating an image of any sample there are two fundamental parameters that describe its utility: resolution and contrast. A new technique should therefore reveal information that was not previously accessible. At the very least, it should be more convenient than existing methods. Any technique with startling spatial resolution will find a ready application since finer detail is always in demand. Techniques such as transmission electron microscopy and, more recently, tunnelling and atomic force microscopy thus have very obvious applications. The acoustic microscope, on the other hand, generally gives resolution close to that of the optical microscope. Its utility therefore lies in the contrast that can be obtained. It is important to understand the interaction of the sample with the ultrasonic radiation.

11.2.1. Velocity

A longitudinal (or compressional) acoustic (or ultrasonic) wave arises from the periodic compression and expansion of the material medium; this stress distribution moves through the material at a (phase) velocity characteristic of the material, which is given by:

$$v_1 = \sqrt{c_{11}/\rho} \tag{11.1}$$

where c_{11} is the bulk modulus (stiffness) of the sample and ρ is the density. The velocity of propagation depends on the mechanical properties of the material medium.

A transverse wave arises from periodic shearing of the material. This wave will travel at a velocity given by equation (11.1) with c_{11} replaced by c_{44}. Longitudinal waves are the only type of wave that can propagate in ideal fluids because they have no resistance to shearing, that is, $c_{44} = 0$ (liquids flow!). From this standpoint water and most soft tissue approximate to ideal fluids. Hard tissue, on the other hand, will allow both shear and longitudinal waves to propagate, which, as we will point out later, can have some significant consequences.

Acoustic waves travel five orders of magnitude slower than electromagnetic waves. This factor means that for a given frequency the wavelength of sound is smaller by this factor. High spatial resolution can therefore be achieved at comparatively low frequencies. To be specific, the velocity of sound in water and most soft tissues is approximately 1500 ms^{-1}, which means that at 2.5 GHz sound has the same wavelength as red light. If diffraction-limited resolution can be achieved it is then possible to achieve resolution comparable with the optical microscope using microwave electronics. The comparatively low frequencies with which it is possible to get short wavelengths means that electronic signal-processing methods can be used that could not be contemplated with optics. It was the realization that optical resolution could be achieved at microwave frequencies while at the same time obtaining image contrast that depended on the mechanical properties of the sample which led to the quest for the modern acoustic microscope.

11.2.2. Acoustic Impedance

To form an image the acoustic waves must be scattered from the sample. Provided the interface between materials is very large compared with the wavelength, the reflection coefficient R of a normally incident wave is given by:

$$\Gamma = \frac{Z_1 - Z_2}{Z_1 + Z_2} \qquad (11.2)$$

where Z refers to the acoustic impedances of the media 1 and 2 as depicted in Figure 11.1.

The acoustic impedance is defined as the product of the acoustic wave velocity and the density, and thus depends on the mechanical properties of the materials. Acoustic impedance may be thought of as analogous to the reciprocal of refractive index in optics. Table 11.1 shows the impedances of a range of biological tissues. The values shown in the table were measured at frequencies below 10 MHz, which is rather lower than the frequencies used in acoustic microscopy, but with certain reservations they will be applicable to acoustic microscopy. We can see immediately that the signal reflected from a soft tissue–water interface is approximately 30 dB smaller than the incident wave. At the other extreme, more than 50% of the sound energy will be reflected from dental enamel. It is interesting that the methods used in materials science are applicable for the examination of hard tissue while distinct methods must be used to look at soft tissue. To emphasize this point, we note that dental enamel has almost precisely the same acoustic properties as aluminium. On the other hand, the impedance of lung tissue is particularly low. This is because the tissue is not continuous and is highly porous. At a low frequency, the tissue may be regarded as continuous with a very low impedance. When the frequency is high the tissue

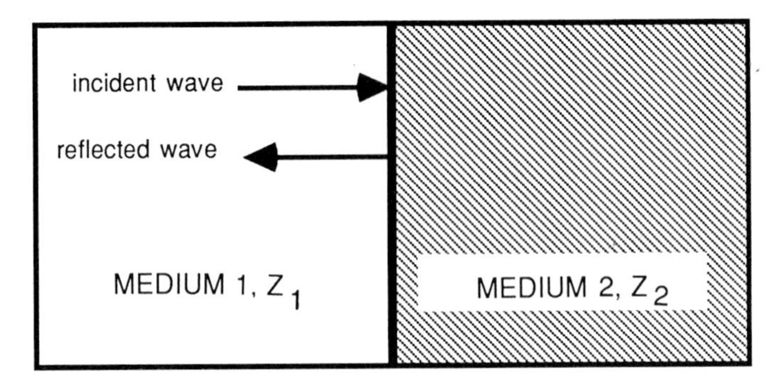

Figure 11.1. Reflection of acoustic wave from an interface of materials with different acoustic impedances.

must be thought of as heterogeneous with areas of comparatively high impedance and areas of virtually zero impedance (air).

It is inappropriate to discuss the theory of the scattering of ultrasound, but let us say that the situation is greatly complicated when the interface is not large compared to the wavelength. The theory of scattering has been considered extensively in the literature of medical ultrasound.[1]

Later in this chapter we will describe how additional effects such as nonlinearity and attenuation of the wave can affect the performance of the instrument and the interpretation of the results. The motivation for the development of the acoustic microscope both in biology and material science is not that the resolution is particularly spectacular but that the interaction of the acoustic waves with the mechanical properties of cells can give information that cannot be readily obtained with other instruments.

Table 11.1. Densities and Acoustic Impedances of Some Tissues

Material	Density ($kg\ m^{-3}$)	Acoustic impedance ($10^6\ kg\ m^{-2}\ s^{-1}$)
Water	1000	1.52
Blood	1060	1.62
Bone	1380–1810	3.75–7.38
Brain	1030	1.55–1.66
Kidney	1040	1.62
Lung	400	0.26
Muscle	1070	1.65–1.67
Dental enamel aluminum	2700	17

11.3. History

The history of the development of the acoustic microscope is particularly interesting if only because the present instrument is so much more simple than many of the ingenious attempts that have fallen by the wayside. In the mid-1930s, Sokolov realized that resolution similar to that of an optical microscope should be obtainable using ultrasound at microwave frequencies. He obtained the first acoustic images somewhat later—the resolution however, was rather poor.[2,3] The technique used in these early experiments was naturally limited by the technology available at the time. Sokolov allowed the acoustic field to build up a charge concentration on a thin piezoelectric plate, which was then read by scanning an electron beam across the surface to detect the charge accumulation.

Much later several attempts were made to create acoustic analogues of the optical microscope. The first of these systems, developed by Dunn and Fry,[(4)] measured the absorption of the acoustic beam using a power-sensitive detector. A phase-contrast system was developed by Suckling and Ben-Zui.[(5)] These early systems were hampered, however, by the difficulties in getting diffraction-limited resolution, the low operating frequencies, and the prohibitively long times to obtain an image (in the case of the Dunn and Fry system, for example, each image point took 1 s). Spatial resolution of around 100 μm was obtained with both these instruments.

Towards the middle to late sixties groups of scientists and engineers with experience in microwaves had developed the technology to produce coherent ultrasonic waves with wavelengths comparable to that of light.[(6)] The acoustic microscope was therefore "ripe" for development. The story of the development has several strands, but since space is limited, we will just follow the two that led to instruments presently in use.

Workers at Zenith Corporation led by Robert Adler, a distinguished microwave engineer, developed a system in which the acoustic field is read optically, the so-called "scanning laser acoustic microscope," or SLAM. The system is shown in Figure 11.2a. The acoustic wave from the transducer passes through the sample and causes the surface to be displaced in sympathy with the acoustic wavefront. Because the wave is travelling obliquely, the surface displacement traces out a sinusoidal movement. The sample is then illuminated with a focused beam of laser light, which is deflected by an angle, say, θ. When the acoustic wave is 180° out of phase the laser beam will be deflected by $-\theta$. The reflected acoustic beam is therefore forced to swing to and fro at the acoustic frequency. This variation of the angle of the reflected beam is turned into an amplitude variation by adjusting the position of the returning beam so that it is partially obscured by a knife edge. When the beam is in position A, the photodetector will receive the minimum signal; at position B the signal will be a maximum (see Figure 11.2b). The signal detected by the photodetector is a sinusoidal oscillation

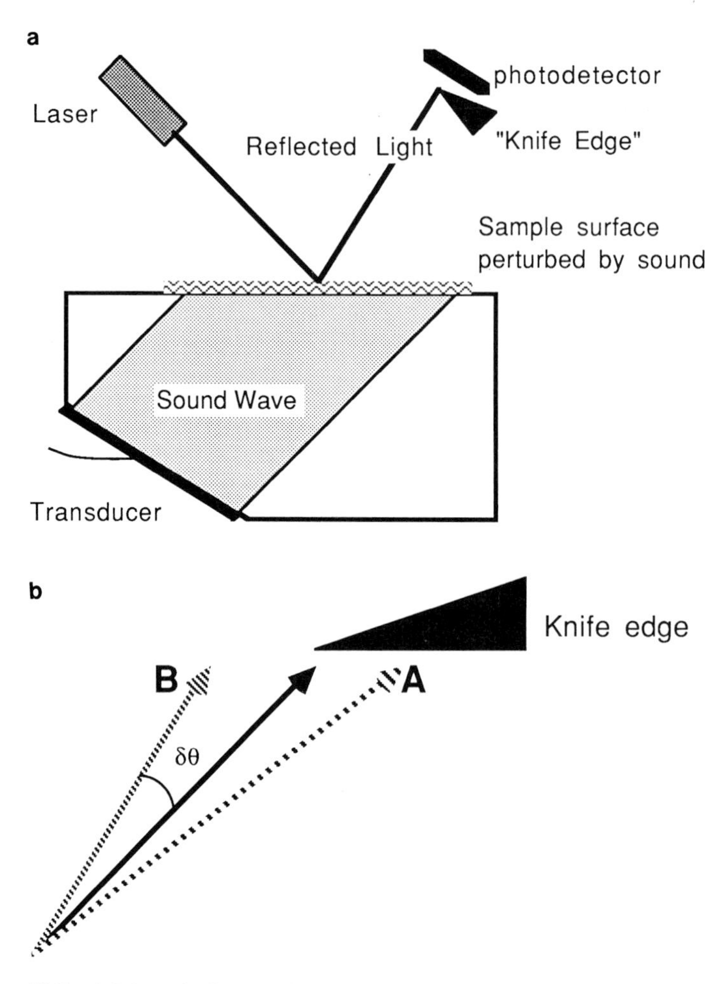

Figure 11.2. (a) Schematic diagram of scanning laser acoustic microscope. (b) Schematic diagram showing knife edge detection; arrowed lines represent reflections from sample surface. As the returning angle changes, the fraction of the beam obscured by the knife edge varies at the acoustic frequency.

at the acoustic frequency whose amplitude is proportional to θ, which is in turn proportional to the acoustic signal incident at the sample surface. The output amplitude from the instrument can therefore be related to the acoustic wave attenuation in the sample. An image is formed by scanning the laser spot across the sample at TV rates, thus allowing observation of living tissue. The instrument is generally operated at frequencies around 100 MHz, and has been used extensively (primarily at the University of Illinois) for determination of tissue properties.

The SLAM became available in the early seventies at about the same time as the other strand in the development of acoustic microscopy was reaching fruition. This development originated primarily from Stanford University where they set themselves the task of achieving optical resolution with acoustic waves. Interestingly, the justification for some of this early work was in terms of the potential application in the biological sciences. The most obvious route to achieving optical resolution with acoustic waves was to make an analogue of an optical lens. This was indeed attempted by the eventual inventor of the modern acoustic microscope, Calvin Quate. The problem was to design lenses where the off-axis aberration was small enough to allow for diffraction-limited resolution over a full field. The lenses were very difficult to align partly because they were opaque and partly because any solid material would generate considerable aberration due to the presence of both shear and longitudinal waves. This line of attack was eventually abandoned.

Several attempts were made to achieve high resolution, notably the system developed by Auld *et al.*.[7] In this instrument, the acoustic shadow on a transducer was read by scanning a laser across the element. The photoconduction induced by the laser reduced the overall piezoelectric signal so that a low-contrast negative of the acoustic field was obtained. Despite the operating frequency of 1.1 GHz the resolution achieved with this system was between 20 and 30 microns, which is much worse than diffraction-limited resolution.

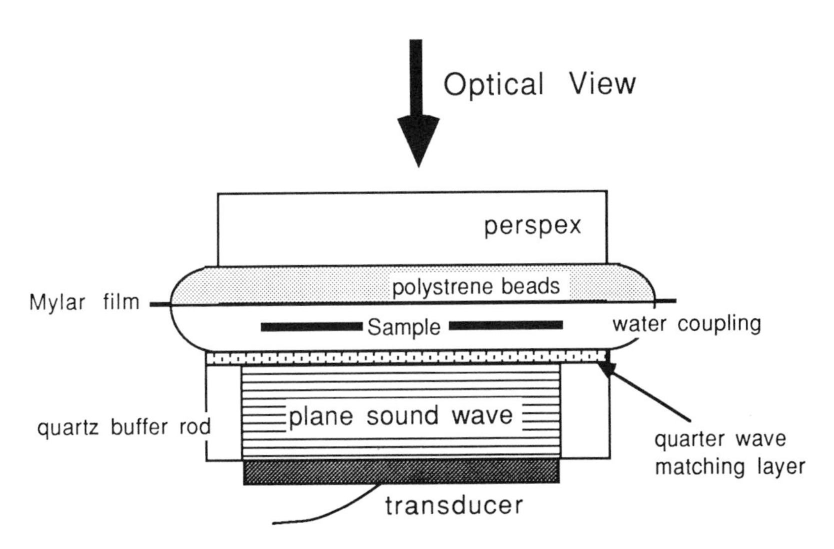

Figure 11.3. Schematic diagram of "contact printing" method developed by Cunningham and Quate.[8]

Perhaps the most interesting precursor to the modern acoustic microscope is that developed by Cunningham and Quate,[8] which is shown in Figure 11.3. A plane beam of acoustic energy is excited from a zinc oxide transducer. This wave passes through a thin block of quartz prior to its passage into a water cell in which the sample is immersed. The wavefield leaving the sample will be a shadow of the sample whose intensity will be directly related to the attenuation of the tissue. The water cell and sample are separated from a second very small cell by a thin sheet of mylar which contains polystyrene beads of 1 μm diameter. The beads are driven away from the regions where the acoustic intensity is high, and accumulate in regions where the intensity is lower. The surface above the sample is examined with a conventional optical microscope, in which the regions of high bead concentration appear reflective and the other regions appear comparatively transparent. The authors describe this method as contact printing, an appropriate term since the success of the method hinges on the fact that the acoustic waves do not diffract significantly. The method is thus a rather interesting example of how to achieve high resolution—either achieve diffraction-limited resolution, which up to that time was proving very difficult, or simply probe the near field before diffraction takes place. As we will mention in the concluding section to this chapter, there is a strong likelihood that near-field acoustic microscopy will be the acoustic microscopy of the future, albeit in a modified form.

The significance of the technique is that the method provides much of the technology for the modern scanning acoustic microscope, notably zinc oxide transducers capable of operating at microwave frequencies and acoustic antireflection coatings made up of materials whose acoustic impedance is intermediate between the buffer rod and the fluid couplant.

11.4. Instrumentation and Operation

11.4.1. The Acoustic Lens

Not surprisingly, the development of the modern acoustic microscope came from Stanford, where much of the precursory work was carried out. There was clearly a formidable problem of obtaining diffraction-limited resolution with a full-field analogue of the conventional optical microscope. However, Quate and his student Ross Lemons realized that a far better approach was simply to use mechanical scanning and to build up the image point by point. The design of an acoustic lens is shown in Figure 11.4 where one can see the zinc oxide transducer used to generate a plane sound wave in the buffer rod, which typically has a velocity at least a factor of four greater than the coupling fluid (usually water). When the wave impinges on the lens surface the concave indentation filled with the slow coupling fluid acts as a converging lens refracting the sound to a sharp focus. The acoustic pulse then hits the sample and is reflected back through the water to the lens.

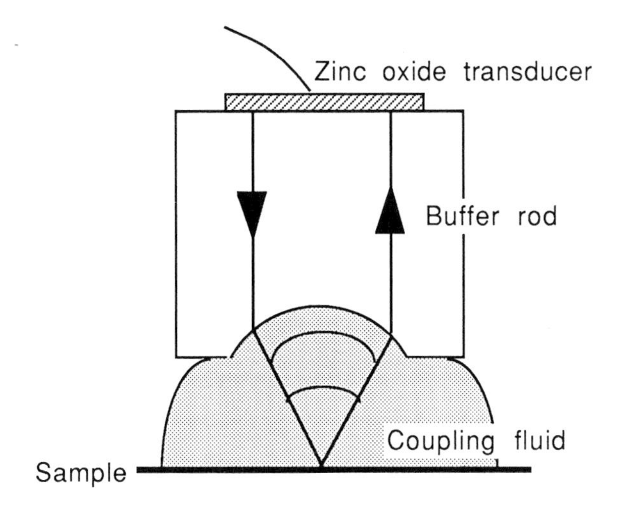

Figure 11.4. Acoustic lens for scanning microscope showing buffer rod and spherical indentation (matching layer now shown).

Simple ray tracing shows that the focus suffers from a negligible degree of geometrical aberration because of the very large velocity ratio between the wave in the buffer rod and the wave in the coupling fluid.[9]

An antireflection coating is often deposited on the lens surface in order to reduce reflections at the lens surface. The layer is virtually essential at high frequencies above 1 GHz where losses in the coupling fluid mean that a high signal-to-noise ratio must be maintained by reducing interface losses to a minimum. At lower frequencies, on the other hand, this layer is desirable but not essential as the signal-to-noise ratio is normally fairly high.

Figure 11.5 shows a photograph showing four typical acoustic lenses. The large lens (C) has an aperture of approximately 4 mm, and is operated around 50 MHz (wavelength in water: -30 μm). It has a large aperture angle of more than 50°. The buffer rod is made of fused quartz and there is no matching layer. The small lens (A) is very similar in construction except that the radius of curvature of the lens is close to 100 μm and is only just visible to the naked eye. The lens is designed for operation around 1 GHz (-1.5 μm). The buffer rod is made of sapphire, and the matching layer is sputtered glass. Lens (D) has a line focus which accounts for its cylindrical shape. The lens is of little use for imaging because the length of the focal line is only about a millimeter. It is very useful, however, for quantitative measurements, as will be explained later. This particular lens has a buffer rod of sapphire with a matching layer of calcogenide glass and is designed for operation centered at 200 MHz (-7.5 μm). Lens (B) is a focused transducer in which the active piezoelectric element also performs the focusing action.

Figure 11.5. Photographs of four different acoustic lenses: (A) is a high-frequency lens for operation at a gigahertz; (B) is a focused transducer for operation at less than 50 MHz; (C) is a low-frequency wide-angle lens operating around 50 MHz; and (D) is a line focus lens for quantitative studies operating around 200 MHz.

11.4.2. Electronics, Scanning, and Image Display

As we mentioned earlier the reflections from the lens–water interface can be problematic. The usual way of separating the various reflections is to operate the microscope in pulsed mode so that these reflections may be separated in time. Figure 11.6 shows an oscilloscope trace of the reflections from a low-frequency lens designed for subsurface imaging. The focal length of this lens is about 16 mm, which is particularly long but necessary to achieve satisfactory subsurface imaging. The acoustic image is formed by using a gated detector which converts the pulse height returning within the appropriate time interval to a DC signal that is used to modulate the brightness of the image at the particular x–y position. The image is usually stored in a computer or a digital scan converter.

Sample scanning is achieved in a number of different ways depending on the scan area required. For high resolution, where image scans of more than a millimeter are rarely required, the fast scan is usually performed within a vibrating coil and the slow scan is performed with a DC motor that moves a stage via a simple pulley. This arrangement allows a frame scan to be achieved in a few

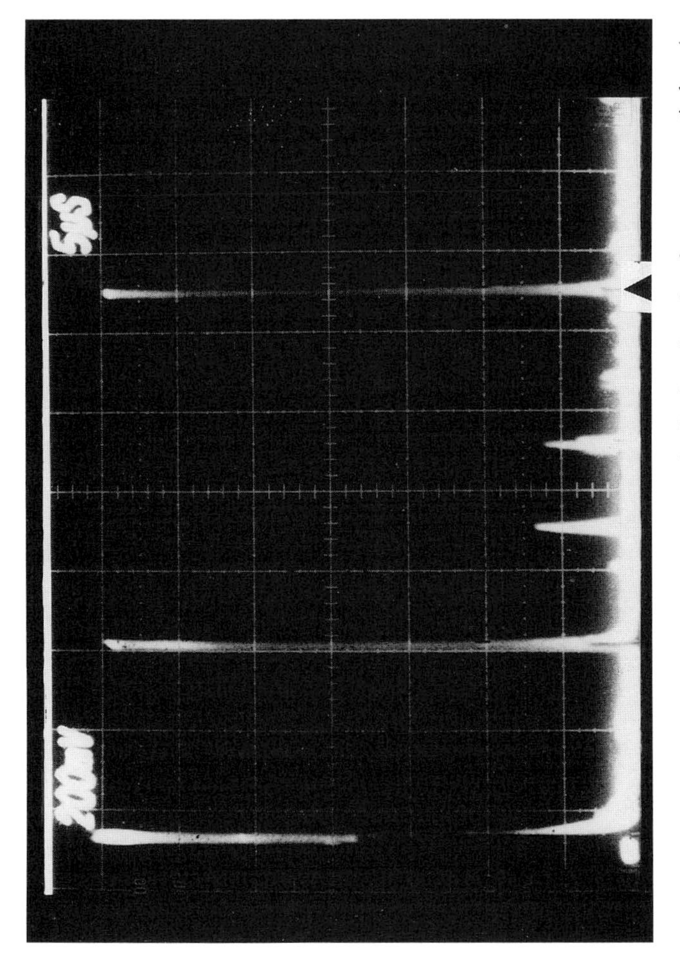

Figure 11.6. Oscilloscope trace showing sequence of reflections from low-frequency acoustic lens (arrowed reflection is from sample surface).

seconds. To scan areas much greater than a millimeter square, one has to sacrifice a considerable amount of speed. Large-area scans at low frequencies normally use two orthogonal stepper motors. In this way very large scans can be achieved, and the resolution of the steppers determines the smallest step size. Several minutes are then needed to build up a full frame.

The acoustic microscope can be configured in either reflection or transmission mode, and much of the early work was performed in transmission.[10] The reason for this was mainly because of the difficulty of achieving sufficiently short pulses to enable separation of the reflection from the lens surface and the sample surface. All the commercial systems are now reflection microscopes, even for applications involving soft tissues. The reflection microscope is considerably easier to operate.

11.4.3. Resolution

The factor that limits the spatial resolution in the acoustic microscope is the operating frequency. To reduce the size of the focal spot it is necessary to increase the operating frequency of the acoustic lens. The attenuation in most fluids, however, increases as the square of the frequency. This means that to maintain a usable signal-to-noise ratio the focal length (i.e., the water path) must decrease as the square of the frequency. A high-frequency lens must not only be smaller in absolute terms but also in terms of wavelength. This places very severe constraints on the pulse electronics because extremely small pulses must be used so that reflections from the sample surface are separated from the spurious reflections. This puts a practical limitation on the maximum operating frequency with which the microscope can be used in an attenuating coupling medium. Most commercial instruments do not operate above 2 GHz (wavelength of 0.75 μm), and the best value obtained using water as the coupling medium is about a factor of 2.5 better than this.[11]

One obvious approach to the improvement of resolution is to use a coupling fluid with reduced attenuation. Reduction of the acoustic velocity also helps because the acoustic wavelength, and hence the attainable resolution, is lower for a given frequency. Two classes of fluid have been used to this end, liquid helium and high-pressure gases.

Liquid helium has vanishingly small attenuation as the temperature approaches absolute zero, so that the need to reduce the path in the coupling fluid with increasing temperature does not exist. This has allowed workers at Stanford to achieve spatial resolution of approximately 250 Å operating at 8 GHz.[12] An effort is presently being made to use operating frequencies approaching 100 GHz where the expected spatial resolution will be competitive with the scanning electron microscope.

High-resolution images of cells have been taken at 8 GHz, but unfortunately there are two main disadvantages with this system. First, the instrumental complexity is formidable. The second disadvantage is the large difference between the acoustic impedance of the coupling fluid and any tissues, which, as equation (11.2) indicates, makes the microscope highly insensitive to sample properties because the specimen will appear almost completely reflecting. This means that the only contrast that can be obtained will result from topographical variations where the signal level varies simply on account of the lens sample separation. The great virtue of using acoustic waves for imaging is that one can map direct interaction with the mechanical properties of the sample, whereas sensitivity to topography alone is of rather more limited use.

The approach of using high-pressure gases has been pursued at University College London.(13,14) This also has the drawback that the contrast is entirely topographic due to the large impedance mismatch between the sample and coupling fluid. The attenuation of a sound wave in an ideal gas decreases as the pressure is increased so that operation at high pressures improves the resolution available. The method is, however, rather easier to apply than cyrogenic microscopy, and the high pressures do not appear to damage the cells. Figure 11.7

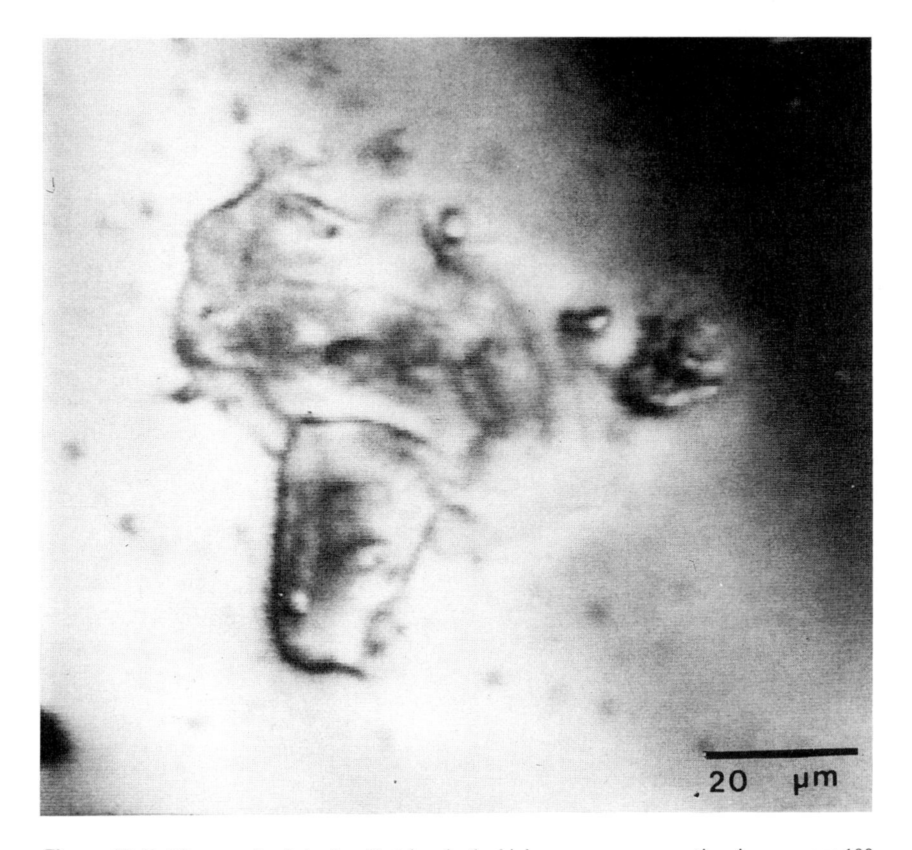

Figure 11.7. Micrograph of cheek cells taken in the high-pressure gas acoustic microscope at 100 atm pressure.

shows a micrograph of cheek cells taken at an operating frequency of 160 MHz which appears to have spatial resolution similar to the value one would expect at five times the frequency in water. Realistically, it does not appear that this method offers many advantages, but it could be useful when one wants to operate in dry environments.

11.4.4. When Is an Acoustic Microscope Not a Microscope?

In the description of the instrumentation above we discuss both high- and low-frequency instruments. Strictly speaking, only instruments operating above 100 MHz may be properly described as microscopes. The patent held by the Stanford group specifically mentions operation above 100 MHz. This point is quite important since low-resolution scanned acoustic images were produced before the invention of the acoustic microscope, but the images after the era of the acoustic microscope were so superior in terms of resolution and quality that it is quite reasonable to attribute the birth of the modern acoustic microscope to the pioneering work of Quate and Lemons. On the other hand, from our present perspective there does not seem to be any point in making artificial distinctions between acoustic microscopy above and below 100 MHz. Technically, if not legally, there is a gradual transition between microscopy and high-resolution C-scans.

There are two or three commercial instruments operating above 100 MHz (usually considerably more) available at the present time. There is also a growing number of lower-frequency instruments available at much lower cost for which the Stanford patent probably does not apply.

11.5. Application to Soft Tissue

From the physical point of view, imaging soft tissue provides a very different challenge from hard tissues, which will be considered in the next section. Soft tissue has acoustic properties very similar to that of water and does not support shear waves.

11.5.1. "Real-Time" Biopsy

The acoustic microscope gives contrast that is determined by the mechanical properties of the tissue. Whether it has a significant role in tissue research remains to be established, but real-time biopsy is one application that should become important in routine diagnosis. The conventional methods of optical microscopy use contrast agents which enable satisfactory diagnosis. However, such stains often take a considerable time to become effective. For instance, collagenous areas show greatly increased optical contrast with trichrome stains.

However, these take several hours to become effective.[15] The acoustic microscope usually provides adequate contrast (most particularly of collagen regions), which enables rapid diagnosis of the tissue, thus enabling an appropriate medical decision to be made while the patient is still in surgery. It appears likely that some of the new techniques of phase-contrast scanning optical microscopy[16] which give greatly increased sensitivity to minute changes in refractive index could also fulfill this important role.

11.5.2. Optical Microscopy and Medical Ultrasonics

The acoustic microscope may be expected to play a linking role between these two important and well-established branches of medical diagnosis. The understanding so gained would undoubtedly increase the precision and power with which medical diagnosis may be performed.

Work to produce this link has not proceeded too far at present, but a useful first methodology has been developed by Daft[17] in an elegant series of experiments. The philosophy behind this work has been to compare optical micrographs of tissue with the corresponding acoustic micrographs and then to use a modified form of the acoustic microscope to get quantitative values for the important acoustic parameters—velocity, density, and attenuation. The eventual aim is that once the acoustic parameters can be measured with microscopic resolution, these values can be used to interpret the images obtained in ultrasonic diagnosis at two orders of magnitude lower resolution. It may then be hoped that features in these medical ultrasound images may be related to structures observable in the optical and acoustic micrographs.

The principle behind Daft's work is extremely simple. The acoustic lens was excited with a very short pulse of about 10 ns duration. The tissue was mounted on a glass slide, and the reflected pulses were received on a sampling scope so that the time delay of the various echoes could be measured with good accuracy. Typical curves showing the time-domain reflected signals are shown in Figure 11.8a and 11.8b which show the reflection from the glass slide only and the reflections from the tissue mounted on the glass slide.

The time delay between the reflection from the glass slide and the surface of the tissue was measured. When this value was divided by the water velocity the thickness of the tissue could be determined. The time delay of the reflection from the top surface of the cell was then compared with that of the reflection from the bottom of the glass slide. Since the thickness of the cell had already been determined, this enabled the velocity in the tissue to be measured. We now recall equation (11.2) in which the reflection from an interface is related to the impedance difference between adjacent media. The magnitude of the reflection therefore enables one to measure the value of tissue impedance. Since the impedance is the product of velocity and density, and the velocity has already been measured, the tissue density may be calculated.

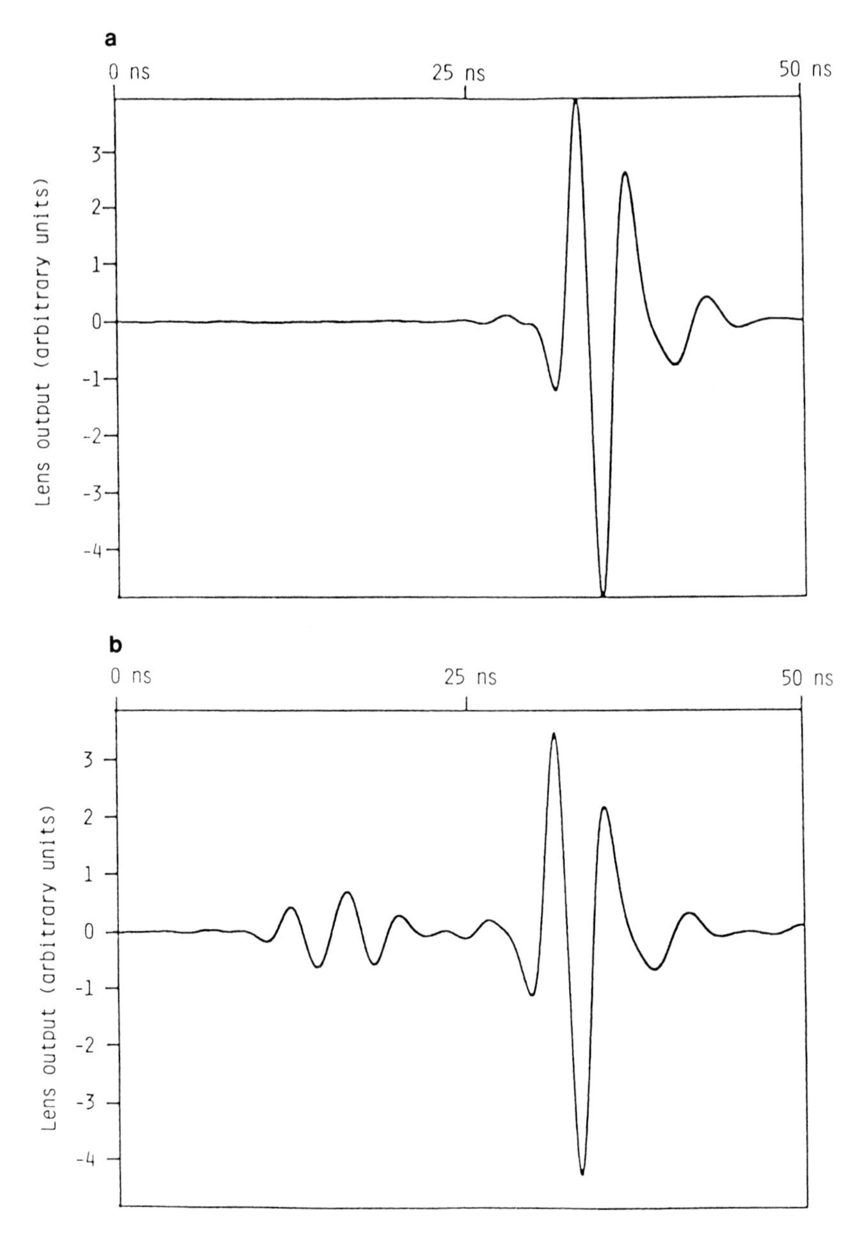

Figure 11.8. (a) Time domain reflection from a glass slide. (b) Time domain reflection from tissue selection mounted on a glass slide. The first small reflection is from the tissue surface. After Daft.[17]

It remains now only to determine the attenuation in the tissue. Because the value of tissue impedance is known, the magnitude of the reflection from the tissue–glass interface in the absence of attenuation can be calculated. Because the actual reflected signal has passed through the tissue twice, the attenuation may be obtained from the ratio of the actual reflection with that expected from the previously calculated impedance values of the media.

Several tissue types have been studied. Here, we will review some results taken on mouse tendon. The scanning acoustic microscope image over which the time-resolved line scans were taken is shown in Figure 11.9a. The tendon is dark and curved, and there is muscle boundary at the top right. A series of time-resolved linescans were taken down the center of the image of Figure 11.9a, thus passing through the bulk of the tendon region. The process described above was then used to compute the values of velocity, thickness, density, and attenuation corresponding to the line through the mid-point of the image shown in Figure 11.9b–e. The work carried out by Daft differs from conventional acoustic microscopy in that the isonifying burst of sound is shorter than the time-of-flight through the tissue. For further details of this work the reader is referred to Refs. 17 or 18.

The method has been further developed to determine the values of the tissue parameters as a function of frequency. Later in this section we will discuss the effects of nonlinearity and their possible limitations on the accuracy of the most simple implementation of the method.

It should also be pointed out that the scanning laser acoustic microscope has also been used for determination of many of the acoustic tissue properties measured with SAM for a much wider range of tissue samples.[19,20] In general, the acoustic frequency used with this instrument is somewhat lower than that used in the SAM, and as such gives very valuable intermediate data in providing the link between microscopic and medical ultrasound frequencies.

11.5.3. Cell Attachment and Examination of Living Cells

The earliest studies of cell attachment in the acoustic microscope have been carried out by Hildebrand[21] using an acoustic microscope operated at 1.7 GHz. The point of attachment between the cell and the substrate can be shown to increase the acoustic impedance compared to other regions in the cell. Hildebrand developed a methodology for determining acoustic impedance and attenuation for various positions within the cell. These results should be compared with those of Daft in which the velocity and density that make up the value of impedance can be independently determined. Daft was not trying to image single cells and was thus able to time resolve the reflections from the substrate and the cell surface. Such time resolution on a single cell would require bandwidths of several gigahertz.

The fact that cell attachment sites give rise to contrast in the microscope has

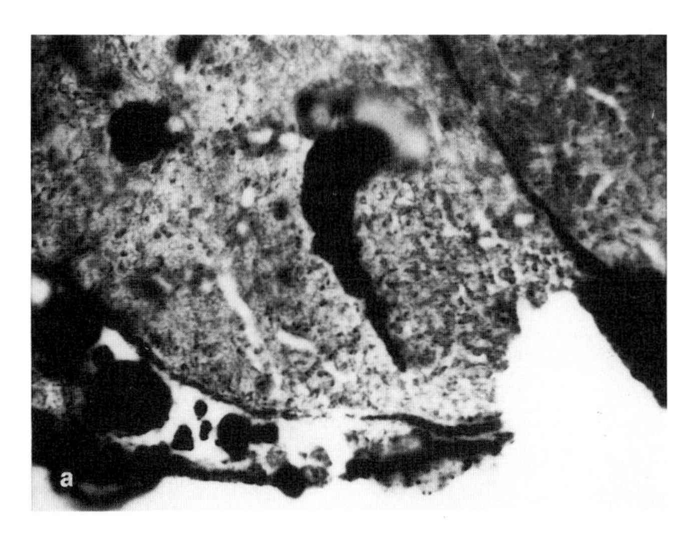

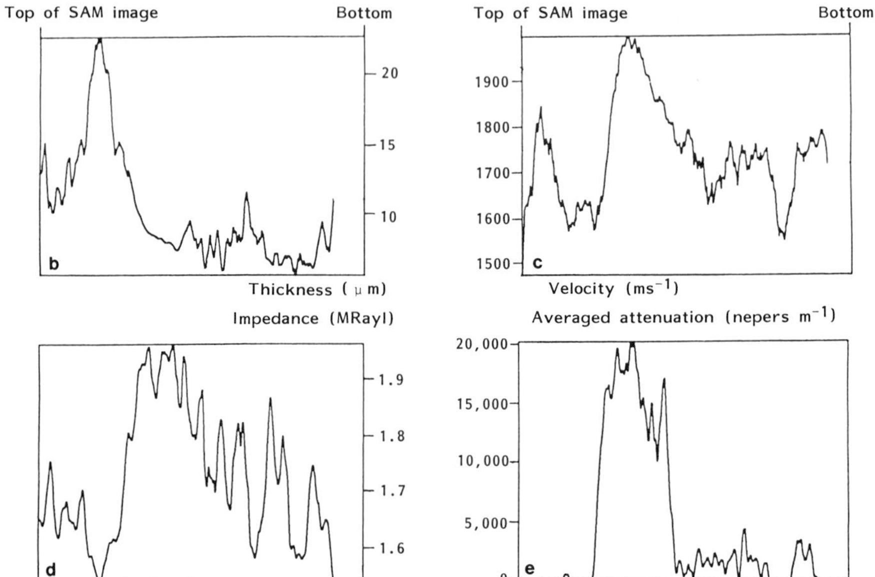

Figure 11.9. (a) Scanning acoustic micrograph of mouse tendon (720 μm field of view). (b) Thickness variation along the midpoint of the image. (c) Velocity variation along the midpoint of the image. (d) Impedance variation along the midpoint of the image (unit MRayl, i.e., kg $m^{-2}s^{-1}$). (e) Average attenuation variation along the midpoint of the image (unit measurement: nepers m^{-1}).

assisted Bereiter–Hahn[22] to make a motion picture, taken at six frames per minute, of moving cells. It is interesting to note that the acoustic beam does not seem to disturb the cells at all, but the scanning motion of the lens can cause detachment of the cells from the substrate. Such studies may ultimately prove useful because release of protein, which inhibits gene expression, will also certainly alter the mechanical properties.

11.5.4. Nonlinear Effects

The basic theory described in Section 11.2 referred to the so-called "linear regime" where Hooke's law applies. Similarly, in the linear region, the local density increase due to the propagation of the wave only varies in proportion to the pressure change. Equation (11.3) gives the approximate equation of state of a liquid that relates the fluid pressure and local density:

$$p = p_0 + A(\rho - \rho_0) + \frac{B}{2}\left(\frac{\rho - \rho_0}{\rho_0}\right)^2 + \text{higher terms} \qquad (11.3)$$

The nonlinear effect is expressed by the parameter B. B/A is an important parameter in diagnostic ultrasound because the value is a good indicator of pathology. Nonlinear propagation causes energy at frequency f to be converted to energy at multiples of the fundamental frequency. If detection occurs at the fundamental frequency, an excess energy loss will be observed. The fractional energy loss will increase with the intensity of the acoustic wave. Nonlinearity can be both a curse and a benefit.

The problem is that nonlinearity can distort any linear measurements. As the power density of a wave increases, a greater proportion of the energy will be transferred from the fundamental. This proportion also increases as the wave propagates. In the acoustic microscope the greatest power density is close to the focus of the lens, so that the greatest rate of harmonic power accumulation is close to the focus. On the other hand, the depth of focus is generally small so that the distance over which this occurs is small. Precise calculations therefore are needed in each case to determine where the bulk of the harmonic generation occurs, and without this information interpretation of results is somewhat difficult. If the harmonic is generated in the focal region, then most of the energy will be generated in the tissue being examined. This case is useful for imaging the nonlinear parameters of the material, but would be a problem if one wants to make an interpretation similar to that given by Daft. If the majority of the harmonic is generated in the coupling fluid prior to focus, most of the information obtained from the sample is linear. In this case Daft's results are probably valid. It is important, therefore, to assess the effects of nonlinearities before quantitative results can be interpreted reliably.

If nonlinearity is a real problem for time-domain tissue measurements (and without the precise details of power levels it is not possible to say whether that is the case), the methodology is still valuable because there are well-known expedients to get around this problem. The large peak powers usually associated with a pulse of short temporal duration, and therefore large bandwidth, may be avoided by using coding techniques well known in radar. Either pulse compression[(23)] or frequency-modulated continuous waves[(24)] may be used. These allow the same equivalent bandwidth while decreasing the peak power for a given signal-to-noise ratio. Pulse compression is actually used in liquid helium microscopy because the large value of the nonlinear parameter in this medium makes the signal-to-noise ratio of an uncoded signal rather poor.

On the other hand, nonlinearity can offer a very effective means of examining tissue. Early images of kidney have been obtained by Kompfner and Lemons with a transmission microscope.[(25)] Two linear images were obtained at f (around 450 MHz) and $2f$. These images showed very similar features with the higher-frequency image giving better resolution, as one might expect. Another image was taken transmitting at f and detecting at $2f$. In addition to good resolution, the contrast is different with the boundaries of individual cells emitting more second harmonic than the centers of the cells.

Nonlinear effects appear to be a powerful and under-exploited means of imaging and offer a largely unexplored possibility for tissue diagnosis. The potential is perhaps brought out even more clearly in some recent work by Germain and Cheeke[(26)] in which signals up to the 14th harmonic have been detected by placing a plane-detecting transducer close to the focal plane.

11.6. Applications to Demineralization

As we mentioned before, hard tissue presents a very different problem from soft tissue. First, the values of acoustic impedance are very much greater so that there is a considerable reflection from a water interface, and, second, shear waves can be generated.

In the conventional imaging mode the acoustic microscope provides a means for examination of the effects of demineralization of hard tissue. The reflection coefficient for waves hitting the sample near to normal incidence is determined by equation (11.2). A simple acoustic image thus gives ready mapping of the product of the density and the velocity. This itself can be useful for assessing the extent of demineralization and monitoring new bone growth. Adaptation to the presence of local stress may also be studied.[(27)] In addition, averaging over a cluster of pixels enables one to eliminate some of the effects of inhomogeneity. Results obtained microscopically also appear to agree with bulk measurements quite well. If the velocity of the material can also be measured a direct mapping of density may be possible by, say, the $V(z)$ method discussed later in this section.

The very simplest imaging mode provides a measure of impedance in tissue such as bone. The acoustic velocity of very hard tissue such as enamel is about a factor of two greater, which allows the so-called "Rayleigh wave mechanism" to operate. For this mechanism to operate, acoustic lenses with high numerical aperture are used. This is not only to achieve the best possible resolution at the appropriate frequency but also to make use of a unique contrast mechanism.

Figure 11.10 is a curve showing the reflection coefficient between an interface of aluminum (dental enamel!) and water. At angles of incidence greater than approximately 30° the modulus of the reflection coefficient is unity. Examination of the phase of the reflection coefficient shows that there is a rapid phase change through 2π rad, just above the shear wave critical angle. This phase change is due to excitation of surface waves across the specimen surface. These waves (sometimes called Rayleigh waves after Lord Rayleigh who first predicted their existence) are unique to acoustics, and result from a coupling of the shear wave in the solid to the longitudinal wave. These waves decay away from the sample surface and only have a significant amplitude two or three wavelengths away from the surface. They travel with a velocity and attenuation unique to the particular material (slightly less than the shear wave velocity), and as such give a great deal of useful information about the material properties of the sample. Rayleigh waves also interact very strongly with discontinuities in the sample surface so they can be used to detect the presence of very thin cracks much smaller than the acoustic wavelength.(28,29)

Figure 11.11 shows the principal ray paths in an acoustic microscope when the sample is above the focal plane. When the sample is at focus, all the rays will return to the transducer in phase, so the received voltage is large. When the

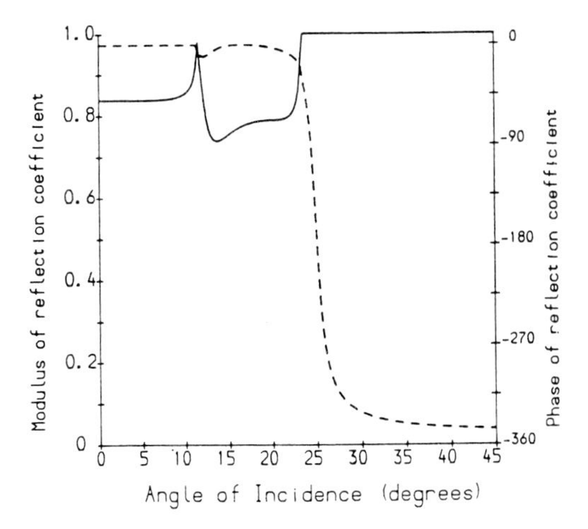

Figure 11.10. Theoretical reflection coefficient at an aluminum (dental enamel)/water interface, showing amplitude and phase variation of the reflected signal as a function of incident angle.

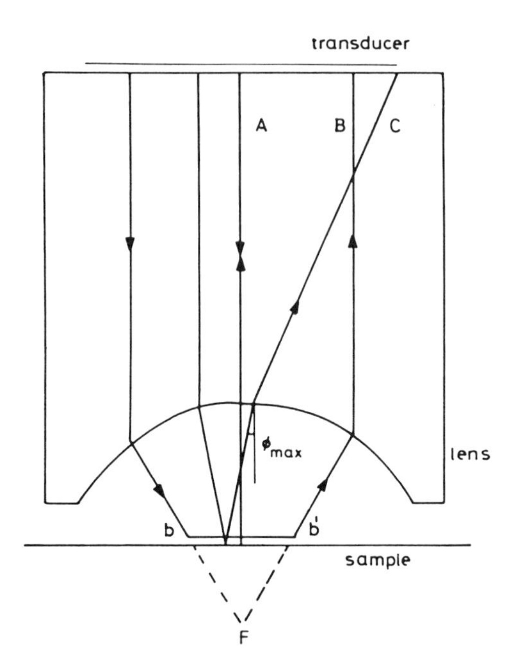

Figure 11.11. Principal ray paths in the acoustic microscope when the sample is above the focal plane.

sample is moved toward the lens, only those rays incident at or near to normal incidence will be reflected so that they intercept the transducer (ray path A). Rays incident at an angle greater than ϕ_{max} will miss the transducer. The received output due to these rays is therefore reduced as the sample is defocused.

If the semi-angle of the lens is greater than the critical angle for excitation of surface waves, the surface waves will be excited on the specimen surface at point b of Figure 11.11. These waves will propagate along the sample surface, continuously leaking out their energy into the coupling fluid; those rays radiated at or close to point b' will appear to come from the focus F. They return to the transducer along ray path B. The modulus of the output voltage is therefore determined by the vector sum of the rays incident or close to normal incidence and those at or close to the Rayleigh wave critical angle. As the focal position z is changed, the relative phase between the rays A and B will change so that the modulus of the output response V will vary periodically with maxima occurring when A and B are in phase and minima when they are in antiphase, thus tracing out the so-called "$V(z)$ curve."

The periodicity of the variation may be easily calculated using simple ray optics,[30] and this may be used to measure the surface wave velocity to around one part in 1000. In the defocused condition the acoustic microscope acts rather like an acoustic interferometer, and is thus able to image very small variations in elastic properties of the sample. Surface waves therefore provide a powerful contrast mechanism for hard tissue and also an accurate means for quantitative determination of surface wave velocities.

One would expect that one effect of defocusing the specimen would be to reduce the lateral resolution. This is only true to a very small extent, as Peck has shown experimentally on dental enamel.[31] The small reduction in resolution may be understood when one realizes that the surface waves are excited on an annulus and the maximum power density of the waves occurs at the axis of the lens.[32,33] Peck has used this mechanism extensively for imaging to study dental caries and has shown how the elastic properties vary within a lesion. He has also shown in a preliminary series of experiments that an acidic fluid coupling can be used to advantage.

In addition to imaging experiments, quantitative $V(z)$ measurements have been taken on teeth by Peck.[31] A so-called "line focus lens" in which the surface waves can only propagate in one direction can be used to determine the velocity and attenuation relative to the prismatic direction.

The $V(z)$ technique has also been used where surface skimming bulk waves rather than Rayleigh waves interfere so that the wave velocity of compressional waves may be assessed. This has been used by Kushibiki and Lees[34] to measure the bulk wave velocity of bone as a function of direction. The values obtained at around 150 MHz agree well with conventional bulk acoustic measurements at low frequencies and Brillouin scattering measurements at 10 GHz.[35] These anisotropic effects may be interpreted using the theoretical framework developed in Somekh *et al.*[36]

11.7. Conclusion and the Future

The acoustic microscope has definite potential as a routine tool for biological studies, and if it were not so expensive would have an excellent chance of replacing the optical microscope for some cytological applications. The sensitivity of the contrast to mechanical properties of tissue is a powerful advantage, and the contrast mechanism can reveal information of real functional importance to the biologist. However, the vast experience with other techniques naturally means that there is resistance to using the method. Unfortunately, the resolution is not high enough for many applications such as cell differentiation and developmental studies, where the scanning motion may disturb the cells.

Overcoming the mechanical scanning problem takes us back full circle, since the early success of the microscope lay in the simple virtues of mechanical scanning. Technologically, scanning in one direction with arrays is possible with resolution approaching a micron while at the same time retaining the mechanical motion in the other direction. While the problem can be solved it is unlikely that such a high level of funding would be made available for very uncertain benefits.

The quest for high resolution, while retaining the advantage of a contrast mechanism that relies on the mechanical properties is a goal that is likely to be addressed very shortly. We described in the brief historical survey the stages in development of the acoustic microscope which can be summarized as follows: full-field, far-field microscope[4,5]; full-field, near-field microscope (i.e., the contact print method of Cunningham and Quate[8]); and the present scanning microscope which may be described as far-field scanning.

The next stage in the progression is a scanning near-field microscope. The concept of such an instrument was first developed by Ash and Nicholls using microwaves.[37] They realized that if a cavity with a very small aperture was scanned very close to the surface of a sample, the local variations in dielectric constant and magnetic permeability could be mapped with resolution far exceeding the wavelength of the radiation, and using microwaves they managed to achieve resolution of a sixtieth of the wavelength.

The essence of the method is the requirement to scan the probe very close to the sample so that the radiation pattern does not spread significantly due to diffraction. Until recently this could be thought to be a barrier against operating at optical wavelengths. The advent of the tunnelling microscope, however, has changed this. Tips are scanned Angstroms above the surface. The near-field optical microscope[38] has recently been developed using technology similar to that of the tunnelling microscope. Resolution of less than 1000 Å has been achieved readily, and values much smaller than this should soon be achievable routinely. The scanning near-field acoustic microscope should therefore soon become available. The principle has been demonstrated by Zieniuk and Latuszek,[39] but the resolution achieved is comparatively poor due to the limited access to advanced technology. The stage is now set, however, for high-resolution, near-field acoustic microscopy, giving SEM resolution and a contrast mechanism depending on local mechanical properties—an enticing prospect!

ACKNOWLEDGMENTS. I would like to thank Chris Daft for making available some of his results prior to publication.

References

1. P. N. T. Wells, in: *Biomedical Ultrasonics*, Academic, New York (1977).
2. S. Sokolov, USSR Patent (1949).
3. S. Sokolov, *Dokl. Akad. Nauk. (USSR)* **64,** 333–336 (1949).
4. F. Dunn and W. J. Fry, "Ultrasonic absorption microscope," *J. Acoust. Soc. Am.* **31,** 632–633 (1959).
5. E. E. Suckling and S. Ben-zui, "Ultrasonic phase microscopy," *J. Acoust. Soc. Am.* **34,** 1277–1278 (1962).
6. T. M. Reeder and D. K. Winslow, "Characteristics of microwave acoustic transducers for volume wave excitation," *IEEE Trans. on Microwave Theory and Techniques* **17** (11), 927–941 (1969).

7. B. A. Auld, R. J. Gilbert, R. J. Hyllested, C. G. Roberts, and D. C. Webb, "A 1.1 Ghz scanned acoustic microscope," *Acoust. Hologr.* **4,** 73–96 (1972).
8. J. A. Cunningham and C. F. Quate, "High-resolution acoustic imaging by contact printing," *Acoust. Hologr.* **5,** 83–102 (1974).
9. R. A. Lemons and C. F. Quate, in: *Physical Acoustics, Vol. XIV,* pp. 1–92, Academic, New York (1979).
10. R. A. Lemons, in: *Acoustic Microscopy by Mechanical Scanning,* M. L. Report No. 2456, Microwave Laboratory, Stanford University (1975).
11. B. Hadimioglu and C. F. Quate, "Water acoustic microscopy at suboptical wavelengths," *Appl. Phys. Lett.* **43,** 1006–1009 (1983).
12. J. S. Foster and D. Rugar, "Low-temperature acoustic microscopy," *IEEE Trans. on Sonics and Ultrasonics, SU-32* **2,** 139–151 (1985).
13. C. R. Petts and H. K. Wickramasinghe, "Acoustic microscopy in gases," *Electron. Lett.* **16,** 9–11 (1980).
14. F. Faridian, "Gas medium acoustic microscopy at 160 MHz microscopy," *Proc. IEEE Ultrasonics Symp.,* pp. 759–762 (1985).
15. R. A. Lemons, in: *Acoustic Microscopy by Mechanical Scanning,* M. L. Report No. 2456, p. 171, Microwave Laboratory, Stanford University (1975).
16. C. W. See, R. K. Appel, and M. G. Somekh, "Differential optical profilometer for simultaneous measurement of amplitude and phase variation," *Appl. Phys. Lett.* **53,** 10–12 (1988).
17. C. M. W. Daft, "Acoustic Microscopy of Biological Tissues," D. Phil. Thesis, University of Oxford (1987).
18. C. M. W. Daft and G. A. D. Briggs, "Wide-band acoustic microscopy of tissue," *IEEE Transactions on Ultrasonics, Ferroelectrics and Frequency Control* **36,** 258–263 (1989).
19. C. A. Edwards and W. D. O'Brien, Jr., "Speed of sound in mammalian tendon threads using various reference media," *IEEE Trans on Sonics and Ultrasonics, SU-32* **2,** 351–354 (1985).
20. W. D. O'Brien, Jr., J. Olerud, K. K. Shung, and J. M. Reid, "Quantitative acoustical assesment of wound maturation with acoustic microscopy," *J. Acoust. Soc. Am.* **69,** 575–579 (1981).
21. J. A. Hildebrand, "Observation of cell-substrate attachment with the acoustic microscope," *IEEE Trans on Sonics and Ultrasonics, SU-32* **2,** 332–340 (1985).
22. Work relating to this film may be found in: J. Bereiter-Hahn, "Scanning acoustic microscopy of living cells," *J. Microsc.* **146,** 29–39 (1987).
23. M. Nikoonahad, G. Yue, and E. A. Ash, "Pulse compression acoustic microscopy using SAW filters," *IEEE Trans. on Sonics and Ultrasonics, SU-32* **2,** 152–163 (1985).
24. F. Faridian and M. G. Somekh, "FMCW techniques in scanning acoustic microscopy," *Proc. IEEE Ultrasonics Symp.,* pp. 769–774 (1986).
25. R. Kompfner and R. A. Lemons, "Nonlinear acoustic microscopy," *Appl. Phys. Lett.* **28,** 295–297 (1976).
26. L. Germain and J. D. N. Cheeke, "Generation and detection of higher-order harmonics in liquids using a scanning acoustic microscope," *J. Acoust. Soc. Am.* **83,** 942–947 (1988).
27. J. L. Katz, reported in: *Ultrasonic Microspectroscopy Workshop,* Sendai, Japan (1988).
28. C. Ilett, M. G. Somekh, and G. A. D. Briggs, "Acoustic microscopy of elastic discontinuities," *Proc. Roy. Soc. Lond.* **A393,** 171–183 (1984).
29. M. G. Somekh, H. L. Bertoni, G. A. D. Briggs, and N. J. Burton, "A two-dimensional imaging theory of surface discontinuities with the scanning acoustic microscope," *Proc. Roy. Soc. Lond.* **A401,** 29–51 (1985).
30. W. Parmon and H. L. Bertoni, "Ray interpretation of the material signature in the acoustic microscope," *Electr. Lett.* **15,** 684–686 (1979).
31. S. D. Peck, "Acoustic Microscopy of Caries in Human Dental Enamel," D. Phil. Thesis, University of Oxford (1986).
32. I. R. Smith, H. K. Wickramasinghe, G. W. Farnell, and C. K. Jen, "Confocal surface acoustic wave microscopy," *Appl. Phys. Lett.* **42,** 411–413 (1983).

33. M. G. Somekh, "Consequences of resonant surface wave excitation in reflection scanning acoustic microscopy," *IEEE Proc., Part A* **134,** 290–300 (1987).
34. J. Kushibiki and S. Lees, reported in: *Ultrasonic Microspectroscopy Workshop,* Sendai, Japan (1988).
35. S. Lees, N.-J. Tao, and S. M. Lindsay, Sonic velocities in bone at GHz frequencies, *Acoustical Imaging,* **17,** 371–380 (1988).
36. M. G. Somekh, G. A. D. Briggs, and C. Ilett, "The effect of anisotropy on contrast in the scanning acoustic microscope," *Phil. Mag.* **49A,** 179–204 (1984).
37. E. A. Ash and G. Nicholls, "Super-resolution aperture scanning microscope," *Nature* **237,** 510–512 (1972).
38. U. Ch. Fischer, V. T. Durig, and D. W. Pohl, "Near-field scanning microscopy in reflection," *Appl. Phys. Lett.* **52,** 249–251 (1988).
39. J. K. Zieniuk and A. Latuszek, Non-conventional pin scanning ultrasonic microscopy, *Acoustical Imaging* **17,** (1988)

12

Scanning Tunnelling Microscopy

M. E. Welland and M. E. Taylor

12.1. Introduction

Since its introduction by Binnig *et al.*,[1] the Scanning Tunnelling Microscope (STM) has engendered much excitement among surface scientists, not only for the atomically resolved surface topography it can achieve, but also for the range of surface spectroscopy possible. In this chapter we discuss scanning tunnelling microscopy in the context of the study of thin films (for example, an organic adsorbate on a metal substrate), highlighting the strengths and weaknesses of the technique. Greater detail may be found in the literature; in particular, the review by Hansma and Tersoff[2] provides a wide survey of the field. For more general reviews of tunnelling and tunnelling spectroscopy the reader is directed to the books by Wolf[3] and Hansma.[4]

12.2. Outline of Scanning Tunnelling Microscopy

The phenomenon of tunnelling, in which a particle may traverse a region through which, classically, it is forbidden to pass, has been of great interest since the early days of quantum mechanics. Tunnelling of electrons through solid-state barriers was first demonstrated convincingly about 30 years ago in semiconductor junctions by Esaki[5] and in metal-insulator-superconductor junctions by

M. E. Welland • Department of Engineering, University of Cambridge, Cambridge CB2 1PZ, United Kingdom. ***M. E. Taylor*** • Interdisciplinary Research Centre in Superconductivity, University of Cambridge, Cambridge CB3 0HE, United Kingdom.

Giaever.[6] Apart from field emission, however, vacuum tunnelling of electrons was generally neglected until the advent of the STM. Some of the ideas of scanning tunnelling microscopy were found in the topographiner of Young *et al.*,[7] but the much greater tip-sample distances used in this device (of order 70 Å rather than 10 Å) prevented attainment of atomic resolution.

In principle, the microscope is very simple (see the block diagram in Figure 12.1 and an example of a commercial UHV STM in Figure 12.2). A sharp metal tip (usually tungsten or platinum-iridium) is brought close to the (conducting) surface of interest, say within 10 Å, and a tunnel current established by application of a suitable bias. A feedback loop adjusts the vertical (Z) position of the tip

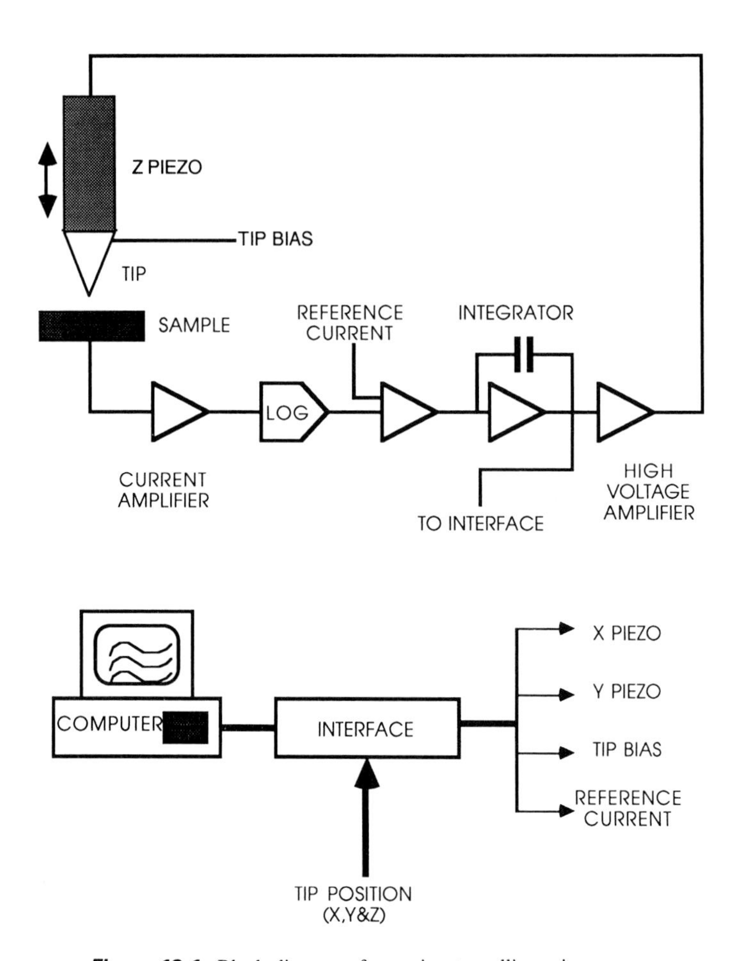

Figure 12.1. Block diagram of scanning tunnelling microscope.

Figure 12.2. Photograph of an STM head for use in UHV. (Courtesy of W. A. Technology.)

to keep the current constant as the tip is scanned across the surface. Thus, by monitoring the output of the feedback loop, it is possible to obtain the surface relief of the specimen [see Figure 12.3(a)]. The spatial resolution available with a good tip is about 2 Å in the sample plane, and 0.1 Å vertically (or even 0.01 Å using a.c. modulation of the tunnel current). Factors affecting resolution are considered below. STMs have been applied to many problems. Figure 12.4 shows the 7×7 reconstruction of the silicon (111) surface in ultrahigh vacuum, first observed by Binnig *et al.*.[(8)] Further examples are: surface reconstruction of other semiconductors and of metals by Kuk *et al.*,[(9)] the effect of adsorbates on surface structure by Baró *et al.*,[(10)] the behavior of charge-density waves in layer compounds by Coleman *et al.*,[(11)] and imaging the flux lattice of a superconductor in a magnetic field by Hess *et al.*.[(12)]

There are several alternatives to the constant current mode of operation. With a sufficiently flat sample the microscope may be run at constant voltage, that is, the scan speed is chosen to be much faster than the time constant τ of the feedback loop, so that changes in the tip-sample distance are revealed as variations in the current, as was shown by Bryant *et al.*.[(13)] Tunnelling spectroscopy is also possible. Repeated scanning over a region with a variety of bias voltages

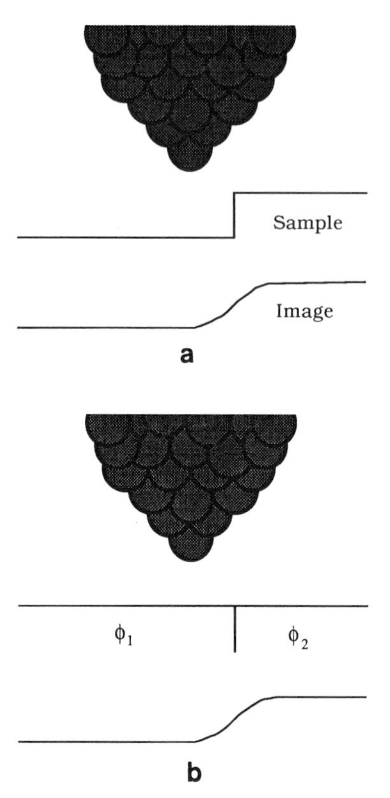

Figure 12.3. Schematic representation of steps in STM images produced (a) by a surface step and (b) by a change in work function.

gives information on the variation of electronic states with energy, and thus provides information on bonding. Alternatively, the feedback loop may be suspended temporarily at a point of interest on the surface, and a current–voltage characteristic taken (see Hamers *et al.*[14]). These measurements give, for instance, changes in the gap parameter of a superconductor, or the distribution of localized electronic traps in a semiconductor or an organic film. Measurement of the current as a function of tip-sample separation s gives the sample work function. Finally, with slight modifications, the STM may be used for potentiometric (see Muralt and Pohl[15]) or for electrochemical measurements (see Sonnenfeld and Hansma[16]).

Recent development has involved combining STMs with other surface tools (for example, electron microscopy, field emission, Auger spectroscopy, etc). Alternative scanning microscopies have also been developed, the most important of which is the Atomic Force Microscope (AFM), originally introduced by Binnig *et al.*[17] in 1986. In this device the force between a surface and tip in close proximity is measured as the tip is scanned over the surface. The tip is mounted on a lever, the deflection of which gives the force. Several means of measuring

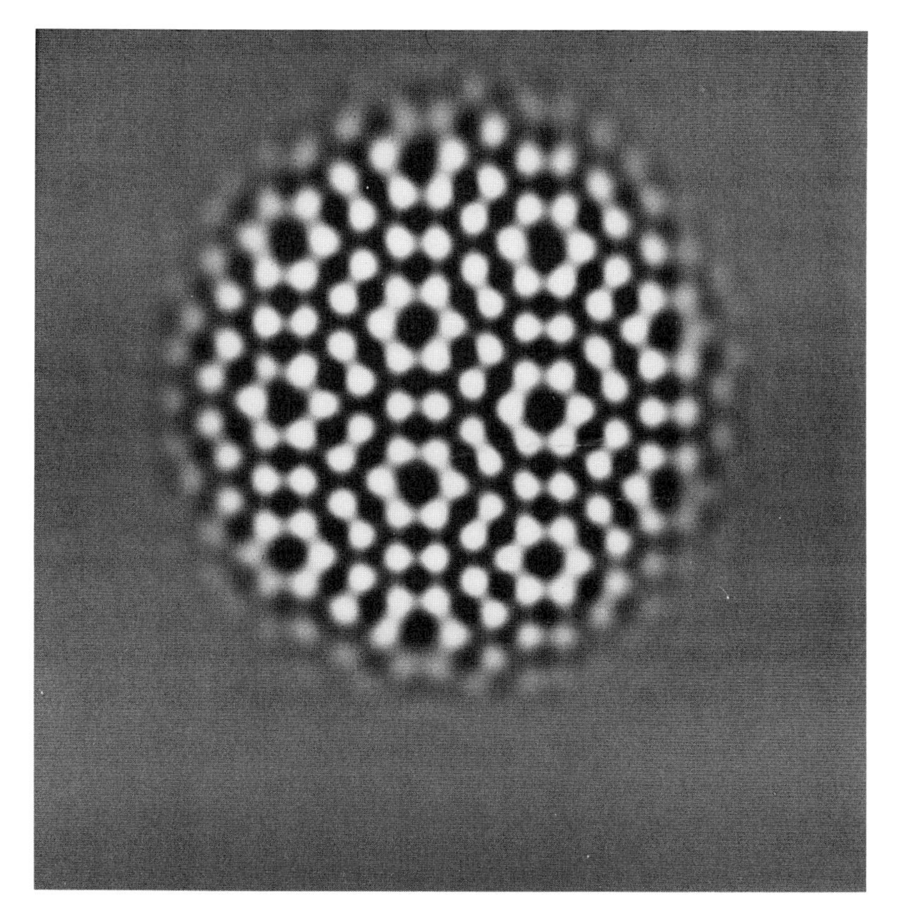

Figure 12.4. Scanning tunnelling micrograph of the 7×7 reconstruction of the silicon (111) surface. The nearest-neighbor spacing is about 4.7 Å and the white-to-black contrast about 1.5 Å.

the deflection have been employed; for example, use of an STM, capacitative techniques, and optical interferometry.

12.3. Theory

The great sensitivity of the scanning tunnelling microscope arises from the approximately exponential dependence of tunnel current on the tip-sample distance. In the first approximation, sufficient for many applications, the tunnelling probability for an electron incident on one side of a potential barrier is proportional to $e^{-2\alpha s}$, where $\alpha = \sqrt{2m\Phi}/\hbar$ is the decay constant of the electron wave

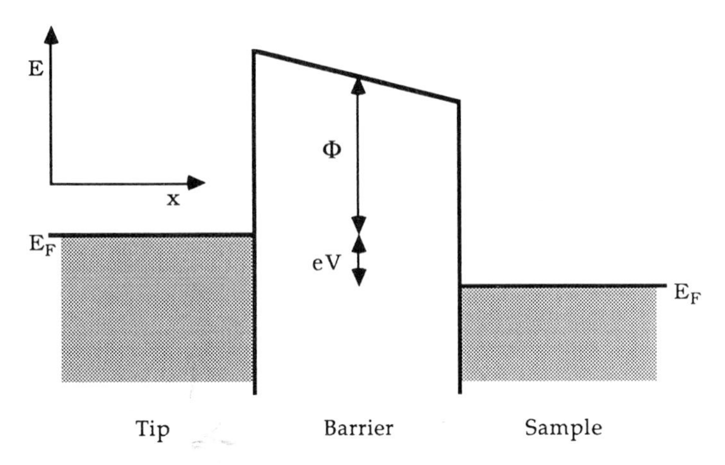

Figure 12.5. Diagram of free-electron metals (tip and sample) separated by a tunnelling barrier.

function in the barrier, and Φ is the barrier height (see Figure 12.5). Strictly, this is only true for planar electrodes separated by a rectangular barrier. As an asymptotic form for other configurations, however, it may be justified by application of the WKB approximation (see, for example, Simmons(18)). Neglecting effects such as the electron image potential (still a controversial issue) the height of a vacuum barrier should be approximately the work function of the electrode material. For a typical metallic work function of about 4 eV, the decay constant α is roughly 1.0 Å^{-1}.

Referring again to Figure 12.4, at absolute zero the Pauli exclusion principle restricts tunnelling to electrons in one side that are opposite empty states in the other, so that the current I is given by:

$$I \propto V \rho_t \rho_s e^{-2\alpha s} \tag{12.1}$$

where ρ_t and ρ_s are the local densities of electronic states in tip and sample. Thus, in constant current mode the STM images surfaces of constant local density of states. It is important to realize that the STM is sensitive to the local density of states at the surface, rather than the surface charge density as is sometimes erroneously stated. This is only true, however, if the barrier height, and thus α, are independent of position. Figure 12.3(b) indicates schematically how a change in Φ can mimic a surface step. The form of equation (12.1) suggests a method for extracting variations in α. If a small modulation, $\delta s \cos \omega t$, is added to the separation, a lock-in amplifier may be used to obtain the logarithmic derivative of the current:

$$\frac{d\ln I}{ds} = -2\alpha$$

In practice, the ohmic region to which equation (12.1) is directly applicable is limited by differences in the tip and sample work functions, and by other inherent asymmetries. Also, above absolute zero, the electronic distributions are no longer cut off sharply at the Fermi energy, but drop to zero over a scale of about kT. Lambe and Jaklevic[19] showed that at finite temperatures tunnelling features are broadened by a width of approximately 5 kT.

The simple, essentially one-dimensional, model we have discussed is sufficient for qualitative understanding of the behavior of an STM. However, treatment of problems such as the lateral resolution of the instrument requires a much more careful consideration of the effect of real surface structure, both of sample and tip, on the tunnel current. Precise simulation of real surfaces requires numerical modelling, and applications to scanning tunnelling microscopy are rendered difficult by the lack of good information on tip structure. Consequently, although the sample surface may be treated more or less accurately, most investigators have restricted themselves to consideration of extremely idealized models of the tip. It is important to be aware of the model dependence of interpretation of STM data, just as with other tunnelling experiments.

Given a particular model of sample and tip, complete calculation of the tunnel current requires first matching the conduction electronic wave functions of sample and tip with a suitable evanescent form in vacuo to find the tunnelling probability between tip and sample states, and then, as in the one-dimensional case, summing over the initial and final states with due regard for the exclusion principle. Several approximate schemes are available for calculating the tunnelling probability, the most common of which is the transfer Hamiltonian method (see Bardeen[20]), applicable to weak coupling (i.e., low tunnelling probability) between electrodes. Using this approximation, and taking the tip to be a hemisphere of free-electron metal, Tersoff and Hamann[21] have derived an expression for the tunnel current that is essentially equation (12.1), but the sample local density of states is evaluated at the center of curvature of the tip. Furthermore, lateral detail in the sample is broadened by a width $\sqrt{(s + r_t)/2\alpha}$, where r_t is the radius of curvature of the tip. If the hemisphere is taken to be a single terminating atom (admittedly an extreme case, in which some of the approximations made become invalid) the width is about 2.5 Å using the value of α given above. Similar conclusions have been reached by other workers.

The current contains information on the energy variation of the local electronic density of states in the sample, but combined with both the density of states of the tip and the tunnelling probability. If we can make the assumption (by no means always justified) that these last two factors are roughly constant on the energy scale of interest, then

$$I \propto \int_{E_F}^{E_F+eV} dE\rho_s(E)$$

[which is equivalent to equation (12.1) at low bias]. The current is therefore a rather insensitive measure of the density of states, unless ρ_s is dominated by a particular sharp feature such as the energy gap in a superconductor. Increased sensitivity to features in the density of states may be achieved in principle by measuring the conductance

$$\sigma = \frac{dI}{dV} \propto \rho_s(E_F + eV)$$

Use of derivative techniques, ubiquitous in solid-state tunnelling, is restricted by noise and vibration in the STM. Good vibration isolation and stability allow the prolonged sampling times required for adequate noise reduction.

The assumption of constant tunnelling probability is invalid in several common situations. If, for example, the barrier includes a thin insulating region such as an oxide layer, or the sample is an organic coating of high resistance on a conducting substrate, tunnelling via localized states may occur. Similarly, a semiconductor surface may have localized surface states or traps, only weakly coupled to the bulk conduction states. In all these cases the phenomenon of resonant tunnelling may occur, where the tunnelling probability is greatly enhanced for electrons having energies close to that of the localized state (see, for example, Payne[(22)]). Furthermore, vibrational and other collective modes can couple to tunnelling electrons to give an *inelastic* contribution to the current. In solid-state structures this contribution has been exploited as Inelastic Electron Tunnelling Spectroscopy (IETS), but similar spectroscopy using STMs has so far been limited by the noise problem.

12.4. Equipment and Instrumentation

The original STMs were designed for use in ultrahigh vacuum systems, but subsequent development has brought microscopes operating in air, water and other solvents, at low temperatures, and also STMs combined with other microscopic and surface techniques. Despite these diverse applications there are some basic design principles vital to successful operation of the STM. The most important features are stability and immunity to vibration. To attain a given vertical resolution (say 0.1 Å) the mutual vibration amplitude of tip and surface must be kept below this level. In principle, the STM head should be extremely light and rigid, with a lowest resonant frequency of at least 10 kHz. Attention to this requirement may well reduce the extent to which the head assembly needs to be vibrationally isolated from its surroundings, but this depends to a great extent on the nature of the environment. In a laboratory with many pumps or other heavy machinery much more isolation will be required than in one without. Similarly, attention should be paid to the frequencies present. Building vibrations are usually in the range 20–25 Hz and can be attenuated successfully by use, for example, of an air table, while sensitivity to sound levels may be obviated by acoustic shielding. More discussion of vibration isolation is given in Hansma and Tersoff.[2]

Precise positioning of the tip with respect to the sample is effected by piezoelectric drives. The original STM used a tripod design with three separate piezos for independent *x*, *y*, and *z* motion. Currently, most groups use tube scanners (see Binnig and Smith[23]), which are simpler in that one piece gives all three movements. In addition, tube scanners have higher resonant frequencies, important both for vibration rejection and for high-frequency scanning. A coarse motion is also necessary to enable initial positioning of sample and tip, and perhaps also to allow in situ modification of the sample by ion bombardment or evaporation. The coarse adjustment may be effected by several techniques. Differential screws, springs or lever mechanisms may be employed to step down the motion of a motor or hand-operated drive. Recently, piezoelectric inchworm linear drives have been used. These give steps in position as fine as a few nanometers, combined with a travel of a few millimeters. UHV-compatible inchworms are available, but at present it is impossible to operate them much below room temperature. The "louse," or walker, used in the original STM has problems both of reliability and also of a low resonant frequency. Several feedback circuits have been published (for example, Fein *et al.*[24] or Park and Quate[25]). It is common to include a logarithmic amplifier in the circuit to linearize the current. Such amplifiers are relatively slow devices, however, and may be omitted when operating at high scan rates with flat samples. The time constant of the feedback circuit can be set either by an integrator or by an RC network; the latter does not follow the surface topography quite as accurately, but may be made less susceptible to long-term drift (see Park and Quate[26]). If I/V or I/s curves are to be taken, a sample-and-hold circuit is required to interrupt the feedback while the bias or distance is changed.

High voltages are required to obtain the full movement of the piezos (up to say $\pm$ 500 V). The output of the feedback loop may be amplified by a good-quality high-voltage amplifier, or alternatively, added to a high-voltage supply. To achieve good resolution it is important to have a very quiet signal, especially for the tip-sample separation.

Finally, computer control is not indispensable, although it may be extremely useful. Output may be generated on a storage oscilloscope, and photographed as required and manual control of most functions is possible. However, the advantages of automation and computer graphics are difficult to resist!

12.5. Tip and Sample Preparation

There are many recipes for manufacturing tunnelling tips as there are groups with STMs. Usually tips are either electrochemically etched, or are ground to the required dimensions. It is possible to form fine tips in vacuo by application of a strong electric field, but this is not always reliable, and may also damage the sample. In general, the rule appears to be that carefully made fine tips (with no curvature visible in an optical microscope) will work well as long as they are not damaged by careless handling or by direct mechanical contact with the sample surface. To avoid this latter fate the initial approach to tunnelling should be judicious. Some quantitative investigations of the relationship between tip morphology and STM image have been reported: for instance, Kuk *et al.*[9] compared field ion microscopy images of various tips with their STM resolution.

Almost any conducting or semiconducting material may be studied by scanning tunnelling microscopy. However, the best results (at least in terms of interpretation) are given by samples with relatively flat regions of surface. With a really rough sample the observed signal may contain as much information on the tip structure as on the sample itself. In UHV work on single crystal surfaces, samples are annealed to remove oxide layers or other surface impurities and defects. Ion guns are also sometimes used to remove surface layers, but subsequent annealing is always required. Good surfaces may often be prepared by cleaving under vacuum or in a glovebox. Layered materials, such as graphite, cleave particularly easily, so that removal of a piece of sticky tape may be sufficient to strip off the old surface.

Scanning tunnelling microscopy of samples with a thin insulating surface layer (such as a native oxide) is possible if the surface layer is less than about 20 Å thick. However, the STM trace will not be simply related to the surface topography. Changes in the layer thickness or its dielectric constant will have a marked effect on the average barrier height, and thus on the tunnelling exponent α. Observation of the surface structure of a bulk insulator is difficult. One method that has been shown to be feasible is to coat the surface with a thin layer ($\sim$ 20–40 Å) of gold, deposited at 77 K or below. As long as the surface is clean, there is a good chance that the gold film will not agglomerate on warming to room temperature, and will in fact give a faithful representation of the real surface (see Jaklevic *et al.*[27]). The only alternative presently available is the atomic force microscope (AFM). Results with this instrument have been obtained on amorphous silica (see Heinzelmann *et al.*[28]), but interpretation of AFM data is still controversial.

12.6. STM of Organic Macromolecules

12.6.1. Introduction

The possibility that the STM can be used to image the internal structure of macromolecules such as proteins, liquid crystals, and DNA is of enormous significance. Techniques such as electron microscopy and X-ray scattering have been successfully used to determine the structure of a variety of molecules through diffraction effects. They have the drawback, however, that the scattering cross sections for organic molecules are small for both techniques, and best results are obtained from molecules that crystallize and thus represent a stronger scattering site for diffraction. A further problem with electron microscopy is that the energy of the electrons, typically 100 KeV, is enough to damage or destroy the molecule in a relatively short period of time. STM has several potential advantages. First, it produces atomic resolution images of surfaces in real space—a distinct advantage over a diffraction technique where the real structure has to be deconvoluted from the diffraction data. Second, the electrons that tunnel to form the image in the STM have energies of a few electron volts—insufficient to damage the molecule in the way that 100-KeV electrons can. Finally, the STM can operate in air or in a liquid, making possible the imaging of molecules in their natural environment.

These advantages appear very attractive, and the inclination to obtain access to an STM and try it on virtually any biochemically interesting molecule or even cell may appear strong. Unfortunately, there are very restrictive conditions on the potential success for a given molecule, and this section will attempt to address the problems of imaging molecules with the STM.

12.6.2. Insulating Molecules: Contact and Substrate Effects

To image an organic molecule that is nominally insulating it is necessary to deposit it on a conducting substrate. The idea then is that the tunnelling current flowing between tip and substrate will in some way be modified by the presence of the molecule, thereby generating contrast in the image. Where this current modification changes from one atomic site to the next within the molecule it is not unreasonable to expect that atomic details will be visible in the image. Of course, the substrate itself will inevitably contribute contrast to the image either through electronic interactions with the molecule or, more simply, through its own surface structure. Ideally, the substrate needs to be atomically flat over large areas so that its contribution to any image is minimized. At first sight highly oriented pyrolitic graphite appears to fulfill this criterion since it cleaves easily to produce extensive, atomically flat regions. There is, however, doubt over the mechanism by which graphite images in the STM are formed since it is possible to obtain atomic images of the surface while the tip is firmly in contact with it. This has been pointed out by Pethica.[29] There are a variety of single-crystal semiconductors that cleave readily on low-index crystallographic planes, silicon being an obvious candidate, but they have the disadvantage that a thin oxide layer develops in air, which may preclude imaging. A good choice of substrate is that of an x-ray mirror, which can be made with an RMS roughness of 2.5 Å over large areas with an oxidation resistant surface of amorphous carbon (see Welland *et al.*[30]).

As stated previously, significant tunnelling will only take place between a metallic tip and sample when their spacing is less than ~ 20 Å. A further important limitation is that for an STM to operate successfully it is of crucial importance that the tip and sample do not come into physical contact; there must be a vacuum gap between the two. To image a molecule adsorbed onto a conducting substrate it is therefore necessary to ensure that the molecule is thinner than ~ 20 Å in the direction normal to the substrate. This is a severe limitation because a substantial number of macromolecules such as proteins exceed this thickness. The size of the gap between tip and substrate is controlled by the choice of tip voltage and tunnelling current; a high tip voltage and low current will maximize the gap. For molecules or thin films with thickness in the range 10–20 Å it is therefore necessary to operate the STM at high tip voltages and low currents.

The situation would be less complicated if it was obvious from an image when contact between molecule and tip was taking place. Unfortunately, as predicted by Coombs and Pethica,[31] it is possible to obtain reproducible images while in contact, and thereby to obtain grossly misleading results. This is illustrated in Figure 12.6 where it is assumed that the tip is in direct contact with the molecule or with some other insulating matter such as oxide or dust. The STM operates by moving the tip normal to the surface so that a constant current is maintained. Here the whole ensemble of tip, molecule, and substrate will be compressed until the required current flows between tip and substrate. The degree to which each is deformed will depend on the size of the contact area and the mechanical compliance of each. In simplistic terms one can assign a spring constant K to each. The higher the spring constant the less that element will be compressed for a given contact force. For a given movement of the piezo, which maintains a constant current, there will be a reduced alteration of the gap between tip and substrate, the actual value of which depending upon the K's of tip, molecule, and substrate. This preserves the exponential dependence of current on separation, but since the actual gap width is changing by less than that of the positioning piezos, a lower apparent electron barrier height is observed. Clearly, if the tunnelling takes place through the molecule or particle and the compressibility of the particle is significantly less than that of the substrate, an exceedingly low apparent work function will be measured.

In the slightly different situation where tunnelling takes place through a vacuum gap a little way away from the contact region Coombs and Pethica[31] showed that the apparent barrier height Φ_a is approximated by

$$\Phi_a = \Phi_0 \, [1 - (a/r)]^2$$

where Φ_0 is the vacuum barrier height, a is the radius of the molecule or particle, and r is the distance between particle and tunnelling region. Once again the measured barrier height will have an arbitrary value less than the true value. If one attempts to take an image in this situation, the measured sizes of any features seen will bear little resemblance to any real sizes. Indeed the features themselves

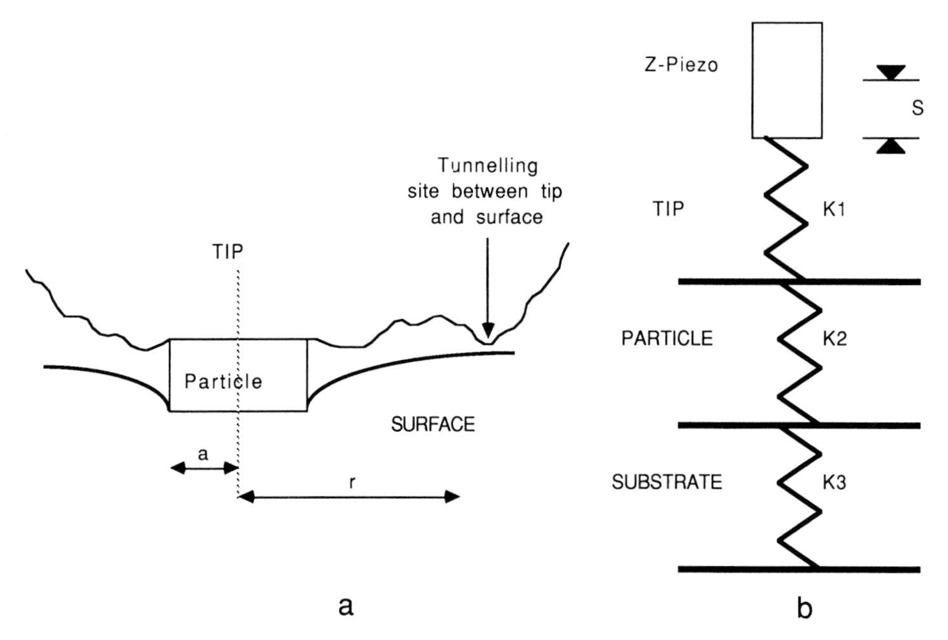

Figure 12.6. Displacements when a contact is made between tip and surface. (a) An insulating particle sandwiched between tip and surface. The tunnelling site is at a distance r from the center of the particle. (b) Simplified spring constants K_1 for tip and mounting structure, K_2 for gap particle, and K_3 for near-particle substrate.

may have little or no correspondence with real features because the image is essentially being generated through a complicated physical compression of the tip, particle, and substrate as well as the constant tunnel current contours of a contactless image.

Since a low apparent barrier height is indicative of contact it is important to measure its value routinely during an experiment. This is simply effected by modulating the gap between tip and sample and measuring the corresponding change in the tunnelling current (or the logarithm of the tunnelling current) with a lock-in amplifier. If the amplitude of modulation is Δs and the corresponding change in the logarithm of the current is $\Delta \ln I$, then

$$\frac{\Delta \ln I}{\Delta s} = -\Phi^{1/2}$$

This result is obtained by differentiating the low-voltage tunnelling equation at constant voltage. A barrier height of greater than 1 eV measured in this way indicates a vacuum gap. In UHV on clean surfaces, values of ~4 eV are not uncommon. Experiments in air in our laboratory often record values of a few millivolts—synonymous with contact. In practice, it is very difficult to obtain

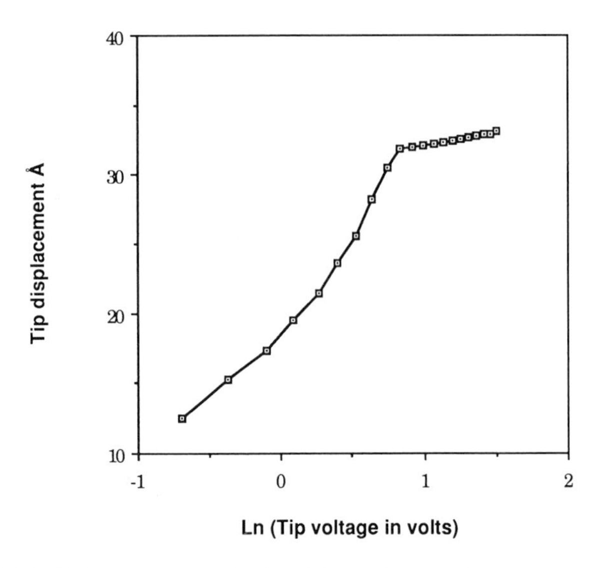

Figure 12.7. Tip displacement versus the logarithm of the tip voltage above a molecule of the protein vicilin.

contactless images in air on the best of surfaces due to surface and/or tip contamination.

A graphic example of contact in an experiment is shown in Figure 12.7. This is a plot of tip displacement in Angstroms versus the logarithm of the tip voltage (s–lnV curve) at constant current above the globular protein vicilin. An STM image of the protein can be seen in Figure 12.8 (see Welland *et al.*[(30)]). The s–ln$V$ curve is seen to consist of two distinct regions, a linear region above ~2.5 V and a nonlinear region below ~2.5 V with a discontinuity in slope between the two. Using the low voltage tunnelling equation

$$I = I_0 \, V \exp(-\Phi^{1/2} \, s)$$

the gradient $ds/d \ln V$ at constant current I is equal to $\Phi^{-1/2}$. The s–ln$V$ curve is linear, as predicted, with gradient corresponding to a barrier height of ~2 eV above the discontinuity at ~2.5 V. This is consistent with a vacuum gap between tip and molecule. Below ~2.5 V, however, the gradient is greater corresponding to a reduced barrier height, and also nonlinear. This indicates that, with the tip in contact with the sample, the barrier height has been reduced as predicted by the contact model. To obtain a noncontact image it was necessary to tunnel at voltages in excess of 3 V.

If we assume that the insulating molecule or film is thin enough so that the STM can be operated without contact taking place, then image contrast will depend on the relative barrier heights of vacuum gap and molecule. We are now

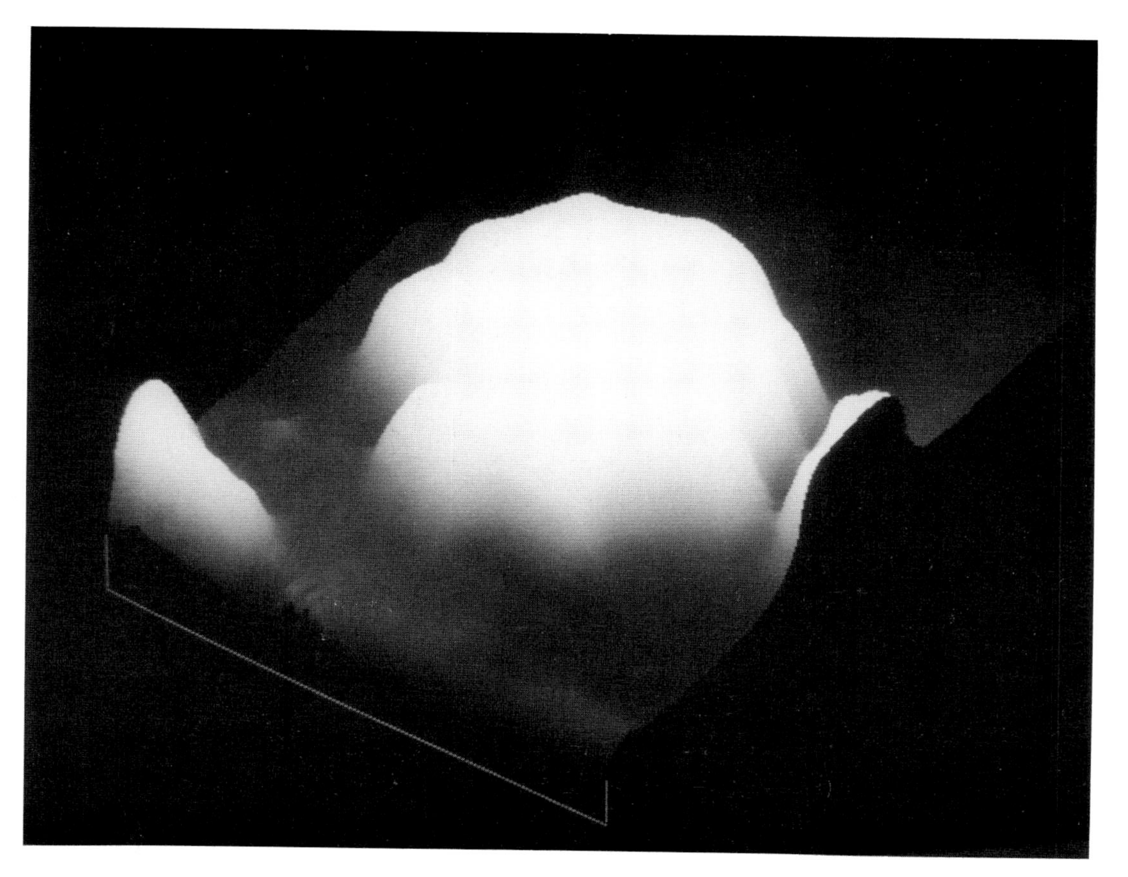

Figure 12.8. STM image of the globular protein vicilin. The width of the protein is ~ 100 Å.

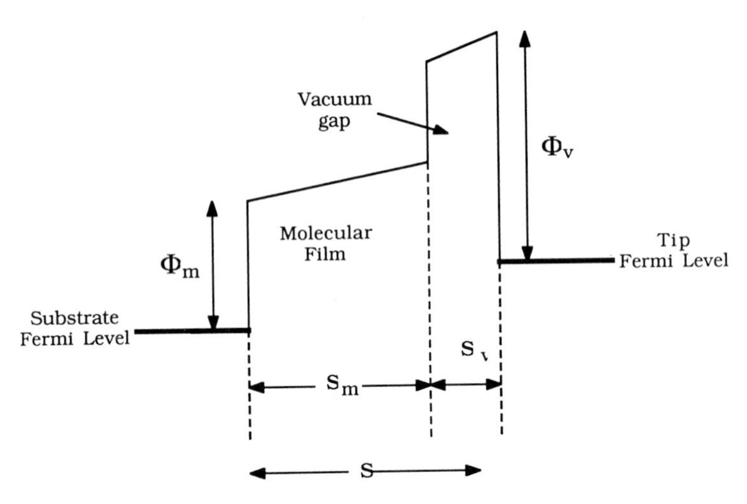

Figure 12.9. Diagram of free-electron metals (tip and substrate) separated by a double tunnelling barrier composed of a vacuum gap and a thin insulating film.

dealing with a two barrier system as shown in Figure 12.9. Considering the case of direct tunnelling through both the vacuum barrier and molecule, the total electron transmission probability D can be approximated by

$$D \propto \exp(-\Phi^{\frac{1}{2}}{}_m\, S_m - \Phi^{\frac{1}{2}}{}_v\, S_v)$$

where the terms are defined in Figure 12.9 (see Coombs[32]). If the substrate is assumed to be flat, then the change in tip position δS for constant transmission probability, and thus constant current, is given by

$$\delta S = \delta S_m\, [1 - (\Phi_m / \Phi_v)^{1/2}]$$

and the observed topographic changes will be smaller than the true film thickness changes by an amount dependent on the relative barrier heights of vacuum and molecule. Only if the barrier height of the molecule is very low will the tip displacement accurately reflect the true thickness of the molecule.

Because the image contrast will depend on the electron barrier height of the molecule, any interaction between the substrate and molecule that alters the barrier height locally will contribute contrast to the image. It is well known (Weber and Peria[33]) that the barrier height of a clean surface Φ is altered by the presence of an adsorbed polar molecule according to

$$\phi = \Phi - e\mu / \epsilon_0$$

where ϕ is the modified barrier height, μ is the dipole moment density of the molecule, e is the electronic charge, and ϵ_0 is the permittivity of free space. The dipole moment of the molecule may either be permanent or be modified by the high electric fields occurring beneath the tip ($\sim 10^9$ V m^{-1}). The electric field will contain components of the image dipole field and depolarizing field of adjacent dipoles as well as the tip-sample field. Thus, μ in the above equation may be some complicated function of field rather than a field-independent quantity. Spong *et al.*[34] have recently ascribed the image contrast in images of liquid crystals to the alteration of the electron barrier height through a dipole interaction with a graphite substrate. Following their calculation, they chose the work function of clean graphite to be 5 eV, and assumed that a benzene ring adsorbed to the graphite surface would have a binding energy of 0.5 eV and an equilibrium distance from the surface of 3.3 Å. The induced dipole moment is then $\sim 2.2 \times 10^{-29}$ cm and the barrier height above the benzene ring is then calculated as being lowered by $\sim$ 2.5 eV. This barrier reduction implies that the tip will move away from the surface as it passes over the molecule by a distance of the order of 1 Å. This was found to agree quantitatively with the contrast in the images of the part of the liquid crystal molecule containing the benzene ring structure. This is an exciting demonstration of the ability of the STM to probe the detailed structure of organic molecules.

Two examples of images of thin molecular films recorded in Cambridge are shown in Figures 12.10 and 12.11 (see Coombs *et al.*[35]). They are both of the same material; a Langmuir–Blodgett (L–B) film of protoporphyrin. L–B films are an attractive proposition for STM because they can be deposited in monolayers with long-range molecular ordering within the film. The first image, Figure 12.10, simply shows that the monolayer of L–B film appears discontinuous in the STM with large rafts of monolayer separated by regions of uncoated substrate. The step height at the periphery of each raft was calculated to be ~13 Å, slightly less than the expected value of 16 Å for the reasons discussed above. A possible arrangement of the disk-shaped porphyrin molecules in an L–B monolayer is with the disks in rows like stacked coins, edge-on to the substrate. Such ordered regions were rarely observed, but a good example is shown in Figure 12.11. The spacing between rows was measured to be ~35 Å. The porphyrin molecule diameter is ~20 Å so it might be expected that the row spacing would be about this value. This apparent discrepancy may be due to structural relaxation through interaction with the substrate.

12.6.3. Semiinsulating and Conducting Molecules

In the case of a highly conducting film or molecule, where the electron transmission occurs within the film, the STM image will be, to a first approximation, that of the surface of the film. This begs two important questions: (1) into or

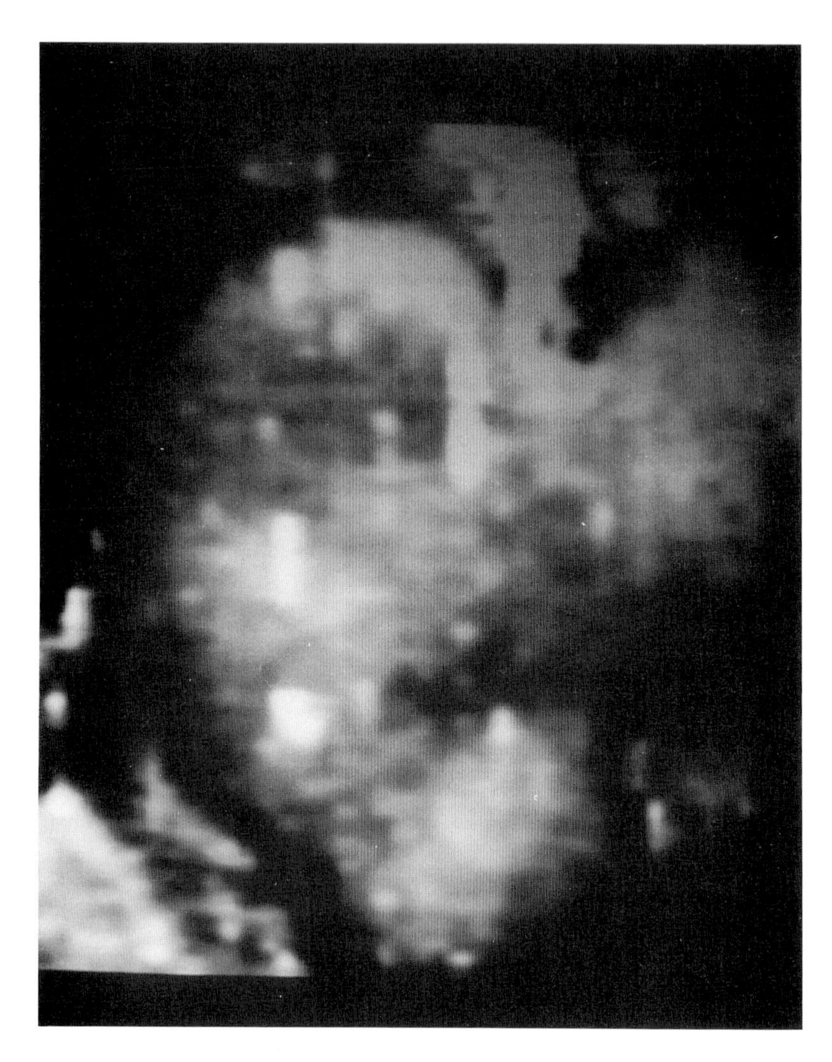

Figure 12.10. STM image of a L–B film of protoporphyrin. The image is ~ 600 Å square.

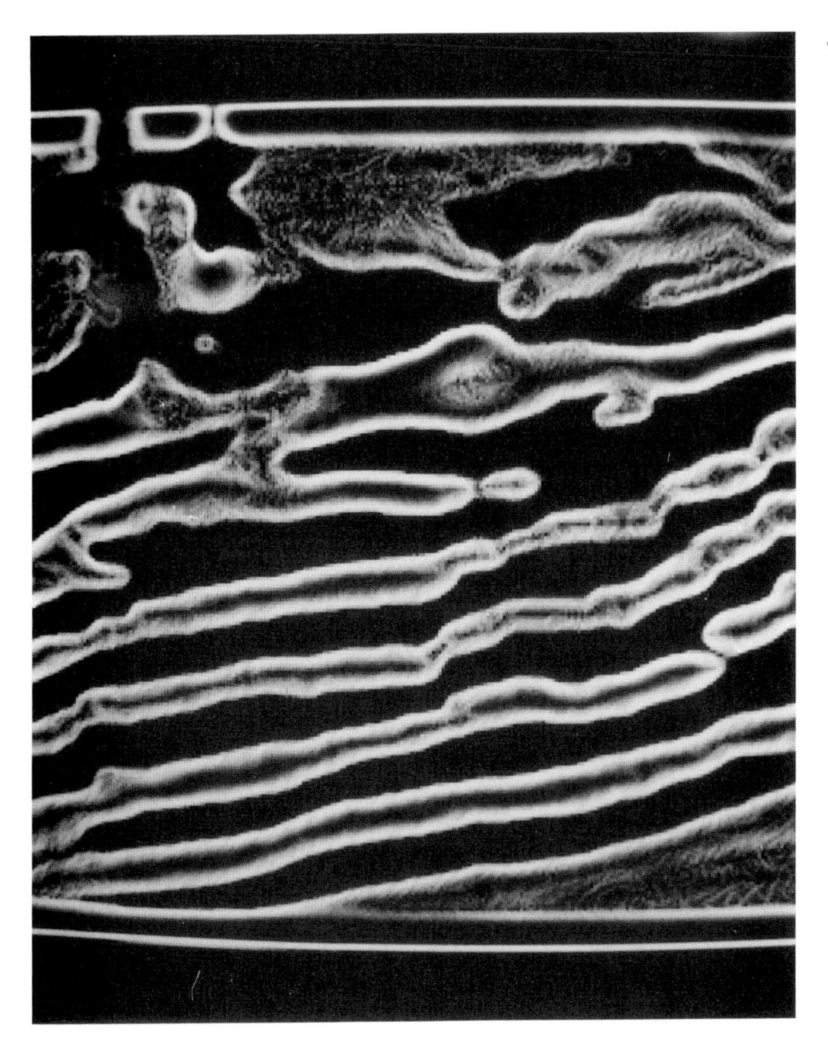

Figure 12.11. STM image of an ordered area of the L–B film of Figure 12.10. The row to row spacing is ~ 35 Å.

from what states do the electrons initially tunnel, and (2) by what conduction mechanism do they traverse the film? The former cannot be answered with any generality and will depend on the surface and bulk states in the particular film. Using the STM as a localized spectroscopic probe may shed some light on this problem. Conduction through the film requires that the effective resistance of the film to electrons tunnelling between tip and conducting substrate is of the order of, or less than, the tunnel gap resistance. This in turn assumes a film "bulk" resistivity several orders of magnitude less than that quoted (see Jones *et al.*[36]) for typical multilayer organic L–B films. Clearly, the use of bulk conductivity values is dubious in this situation; indeed, transferring any bulk parameter based on an assumed continuum to a situation involving such a highly localized electronic probe as the STM is dangerous. Although the electrons tunnelling from the tip to a molecule or film that is conducting will do so at one site immediately below the tip apex, the current dissipation into the substrate can take place anywhere within the film so long as the rate of conduction away from the tunnelling site is greater than the arrival rate of the tunnelling electrons. One can visualize tunnelling into a molecule or chain of molecules with high conductivity along the length of the chain. If the chain is parallel to the surface, then the tunnelling electrons arriving from the tip can move along the chain and subsequently tunnel or conduct directly into the substrate at considerable distance from the position of the tip.

Where electron states are present within a semiinsulating film or molecule that are accessible to tunnelling electrons, two possible effects on the transmission probability through the barrier between tip and substrate may be noted.[3]

The first is known as resonant tunnelling in which the barrier contains an attractive potential of finite width. When the bias voltage is such that the electron tunnelling from the tip to substrate, or vice versa, has an energy matching the barrier level, the electron transmission through the barrier can be increased by orders of magnitude. An important example of transmission resonance occurs in field emission from single-crystal tips that have adsorbed atoms. These atoms provide a resonant level for electrons and enhance the field emission current (see Plummer and Young[37]). Although no reports of this effect in STM have appeared to date, it is possible that a molecule with states at the appropriate resonance energy could indeed enhance the transmission of electrons through the molecule. Because the resonant transmission will be a function of tip voltage it would be necessary to measure the conductance versus voltage at constant separation between tip and substrate. Resonant tunnelling would appear as a peak in this plot.

The second effect of having an electron state in the barrier is where the tunnelling current is substantially reduced when an electron is momentarily trapped in the state. The importance of electron traps in semiconductors is well known since their presence can substantially reduce the electron or hole conductivity. In the STM the trapping of a single electron within the gap between tip and

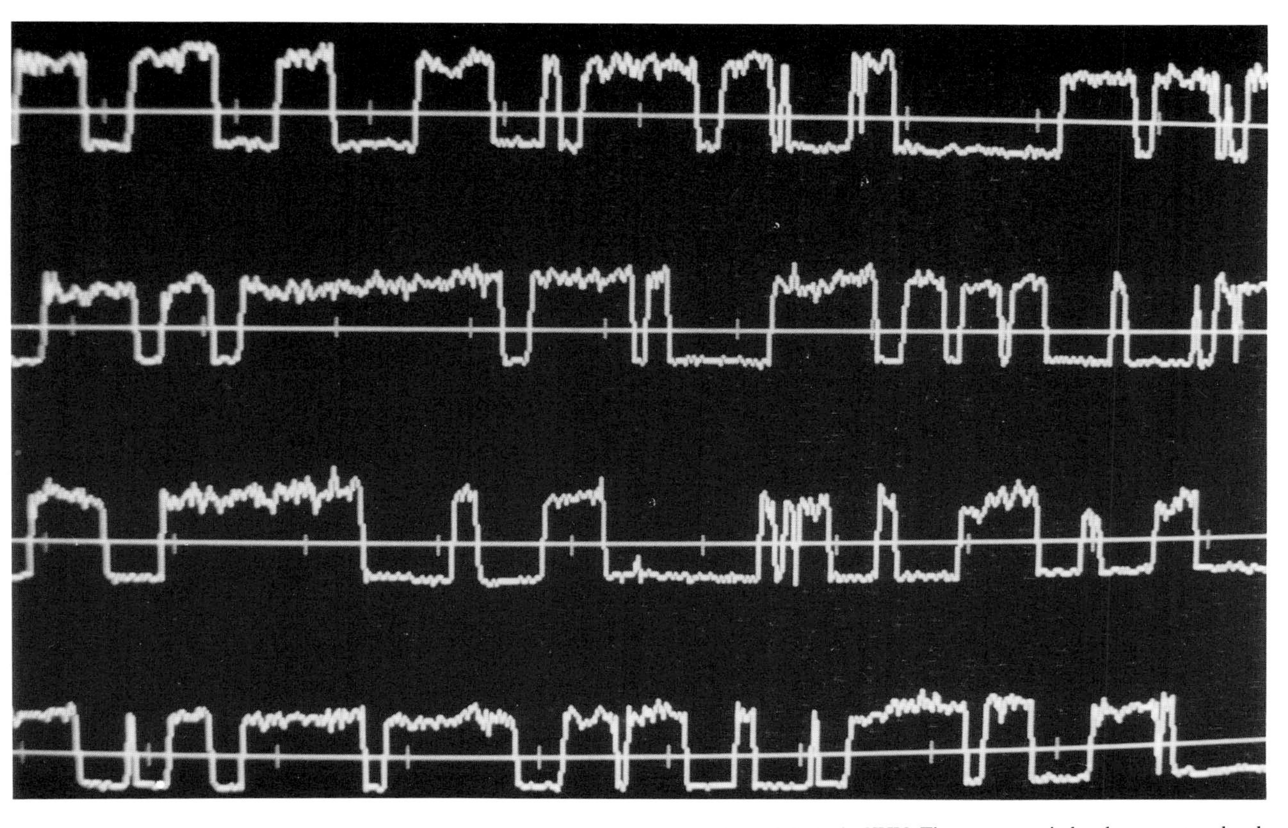

Figure 12.12. A trace of the tunnelling current above an electron trap on the surface of GaAs in UHV. The current switches between two levels corresponding to a single electron hopping onto and off the trap.

substrate is dramatic; the Coulomb repulsion of the electron can prevent any further current flow (Welland and Koch[38]). The trapping time τ is typically thermally activated and is given by

$$\tau = \tau_0 \exp\left(\frac{E_{act} - \gamma e V_{tip}}{kT}\right)$$

where E_{act} is the activation of energy for trapping, γ is the change in energy of the trapping site per unit change in V_{tip}, e is the charge on an electron, k is Boltzmann's constant, and T is the temperature. The range of possible trapping times extends from microseconds to seconds. If the tip of the STM is held stationary over a trap the current will jump between two levels corresponding to the empty and filled states. Since both emptying and filling is thermally activated, the time variation of the current will be that of a random telegraph signal. An example of the effect on the tunnel current of a trap on a GaAs surface is shown in Figure 12.12. Inspection of the above equation shows that the probability of trapping is related to the tip voltage. If a series of measurements on the fractional time that the trap is filled to it being empty are made as a function of tip voltage, both γ and E_{act} can be determined from a plot of ln(occupation probability) versus tip voltage, which will give a straight line with gradient proportional to γ and intercept proportional to E_{act}. An example of such a plot is shown in Figure 12.13 for the trap of the previous figure. Since γ is a measure of the fractional distance of the trap relative to the tip and the substrate, both the activation energy and precise position of the trap can be located with the STM.

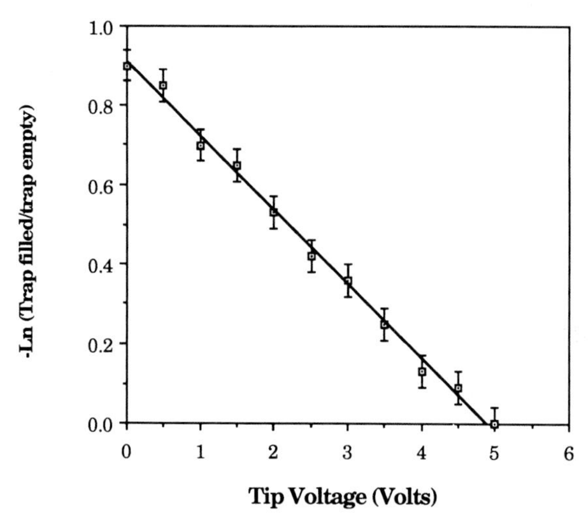

Figure 12.13. Graph of fractional trap occupancy versus tip voltage for the electron trap of Figure 12.12.

12.7. Conclusions

The scanning tunnelling microscope is a uniquely surface-sensitive tool, thus its widespread exploitation by surface scientists. It is, however, essentially restricted to conducting samples or thin insulating films; bulk insulators have to be coated with metal. Imaging of thin insulting films is possible provided that the film thickness does not exceed about 20 Å. Interpretation of STM images of both conducting surfaces and insulating films is not straightforward. There may well be several mechanisms contributing to the observed contrast in addition to the topographical variation. Comparison of STM data with information from other surface techniques is therefore extremely important.

References

1. G. Binnig, H. Rohrer, C. Gerber, and E. Weibel, "Surface studies by scanning tunneling microscopy," *Phys. Rev. Lett.* **49,** 57–61 (1982).
2. P. K. Hansma and J. J. Tersoff, "Scanning tunneling microscopy," *App. Phys.* **61,** R1–R23 (1987).
3. E. L. Wolf, *Principles of Electron Tunnelling Spectroscopy,* Oxford University Press, New York (1985).
4. P. K. Hansma, in *Tunnelling Spectroscopy: Capabilities, Applications, and New Techniques* (P. K. Hansma ed.), Plenum, New York (1982).
5. L. Esaki, "New phenomenon in narrow germanium p–n junctions," *Phys. Rev.* **109,** 603–604 (1958).
6. I. Giaever, "Energy gap in superconductors measured by electron tunneling," *Phys. Rev. Lett.* **5,** 147–148 (1960).
7. R. Young, J. Ward, and F. Scire, "Observation of metal–vacuum–metal tunnelling, field emission, and the transition region," *Phys. Rev. Lett.* **27,** 922–924 (1971); and "The Topographiner: An instrument for measuring surface microtopography," *Rev. Sci. Instr.* **43,** 999–1011 (1972).
8. G. Binnig, H. Rohrer, C. Gerber, and E. Weibel, "7 × 7 reconstruction on Si(111) resolved in real space," *Phys. Rev. Lett.* **50,** 120–123 (1983).
9. Y. Kuk, P. J. Silverman, and H. Q. Nguyen, "Study of metal surfaces by scanning tunneling microscopy with field ion microscopy," *J. Vac. Sci. Technol.* **A6,** 524–528 (1988).
10. A. M. Baró, G. Binnig, H. Rohrer, C. Gerber, E. Stoll, A. Baratoff, and F. Salvan, "Real-space observation of the 2 × 1 structure of chemisorbed oxygen on Ni(110) by scanning tunneling microscopy," *Phys. Rev. Lett.* **52,** 1304–1307 (1984).
11. R. V. Coleman, B. Drake, P. K. Hansma, and C. G. Slough, "Charge density waves observed with a tunneling microscope," *Phys. Rev. Lett.* **55,** 394–397 (1985).
12. H. F. Hess, R. B. Robinson, R. C. Dynes, J. M. Valles, and J. V. Waszczak, "Scanning tunneling microscopy observation of the Abrikosov flux lattice and the density of states near and inside a fluxoid," *Phys. Rev. Lett.* **62,** 214–216 (1989).
13. A. Bryant, D. P. E. Smith, and C. F. Quate, "Imaging in real time with the tunneling microscope," *Appl. Phys. Lett.* **48,** 832–834 (1986).
14. R. J. Hamers, R. M. Tromp, and J. E. Demuth, "Surface electronic structure of Si(111)—(7 × 7) resolved in real space," *Phys. Rev. Lett.* **56,** 1972–1975 (1986).
15. P. Muralt and D. W. Pohl, "Scanning tunneling potentiometry," *Appl. Phys. Lett.* **48,** 514–516 (1986).

16. R. Sonnenfeld and P. K. Hansma, "Atomic resolution microscopy in water," *Science* **232,** 211–213 (1986).
17. G. Binnig, C. Quate, and C. Gerber, "Atomic force microscope," *Phys. Rev. Lett.* **56,** 930–933 (1986).
18. J. G. Simmons, "Generalized formula for the electric tunnel effect between similar electrodes separated by a thin insulating film," *J. Appl. Phys.* **34,** 1793–1803 (1963).
19. J. Lambe and R. C. Jaklevic, "Molecular vibration spectra by inelastic electron tunneling," *Phys. Rev.* **165,** 821–832 (1968).
20. J. Bardeen, "Tunneling from a many-particle point of view," *Phys. Rev. Lett.* **6,** 57–59 (1961).
21. J. Tersoff and D. R. Hamann, "Theory and application for the scanning tunneling microscopy," *Phys. Rev. Lett.* **50,** 1998–2001 (1983).
22. M. C. Payne, "Transfer Hamiltonion description of resonant tunnelling," *J. Phys. C: Solid State Phys.* **19,** 1145–1155 (1986).
23. G. Binnig and D. P. E. Smith, "Single-tube three-dimensional scanner for scanning tunneling microscopy," *Rev. Sci. Instrum.* **57,** 1688–1689 (1986).
24. A. P. Fein, J. R. Kirtley, and R. M. Feenstra, "Scanning tunneling microscopy for low-temperature, high magnetic field, and spatially resolved spectroscopy," *Rev. Sci Instrum.* **58,** 1806–1810 (1987).
25. S-i. Park and C. F. Quate, "Scanning tunneling microscopy," *Rev. Sci. Instrum.* **58,** 2010–2017 (1987).
26. S-i. Park and C. F. Quate, "Theories of the feedback and vibration isolation systems for the scanning tunneling microscopy," *Rev. Sci. Instrum.* **58,** 2004–2009 (1987).
27. R. C. Jaklevic, L. Elie, W. D. Shen, and J. T. Chen, "Application of the scanning tunneling microscope to insulating surfaces," *J. Vac. Sci. Technol.* **A6,** 448–453 (1988).
28. H. Heinzelmann, E. Meyer, P. Grütter, H.-R. Hidber, L. Rosenthaler, and H.-J. Güntherodt, "Atomic force microscopy: General aspects and application to insulators," *J. Vac. Sci. Technol.* **A6,** 275–278 (1988).
29. J. B. Pethica, "Comment on 'Interatomic forces in scanning tunneling microscopy: Giant corrugations of the graphite surface' " *Phys. Rev. Lett.* **57,** 3235 (1987).
30. M. E. Welland, M. J. Miles, N. Lambert, V. J. Morris, J. H. Coombs, and J. B. Pethica, "Structure of the globular protein vicilin revealed by scanning tunneling microscopy," *Int. J. Biol. Macromol.* **11,** 29–32 (1989).
31. J. H. Coombs and J. B. Pethica, "Properties of vacuum tunneling currents: Anomalous barrier heights," *IBM J. Res. Dev.* **30,** 455–459 (1986).
32. J. H. Coombs, "Scanning tunnelling microscopy: Design and tunnelling characteristics," PhD Thesis, University of Cambridge (1987).
33. R. E. Weber and W. T. Peria, "Work function and structural studies of alkali-covered semiconductors," *Surf. Sci.* **14,** 13–38 (1969).
34. J. K. Spong, H. A. Mizes, L. J. LaComb, Jr., M. M. Dovek, J. E. Frommer, and J. S. Foster, "Contrast mechanism for resolving organic molecules with tunneling microscopy," *Nature* **338,** 137–139 (1989).
35. J. H. Coombs, J. B. Pethica, and M. E. Welland, "Scanning tunneling microscopy of thin organic films," *Thin Solid Films* **159,** 293–299 (1988).
36. R. Jones, R. H. Tredgold, and P. Hodge, "Langmuir–Blodgett films of simplified esterified porphyrins," *Thin Solid Films* **99,** 25–32 (1983).
37. E. W. Plummer and R. D. Young, "Field-emission studies of electronic energy levels of adsorbed atoms," *Phys. Rev.* **B1,** 2088–2109 (1970).
38. M. E. Welland and R. H. Koch, "Spatial location of electron trapping defects on silicon by scanning tunneling microscopy," *Appl. Phys. Lett.* **48,** 724–726 (1986).

13

Resolution

A Biological Perspective

Murray Stewart

13.1. Overview

This chapter aims to identify the types of biological structural problems that might benefit from new microscopic methods for investigating their structure. Light and electron microscopy have already provided a wealth of information about many structural aspects of biological systems. It follows that the new techniques that are most likely to be valuable tools for biologists are those that allow different types of structure to be investigated, either by achieving higher resolution or detection efficiency than existing microscopies, or by allowing specimens to be investigated in different conditions. It is also important to realize that structural work in many biological systems is limited by detection efficiency and not resolution, and, moreover, because of the unique structure of many of these systems, structures that have physical dimensions orders of magnitude below the diffraction limit of the imaging system can be successfully analyzed provided adequate detection methods are available. Light microscopy functions well for investigating cells and many organelles, often in living material. In particular, new methods involving confocal techniques offer exciting possibilities for examining thick specimens and for detecting features of interest that are well below the diffraction limit of the imaging system. Because of the manner in which cells are organized, the next highest structural level to be addressed (macromolecular and membrane organization) requires a resolution of about 10

Murray Stewart • MRC Laboratory of Molecular Biology, Cambridge CB2 2QH, United Kingdom.

nm, and there are not very many interesting biological structural features between this level and the resolution obtainable by light microscopy. Further structural information (mainly about molecular arrangement and internal molecular structure) requires resolution in the range of 0.1 to 10 nm, but critical thresholds seem to be at about 2 nm and 0.3 nm. Electron microscopy is a powerful tool for examining biological structure in these resolution ranges, but suffers from having generally low inherent contrast and, moreover, cannot be applied easily to living or to dynamic systems. New microscopies (such as, for example, scanning tunnelling microscopy) which enhance the contrast (and thus detectability) when viewing biological material, or which enable living and dynamic systems to be investigated at resolutions better than 10 nm, would be a significant advance, as would microscopies that enabled specimens as thick as whole cells to be examined to this sort of resolution. Another fertile area for the development of new microscopies would be the precise location of low concentrations of specific elements, such as calcium, which perform important effector roles in many biological systems.

13.2. Introduction

Biology is probably distinguished from other areas of science more by the fact that living entities are studied than by any other feature. The principal feature that distinguishes a living system from a mixture of molecules in a test tube is the way in which molecules are organized: one of the secrets of life is macromolecular organization. It is for this reason that biologists are so concerned with structure, macromolecular assembly, and the location of molecules and ions in living systems.

There is a belief, perhaps more widespread amongst nonbiologists, that new methods of examining cells are needed to gain deeper insights into their structure and internal organization than can be obtained by light and electron microscopy and x-ray diffraction. It is, of course, true that these traditional techniques are not without their problems and limitations, but it is perhaps worth pausing to consider the different levels of structural information in cells and biological macromolecules (see Chapter 2, this volume), how well these are currently able to be investigated, and what advantages any new technique might have. In particular, in developing new microscopic techniques for use in biological systems, it is important to define what sort of information it is currently difficult or impractical to obtain using traditional methods. New techniques that are able to provide this sort of information are likely to have a much greater impact on biological structural research than methods that provide information that can already be obtained by light or electron microscopy, particularly in view of the development of such techniques as confocal light microscopic methods[1] and of methods to examine frozen hydrated biological material by electron microscopy (see Chapter 2, this volume).[2]

13.3. Biological Structural Problems

When trying to assess the likely usefulness of a particular method for investigating biological structural problems, it is important to distinguish between resolution and detection. In a great number of instances, biologists are more interested in the latter and so, for example, can gain very comprehensive information about the fibrous components of cell cytoskeletons (which typically have diameters of 10–30 nm) using fluorescent labels and light microscopy.[3] This has become even more powerful as a detection method with the introduction of confocal light microscopy techniques which result in a large reduction in the background fluorescence deriving from material above and below the plane of focus and so greatly enhances the signal-to-noise ratio of the image of the fluorescent label.[2] Figure 13.1 shows an example of this sort of imaging in which the individual microtubules (rodlike structures about 30 nm diameter) in a spindle from a divided sea urchin embryo can all be easily seen when they are well separated. Dark-field light microscopy can also often detect particles with diameters not much greater than 10 nm, although formally the resolution of a light microscope is usually considered to be about 200 nm. Thus, one can learn a great deal about biological structures that have dimensions of the order of nanometers by using light microscopy, provided that the structures being investigated are spaced sufficiently far apart. This is often true for cellular components such as cytoskeletal filaments and many membrane-bound organelles, including even the endoplasmic reticulum.

Biologists usually consider resolution more in terms of the highest spatial frequency present in the image than the strict Rayleigh criterion. This is particularly true with electron microscopy and x-ray diffraction, where data are commonly analyzed by Fourier-based methods, although light microscopes (and particularly their objective lenses) are also usually evaluated in terms of the Abbe theory of image formation by their numerical aperture.[4] Resolution is usually assessed using objects that have regular lattices or objects that approximate white noise. With regular objects, one simply examines the regular pattern of reflections in their Fourier transform (often produced as an optical diffraction pattern either directly from the object or from a photograph of it through the microscope imaging system) to determine the highest spatial frequency present, whereas with white noise the attenuation of the power spectrum gives a good measure of the highest spatial frequencies likely to be present.[5] In some instances, such as electron microscopy, the contrast transfer function at high spatial frequencies is rather complicated, and so it is important to know to what level images can be

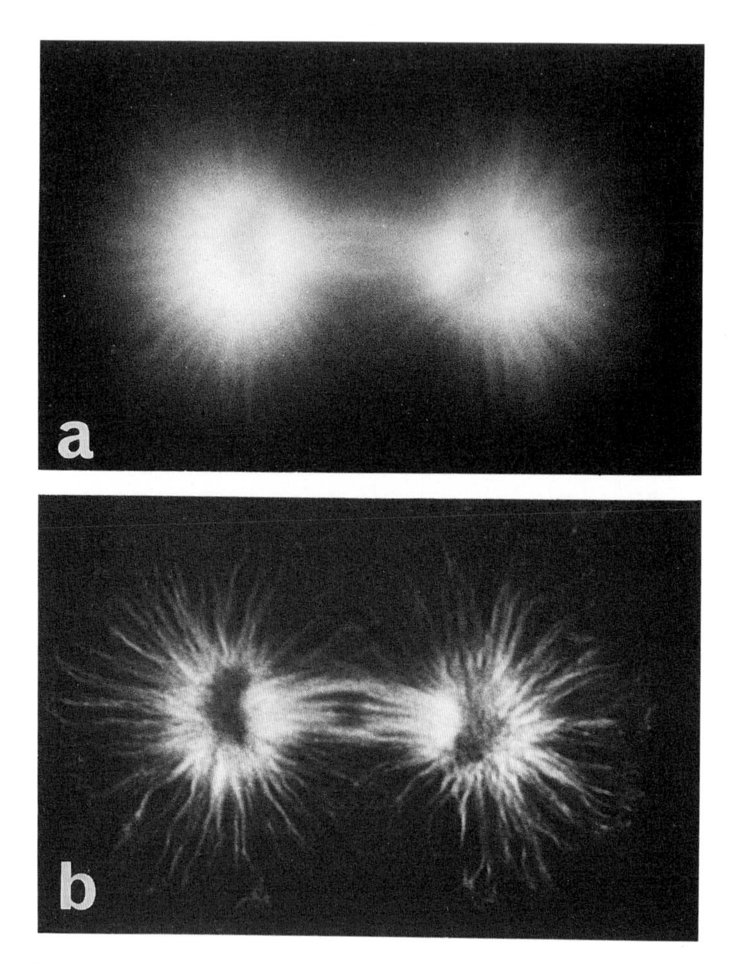

Figure 13.1. Illustration of the improvement produced in a fluorescence image of a sea urchin spindle by using a confocal microscope. In both images the microtubules (long rods about 30 nm diameter) have been labelled with an antibody to which a fluorophore has been attached. In the conventional fluorescence light micrograph (a) the broad outline of the spindle can be made out and there is a hint of finer fibrous material. However, (b) in the image obtained using a confocal microscope (in this case the Biorad-MRC 500 instrument) the fibrous components are very much more clearly delineated and one has no difficulty at all in locating fibers that are roughly an order of magnitude smaller than the diffraction resolution limit of the optical system. (Micrograph courtesy of Dr. W. B. Amos, MRC Laboratory of Molecular Biology, Cambridge.)

naively interpreted (see Chapter 2, this volume).[2] At the moment, light microscopy can attain resolutions of about 200 nm (with a confocal system) and appears to be limited primarily by the numerical aperture of the optical system. Electron microscopes can easily attain resolutions of the order of 0.1 nm on nonbiological material, but low contrast and radiation damage usually limit resolution to the order of 1–2 nm, although in exceptionally favorable cases, resolutions of about 0.3 nm can be achieved.

It is also helpful to consider the different levels of structure present in biological objects and the resolution needed to examine each level in any detail (see Chapter 2, this volume). Although most higher eukaryotic organisms are composed of defined arrangements of cells, the details of this arrangement are usually easily seen by eye or using a conventional light microscope. Cells themselves range from about 1 μm to many centimeters in size, and can usually be easily seen with the naked eye or with a light microscope. Using these methods, cells can usually be observed not only hydrated but often living. The next level of structure in eukaryotic cells is that of organelles, which have dimensions of approximately the order of prokaryotic cells or about 1 to 10 μm. Again these can usually be seen with light microscopes and often in the living state. Three-dimensional views of many of these objects can be built up by varying the level of the focal plane in the object, and this is proving a particularly powerful application of confocal light microscopes (since their depth of focus is very small, of the order of 0.5 μm). Although the intrinsic contrast of many of these objects is quite low, they can often be visualized using dyes (many of which are "vital" dyes and which can be used in living cells) or by using specialized imaging methods, such as phase or differential interference contrast.

Virus particles and most macromolecular assemblies within cells usually have dimensions of the order of 100 nm. These particles can often be detected by high-contrast light microscopy methods (e.g., fluorescence). In some instances this can yield spectacular results, particularly with isolated systems such as the assays now commonly employed to examine elements of motile systems.[6] Sometimes very small movements of particles can be detected by these methods,[7] and they are proving invaluable for investigating cellular molecular motors. However, details of the substructure of biological macromolecular assemblies can usually only be obtained by electron microscopy or x-ray diffraction, which often are used in combination. With electron microscopy, it is often necessary to dehydrate and stain specimens, although recent advances have made it possible to examine a broad range of specimens preserved in amorphous (vitreous) ice, which appears to be a very close approximation to their usual hydrated environment.[2] To even detect substructure in these assemblies, however, requires resolution better than about 50 nm. For example, the repeat distance in collagen (probably the longest spacing commonly found) is 63 nm, whereas the spacings in muscle are 43 and 14 nm. Membranes are a principal organizing feature of cells and generally are about 10 nm thick. Detection of membranes can be a key structural level in cell architecture, and the ability to do this would be a vital feature of any new microscopic technique.

It is important to realize, therefore, that there are very few biological specimens whose internal structure can be investigated in any detail until resolutions of about 50 nm have been reached, and that most objects require a resolution of at least 10 nm. There are very few important biological structural problems in the resolution range of about 10–200 nm. Therefore, it seems that unless resolutions of the order of 10 nm or better can be obtained by a new imaging method, there will only be a very limited number of biological structural problems that it can address. Furthermore, unless resolutions of better than 200 nm can be obtained, it is unlikely that the method will offer significant advantages over light microscopy.

Any details about the fine structure of molecules generally requires a resolution of better than 5 nm. Fine structure usually only becomes clear at about 2.5 nm. The sensitivity of biological material to electron irradiation often presents severe difficulties at spatial frequencies higher than these, although the use of low-dose techniques[(8)] and computer image processing (Chapter 2, this volume) can often alleviate these problems with crystalline specimens. Most biological macromolecules do not then yield a great deal more information until resolutions of about 0.7 nm are obtained, at which time elements of their secondary structure start to become visible. Atomic resolution, which is usually only obtainable by x-ray diffraction, generally requires resolutions of about 0.3 nm or better. However, the solution of the structure of a large biological macromolecule by x-ray diffraction is usually a formidable task, often requiring man-years of effort, and so direct imaging methods, such as scanning tunnelling microscopy, which have the potential to resolve atoms, offer an exciting potential provided problems with specimen thickness and radiation damage do not prove too limiting.

13.4. Location of Specific Elements

Because biologists are generally interested primarily in locating particular elements within cells, within molecules, or with macromolecular assemblies, methods for locating specific elements (such as, for example, calcium) need to achieve resolutions better than about 10 nm to be very useful. Moreover, location of elements to this precision is often very difficult to obtain with the very low concentrations (usually less than 1 m*M*) present in living cells. This is not to discourage the development of new methods for examining biological material, but rather to point out the rather severe criteria they will have to satisfy to become widely applicable.

Electron probe microanalysis can be used in a wide range of applications to locate elements to resolutions of 10–100 nm, but the precision of the location is usually influenced greatly by the absolute concentration of the element being investigated.[9] Unfortunately most biological processes where specific elements (usually metal cations) play a key role, operate either with very low concentrations of the ion (for example, calcium typically functions as a switch for a broad range of biological processes at free concentrations in the micromolar to millimolar range), or with very localized changes in an ion present at high concentration (for example, in nerve impulses). Therefore, although electron probe microanalysis can be very useful in determining bulk ion concentrations in cells or cellular compartments, a considerable improvement in sensitivity would be required to provide information about most biologically relevant ion transport systems. Electron energy loss spectroscopy (which can be used in a scanning mode to produce images analogous to those obtained by electron probe x-ray microanalysis) may offer some advantages in this respect,[10] but still seems beset by severe detection limit problems in many instances.

13.5. Areas of Structural Biology Where New Techniques Might Be Helpful

Although electron and light microscopy have provided a wealth of information about the structure and arrangement of cells and their components, there are a number of areas where these techniques do have severe limitations and in which the development of new microscopical techniques would be of considerable importance. Such areas would include the study of living material at resolutions better than can be obtained by light microscopy, the study of dynamic processes at high resolution, the location of low concentrations of selected elements, determination of the structure of large molecules such as proteins and nucleic acids at atomic resolution, and the study of comparatively thick specimens (such as whole cells). Each of these areas has its own special problems, which are discussed briefly below.

13.5.1. Living Systems

Although light microscopy can be used very effectively to examine living systems (such as intact cells or tissues), these studies have been most powerful where subcellular components can be delineated using special imaging techniques (e.g., dark-field, fluorescence, etc.), often in conjunction with staining with vital dyes. Moreover, although quite spectacular results have been obtained on components such as the cytoskeleton or the endoplasmic reticulum,[3] where the key is actually detection of the structure rather than resolving closely spaced components of it, other, more densely packed, cellular components have not yielded as much information. Although electron microscopy has potentially much higher resolution than light microscopy, it suffers from having to examine specimens that are fundamentally dead: either material is "fixed" and embedded in plastic and/or stain or, at best, is frozen. And even if the specimen was not dead to start with, the pronounced sensitivity of biological material to electron irradiation would ensure that it was dead long before an image could be recorded.

Therefore, because material examined by electron microscopy is clearly not living, some form of new microscopy for examining living material at resolutions higher than can be obtained by light microscopy would be a great advantage. As noted above, however, such a technique would probably have to have a resolution greater than about 10 nm, and preferably a resolution near 2–5 nm, to make a great impact on biology, because it is only at these resolutions that reliable information can be obtained about macromolecular assemblies. Currently this sort of structural level in living material can only be accessed easily by techniques such as low-angle x-ray diffraction, but this method suffers from the fact that phase information is lost and so an image cannot be obtained directly.

This is not to say, of course, that a great deal of very useful biological information cannot be obtained simply from diffraction patterns (as seen, for example, with the elegant studies on muscle by Huxley[11,12] and his colleagues), but that a whole wealth of new and more precise information could become available if it was also possible to form images at this resolution. A second related problem is of contrast, which with most biological material is intrinsically low. New microscopic methods that enhanced the contrast of biological material (probably with respect to the water in which it is invariably embedded) would be a considerable advantage, particularly for work on living material where one must always worry about the possible effects of any staining method employed.

13.5.2. Dynamic Processes

Although some spectacular results have been obtained by light microscopy on *in vitro* motile systems,[6,7] it is generally difficult to study dynamic biological systems at high resolution. This is primarily because of problems associated with electron microscopy and the need for the specimen to be fixed or frozen as part of the preparative protocol. Although in some instances a time-lapse series can be prepared for electron microscopy by freezing specimens at different time intervals during some dynamic process,[13] there are clearly great advantages in following the same specimen throughout the change. Because of the significant damage produced by electron bombardment of biological material, it would seem unlikely that electron microscopy could be used to study a dynamic biological process even if it could be reconstituted within the microscope. Currently most dynamic biological processes are investigated using spectroscopic techniques or low-angle x-ray diffraction, but these techniques do not lend themselves easily to the production of images. It would therefore be a significant advance if a new technique could be developed that would enable images of dynamic biological systems to be recorded at resolutions better than about 10 nm.

13.5.3. Location of Selected Elements

As discussed above, existing methods for detecting selected elements (such as calcium) generally have such low efficiency that the elements cannot be located precisely in biological specimens. Moreover, these methods have been used mainly for examining the distribution of cations and not for anions or atoms specifically attached to molecules. The use of specific labels would be attractive both for locating particular molecules within cells and assemblies and for locating sites on molecules in order to establish the location of active sites or orientations in assemblies. The use of colloidal gold is now a powerful method in conjunction with electron microscopy at the cell biology level (such as locating molecules in organelles or macromolecular assemblies), but the use of labels at higher resolution has only been successful in a very limited number of applications. Heavy metals attached to proteins have been visualized[14] as has the distribution of phosphorus (and so presumably of RNA) in ribosomes,[15] but clearly it would be very helpful to have a precise method for locating low concentrations of specific elements at resolutions better than 10 nm and preferably at at least 2 nm resolution.

13.5.4. Structure of Large Molecules at Atomic Resolution

Although the structure of a considerable number of large biological molecules, such as proteins and nucleic acids, has been solved to atomic resolution using x-ray diffraction, this is generally an extremely laborious technique particularly since phase information is not recorded in x-ray diffraction patterns. Instead, phases have to be determined by indirect methods, such as isomorphous replacement (measuring the change in intensities that results from incorporation of heavy metal atoms at specific sites). Ideally, electron microscopy should be able to form images of biological material at atomic resolution, but sensitivity to radiation damage and low intrinsic contrast combine to make this a dauntingly difficult undertaking. Moreover, both x-ray diffraction and electron microscopy require the production of crystals. A major limiting step in molecular structure determination to high resolution has been the actual production of crystals, and clearly any technique that enabled an atomic-resolution molecular structure to be determined using isolated molecules would be a very significant advance. In this context, the potential of scanning tunnelling microscopy to achieve atomic resolution at high contrast is exciting. It is not yet clear, however, to what extent scanning tunnelling microscopy may be limited by radiation damage, the low conductivity of biological material (which might necessitate the use of specimens as thin as 1–2 nm), or by the presence of large numbers of charged groups on the surface of most biological macromolecules. Moreover, it may not be possible to determine atomic positions in the interiors of molecules in this way.

13.5.5. Thick Specimens

Although one can often use light microscopy to examine whole cells (and sometimes, in favorable circumstances, tissues), in many applications one has to employ sections of material embedded in plastic or paraffin. Certainly building up a three-dimensional picture of most biological objects is usually a very laborious process, although for light microscopy the low depth of focus and availability of sophisticated computer software for confocal microscopes is making this very much easier and straightforward.

Problems related to sectioning (and related dehydration and embedding) and the difficulty in producing three-dimensional representations are much more acute with electron microscopy, where one usually has to employ specimens thinner than about 100 nm. Thus, with the exception of thin cellular processes, most cells examined by electron microscopy have to be sectioned or the majority of their contents removed. Clearly it would be desirable to look at thicker specimens, particularly using electron microscopy, but unfortunately the resolution obtained falls off rapidly with increasing thickness. Although this can be at least partially offset by using much higher accelerating voltages, a further problem results from the large depth of focus in an electron microscope, which makes it difficult to interpret the image obtained because of the superposition of information from different levels. Therefore, a new technique that enabled specimens of the thickness of cells (say 1–50 μm) to be examined, preferably in a way that made interpretation straightforward (perhaps by generating a three-dimensional model of the specimen) would offer considerable advantages to biologists.

13.6. Conclusions

Light and electron microscopy have provided a wealth of information about many different structural levels in biological specimens. Recent new developments in these techniques have greatly extended the range of problems that they can address. Confocal light microscopy, particularly in conjunction with the use of fluorescent labels, is revolutionizing many areas of cell biology. The spectacular detection efficiency of this method, combined with the ability to examine thick objects and generate three-dimensional models by means of computer methods has opened a great range of different problems to investigation and has already provided a wealth of new information about macromolecular and subcellular organization. The development of methods to examine frozen hydrated biological material by electron microscopy, and the development of powerful computer methods for analyzing structural information to high resolution, has also extended significantly the number of biological problems that can be addressed at the level of molecular structure and macromolecular assembly.

For new microscopic techniques to make a significant impact in biological research, they must enable structural features to be investigated that are currently not accessible by light or electron microscopy. Moreover, they should have a resolution sufficient to address questions about a defined level of biological structure. As light microscopy gives adequate results for most questions at the levels of whole cells and organelles, new microscopies will probably be most effective if they can exceed substantially either the resolution or the sensitivity of light microscopy. Moreover, because in many instances, light microscopy can give meaningful information well below its diffraction limit, new techniques will probably have to achieve resolutions of the order of 10 nm before they make a significant impact on biology.

It seems most likely therefore that new microscopic techniques will be most effective when addressing those areas in which electron microscopy has limitations—specifically, in living and dynamic systems, with thicker specimens, or in locating specific elements. At higher resolutions, new methods that could overcome the low inherent contrast of biological material or its sensitivity to electron irradiation would also be significant advances. In terms of analyzing macromolecular structure to atomic resolution, scanning tunnelling microscopy may develop into a powerful new method and has the exciting potential for examining isolated molecules rather than requiring the production of large perfect crystals needed for x-ray diffraction.

It may, perhaps, be sobering to reflect that biologists already have very powerful microscopic techniques for attacking a very broad range of structural problems at the levels of structure that are most meaningful in a biological context. This is not to say that biological microscopy could not benefit substantially from new methods, nor to discourage the development of these methods. Instead it is to indicate the somewhat demanding criteria that may need to be satisfied if new methods are to find broad applicability in biology.

References

1. J. G. White, W. B. Amos, and M. Fordham, "An evaluation of confocal versus conventional imaging of biological structures by fluorescent light microscopy," *J. Cell Biol.* **105,** 41–53 (1987).
2. M. Stewart and G. Vigers, "Electron microscopy of frozen-hydrated biological material," *Nature (London)* **319,** 631–636 (1986).
3. See, for example, *Organization of the Cytoplasm,* Cold Spring Harbor Symposium on Quantitative Biology, Vol. XLVI, Cold Spring Harbor Laboratory, Cold Spring Harbor, New York (1981).
4. E. Hecht and A. Zajak, *Optics,* p. 465, Addison-Wesley, New York (1974).
5. F. Thon, "Zur Defokussierungsabhängigkeit des Phasenkontrastes bei der elektronmikroskopischen Abbildung, *Z. Naturforsch.* **21a,** 476–478 (1966).

6. Y. Y. Toyoshima, S. J. Kron, E. M. McNally, K. R. Niebling, C. Toyoshima, and J. A. Spudich, "Myosin subfragment-1 is sufficient to move actin filaments *in vitro,*" *Nature (London)* **328,** 536–539 (1987).
7. J. Gelles, B. J. Schnapp, and M. P. Sheetz, "Tracking kinesin-driven movements with nanometer-scale precision," *Nature (London)* **331,** 450–452 (1988).
8. P. N. T. Unwin and R. Henderson, "Molecular structure determination of unstained crystalline specimens," *J. Mol. Biol.* **94,** 425–440 (1975).
9. A. P. Somlyo, "Compositional mapping in biology: X rays and electrons," *J. Ultrastruct. Res.* **88,** 135–142 (1984).
10. H. Schuman and A. P. Somlyo, "Electron energy loss analysis at near trace element concentrations," *Ultramicroscopy* **21,** 23–32 (1987).
11. H. E. Huxley and A. R. Faruqi, "Time-resolved x-ray diffraction studies on vertebrate striated muscle," *Ann. Rev. Biophys. Bioeng.* **12,** 381–417 (1983).
12. M. Kress, H. E. Huxley, A. R. Faruqi, and J. Hendrix, "Structural changes during activation of frog muscle studied by time-resolved x-ray diffraction," *J. Mol. Biol.* **188,** 325–342 (1986).
13. S. Tsukita and M. Yano, "Actomyosin structure in contracting muscle detected by rapid freezing," *Nature (London)* **317,** 182–184 (1985).
14. M. Stewart and J. Lepault, "Cryo-electron microscopy of tropomyosin magnesium paracrystals," *J. Microscopy,* **138,** 53–60 (1985).
15. W. Kuhlbrandt and P. N. T. Unwin, "Distribution of RNA and protein in crystalline eukaryotic ribosomes," *J. Mol. Biol.* **156,** 431–448 (1982).

Index